CHEMISTRY
STUDENTS' BOOK II
TOPICS 12 to 18

Revised Nuffield Advanced Science

General editor
Revised Nuffield
Advanced Chemistry
B. J. Stokes

Associate editor
inorganic chemistry
A. J. Furse

Associate editor
organic chemistry
M. D. W. Vokins

Associate editors
physical chemistry
**D. H. Mansfield,
Professor
E. H. Coulson,
Jon Ogborn**

Editor of this book
B. J. Stokes

Contributors to this book
A. W. B. Aylmer-Kelly
Professor E. H. Coulson
B. E. Dawson
A. J. Furse
John Holman
A. J. Malpas
D. H. Mansfield
Professor D. J. Millen
Jon Ogborn
J. G. Raitt
B. J. Stokes
M. D. W. Vokins

Authors of Background reading
Dr R. Barnes
N. J. Cavaghan, British Steel Corporation
G. H. James
Dr A. L. Mansell
E. J. Millett, Philips Research Laboratories,
Solid State Electronics Division
A. R. Milner, Imperial Chemical Industries plc,
Mond Division
Dr J. Stuart-Webb, Imperial Chemical Industries
plc, Petrochemicals and Plastics Division
Dr A. Wiseman
H. Wiseman

Consultant on safety matters
Dr T. P. Borrows

CHEMISTRY STUDENTS' BOOK II

Topics 12 to 18

Revised Nuffield Advanced Science
Published for the Nuffield–Chelsea Curriculum Trust
by Longman Group Ltd

Longman Group Limited
London
Associated companies, branches, and representatives
throughout the World

First published 1970
Revised edition first published 1984
Copyright © 1970, 1984, The Nuffield–Chelsea Curriculum Trust

Design and art direction by Ivan Dodd
Illustrations by Oxford Illustrators Limited

Filmset in Times Roman and Univers
and made and printed in Great Britain
by Hazell Watson and Viney Limited, Aylesbury

Note
All references to the **Book of data** are to the **revised** edition
which is part of the present series.

Cover picture
Ascorbic acid (Vitamin C). Photomicrograph, taken in
polarized light, at × 75.
Copyright, Paul Brierley.

Contents

Foreword   page   *vi*

Acknowledgements   *viii*

Topic 12   Equilibria: gaseous and ionic   *1*

Topic 13   Carbon compounds with acidic and basic properties   *62*

Topic 14   Reaction rates – an introduction to chemical kinetics   *125*

Topic 15   Redox equilibria and free energy   *163*

Topic 16   The Periodic Table 4: the transition elements   *218*

Topic 17   Synthesis: drugs, dyes, and polymers   *262*

Topic 18   The Periodic Table 5: the elements of Groups III, IV, V, and VI   *352*

Index   *403*

Foreword

When the Nuffield Advanced Science series first appeared on the market in 1970, they were rapidly accepted as a notable contribution to the choices for the sixth form science curriculum. Devised by experienced teachers working in consultation with the universities and examination boards, and subjected to extensive trials in schools before publication, they introduced a new element of intellectual excitement into the work of A-level students. Though the period since publication has seen many debates on the sixth form curriculum, it is now clear that the Advanced Level framework of education will be with us for some years in its established form. Although various proposals for change in structure have not been accepted, the debate to which we contributed encouraged us to start looking at the scope and aims of our A-level courses and at the ways they were being used in schools. Much of value was learned during those investigations and has been extremely useful in the planning of the present revision.

The revision of the chemistry series under the general editorship of B. J. Stokes has been conducted with the help of a committee under the chairmanship of Malcolm Frazer, Professor of Chemical Education, University of East Anglia. We are grateful to him and to the committee. We also owe a considerable debt to the London Examinations Board which for many years has been responsible for the special Nuffield examinations in chemistry and to the subject officer, Peter Thompson, who has been an invaluable adviser on these matters.

The Nuffield–Chelsea Curriculum Trust is also grateful for the advice and recommendations received from its Advisory Committee, a body containing representatives from the teaching profession, the Association for Science Education, Her Majesty's Inspectorate, universities, and local authority advisers; the committee is under the chairmanship of Professor P. J. Black, academic adviser to the Trust.

Our appreciation also goes to the editors and authors of the first edition of Nuffield Advanced Chemistry, whose work, under the direction of E. H. Coulson, the project organizer, made this one of our most successful and influential ventures into curriculum development. Ernest Coulson's team of editors and writers included A. W. B. Aylmer-Kelly, Dr E. Glynn, H. R. Jones, A. J. Malpas, Dr A. L. Mansell, J. C. Mathews, Dr G. Van Praagh, J. G. Raitt, B. J. Stokes, R. Tremlett, and M. D. W. Vokins. A great part of their original work has been preserved in the new edition, on which several of them have acted as consultants.

I particularly wish to record our gratitude to Bryan Stokes, the General Editor of the revision. As a member of the original team he has an unrivalled understanding of the aims and scope of the first edition and as a practising teacher he possesses a particular awareness of the needs of pupils and teachers which has enriched the work of the revision. To him, to the editors working with him, A. J. Furse (Inorganic Chemistry), M. D. W. Vokins (Organic Chemistry), J. A. Hunt who is responsible for the Special Studies, and to the team responsible for the Physical Chemistry sections, Professor P. J. Black, J. Holman, D. H. Mansfield, Professor D. J. Millen, and Jon Ogborn, we offer our most sincere thanks.

I would also like to acknowledge the work of William Anderson, publications manager to the Trust, his colleagues, and our publishers, the Longman Group, for their assistance in the publication of these books. The editorial and publishing skills they contribute are essential to effective curriculum development.

K. W. Keohane,
Chairman, Nuffield–Chelsea Curriculum Trust

Acknowledgements

Many people have contributed to this book. Final decisions on the content, and method of treatment used in the first edition were made by the Headquarters team, who were also responsible for assembling and writing the material for the several draft versions that were used in school trials. The Headquarters team consisted of E. H. Coulson (organizer), A. W. B. Aylmer-Kelly, Dr E. Glynn, H. R. Jones, A. J. Malpas, Dr A. L. Mansell, J. C. Mathews, Dr G. Van Praagh, J. G. Raitt, B. J. Stokes, R. Tremlett, and M. D. W. Vokins.

The revision has been undertaken largely by three working groups, whose members were:

Inorganic chemistry: A. J. Furse (chairman), K. W. Badman, M. C. V. Cane, C. Nicholls, and D. Russell.

Organic chemistry: M. D. W. Vokins (chairman), J. J. Eggleton, G. H. James, and Professor D. J. Waddington.

Physical chemistry: Professor M. J. Frazer (chairman), Professor P. J. Black, Dr T. P. Borrows, John Holman, D. H. Mansfield, Professor D. J. Millen, and Jon Ogborn.

Advice on safety matters has been given by Dr T. P. Borrows, Chairman of the Safety Committee of the Association for Science Education.

The authors of Background reading are listed earlier in these preliminary pages, but we should also like to thank those who assisted as consultants and helped to bring the material up to date. They are G. N. Budd, British Alcan Aluminium Ltd; Dr I. P. Freeman, Unilever Research, Colworth Laboratory; Professor G. W. Gray; A. A. Harness, Imperial Chemical Industries plc, Petrochemicals and Plastics Division; Dr T. J. King; Dr R. B. Leslie, Unilever Research, Colworth Laboratory; Dr F. Long, Monsanto plc; Dr A. R. Macrae, Unilever Research, Colworth Laboratory; Dr C. B. Marenah; Dr P. J. Rees; Dr Richard Smith; Dr C. V. Stead, Imperial Chemical Industries plc, Organics Division.

All infra-red spectra were kindly supplied by Reuben B. Girling, University of York.

This book has benefited greatly from the valuable help and advice that have been generously given by teachers in schools and in universities and other institutions of higher education. In particular, the comments and suggestions of teachers taking part in the school trials, both of the original course, and of the revised Topics, have made a vital contribution to the final form of the published material.

Finally, as editor, I should like to record my thanks to the Publications Department of the Nuffield–Chelsea Curriculum Trust for their help, and particularly to Mary de Zouche for her meticulous and painstaking attention to detail in the preparation of the manuscripts for publication and to her colleagues, Deborah Williams, Hendrina Ellis, Nina Konrad, and Sarah Codrington.

B. J. Stokes

Equilibria: gaseous and ionic

12.1
GENERAL INTRODUCTION. THE EQUILIBRIUM LAW

You will know of a number of changes which, when allowed to start, do not proceed to completion. That is, all the starting materials (called the reactants) are not changed into new forms or into new substances (called the products). Some examples are:

1 If some liquid bromine is poured into a bottle which is then stoppered, some of the liquid evaporates to form a brown gas but, unless the volume of liquid used is small, some liquid remains. We say that a *state of equilibrium* has been reached between bromine liquid and bromine gas. This state of equilibrium is represented by

$$Br_2(l) \rightleftharpoons Br_2(g)$$

The symbol $\rightleftharpoons$ indicates that we are considering a system in which equilibrium has been reached and also that the equilibrium state can be reached from either direction. We could start with bromine gas only, at a higher temperature, and cool it, some would then liquefy and give an equilibrium between liquid and gas. A similar equilibrium state is reached by any system of liquid and gas in a *closed* container.

2 When a mixture of water and sodium nitrate is shaken, if there is more sodium nitrate than is required to saturate the water at the temperature concerned, some solid will remain undissolved and an equilibrium state represented by

$$NaNO_3(s) \rightleftharpoons NaNO_3(aq)$$

will be reached. We can approach this equilibrium state from either direction, by dissolving sodium nitrate in water at a given temperature, or by cooling a hot saturated solution of sodium nitrate to the same temperature. If the system of solid sodium nitrate and saturated solution is kept in a closed container at a constant temperature after equilibrium has been attained, no further change in bulk properties can be detected. By 'bulk properties' in this case we mean the concentration of the solution and the amount of undissolved solid. Since, however, the equilibrium state can be approached from either direction it is not unreasonable to suppose that some solid sodium nitrate may dissolve whilst at the same time some dissolved sodium nitrate crystallizes. If this is so, the two processes must go on at the same rate, so that the bulk, or *macroscopic* properties do not alter.

This possibility can be investigated by using radioactive isotopes, which can be detected by a Geiger tube. This involves adding some solid sodium nitrate, in which some of the sodium ions are of a radioactive isotope of sodium (for example, $^{24}_{11}Na$), to an equilibrium mixture of solid sodium nitrate and its solution. It is then possible to detect radioactivity in the solution after some time has elapsed. Thus there must be an interchange of solute between solution and solid. This means that equilibrium must be a dynamic state (and not a static state) in chemical systems.

3 $CaCO_3(s) \rightleftharpoons CaO(s) + CO_2(g)$

When calcium carbonate is heated in a sealed vessel at a fixed temperature, an equilibrium state is reached in which calcium carbonate, calcium oxide, and carbon dioxide are all present. (In an open vessel carbon dioxide will escape and the reaction can then go to completion.)

4 An equilibrium state in which hydrogen iodide, hydrogen, and iodine are all present is reached by heating hydrogen iodide gas in a sealed vessel at a fixed temperature.

$$2HI(g) \rightleftharpoons H_2(g) + I_2(g)$$

5 The formation of esters from carboxylic acids and alcohols, for example the formation of ethyl ethanoate from ethanol and ethanoic acid, involves an equilibrium state.

$$CH_3CO_2H(l) + C_2H_5OH(l) \rightleftharpoons CH_3CO_2C_2H_5(l) + H_2O(l)$$
$$\text{ethanoic acid} \qquad \text{ethanol} \qquad \text{ethyl ethanoate} \qquad \text{water}$$

Equilibrium is attained very slowly indeed at ordinary temperatures in this system. It is attained more rapidly by heating the reagents or by using a catalyst (hydrogen ions).

Characteristics of the equilibrium state

1 A stable state of equilibrium can only be attained in a *closed* system – one that cannot exchange matter with its surroundings. In an open system, which permits matter to enter or leave, stable equilibrium is not possible. (*Note*. A state of equilibrium can also be achieved in an *isolated* system – one that cannot exchange either matter *or energy* with its surroundings. An example of such a system is the contents of a sealed and insulated vessel. We shall not be concerned with such systems in this Topic.)

2 The equilibrium state can be approached from either direction; that is, reactants and products can be interchanged. Changes of this kind are said to be *reversible* and the reactions which take place as equilibrium is approached are called *reversible reactions*.

3 Under given conditions of temperature, pressure, and initial concentrations of reactants, properties such as density and concentration of the various parts of the system do not change. These are *intensive properties* (not dependent on the total quantity of matter present) as measured on the macroscopic scale. Hence we can say that the equilibrium state is characterized by constancy of intensive properties. (Properties which do depend on the total quantity of matter in a system are called *extensive properties*. These include mass, volume, and energy content.)

4 Equilibrium is a dynamic state in that opposing changes on the molecular level are continually taking place. The net result of these is that the macroscopic properties of a system in equilibrium do not change. This dynamic aspect of equilibrium means that it is stable under fixed conditions but sensitive to alteration in these conditions. The existence of an equilibrium state can be recognized by taking advantage of its sensitivity to changes in conditions. For example, if a change in temperature, pressure, or concentration leads to an obvious change in a system it is likely to have been at equilibrium before the change took place.

Relative concentrations under conditions of equilibrium

It has been stated above that the equilibrium state is characterized by constancy of intensive properties, which include concentration. For some equilibria the concentrations of substances in the various parts of the equilibrium mixture do not depend on the relative proportions of substances used initially, that is before equilibrium is approached, provided that other conditions (temperature and pressure) are constant.

Examples of this type are the equilibria

$$Br_2(l) \rightleftharpoons Br_2(g)$$
and
$$NaNO_3(s) \rightleftharpoons NaNO_3(aq)$$

discussed earlier.

For other types of equilibria we would not expect this to happen. For example, in the equilibrium

$$CH_3CO_2H(l) + C_2H_5OH(l) \rightleftharpoons CH_3CO_2C_2H_5(l) + H_2O(l)$$

it might be expected that the concentrations of ethanoic acid, ethanol, ethyl ethanoate, and water would depend in some way on the relative proportions of ethanoic acid and ethanol taken originally. It thus becomes of interest to find out whether there is any relationship between the equilibrium concentrations in systems of this kind when different proportions of starting materials are used. Three investigations for this purpose are described below.

1 The distribution of a solute between two immiscible solvents

Ammonia gas is extremely soluble in water and also dissolves, but to a lesser extent, in 1,1,1,-trichloroethane. Water and 1,1,1-trichloroethane are immiscible liquids, that is, they do not mix together. If an aqueous solution of ammonia is shaken with 1,1,1-trichloroethane, some ammonia is transferred to the 1,1,1-trichloroethane layer. As the ammonia can transfer freely between the layers, an equilibrium is soon established

$$NH_3(CH_3CCl_3) \rightleftharpoons NH_3(aq)$$

We can find the concentrations of ammonia in the two layers by titration, and thus investigate how these concentrations are related.

EXPERIMENT 12.1
The distribution of ammonia between water and 1,1,1-trichloroethane

Caution: Do not mouth-pipette any mixtures containing 1,1,1-trichloroethane. Use a safety pipette filler.

Procedure

Put 75 cm^3 of 6M ammonia and 75 cm^3 of 1,1,1-trichloroethane in a 250-cm^3 bottle with a well fitting glass stopper. Shake the bottle and its contents thoroughly, and allow the layers to separate. Using a pipette *and pipette filler*, carefully remove 25 cm^3 of the upper aqueous layer, and run it into a conical flask. Using a second pipette, *again with pipette filler*, carefully remove 25 cm^3 of the lower 1,1,1-trichloroethane layer, taking care not to collect any of the aqueous layer, and run this into a second conical flask. Add two drops of methyl orange indicator to each flask. Titrate the aqueous layer against 4.0M hydrochloric acid and the 1,1,1-trichloroethane layer against 0.10M hydrochloric acid. In the case of the 1,1,1-trichloroethane titration it will be necessary to shake the contents of the conical flask thoroughly during the titration to ensure that all the ammonia is transferred to the acid layer.

After the titration, transfer all the liquid in the second conical flask to a separating funnel, and allow the layers to separate. Carefully run out the 25 cm^3 of 1,1,1-trichloroethane and return it to the stoppered bottle, and then add another 25 cm^3 of water. Shake the bottle as before, and then remove 25 cm^3 of each layer for titration. Repeat this procedure until the titration value for one layer falls below 10 cm^3; about 4 titrations will be needed. Copy the following table into your notebook and use it to record your results.

Aqueous layer		1,1,1-trichloroethane layer		$\dfrac{[NH_3(aq)]_{eqm}}{[NH_3(CH_3CCl_3)]_{eqm}}$
Volume of 4.0M HCl/cm^3	$[NH_3(aq)]_{eqm}$ /mol dm^{-3}	Volume of 0.1M HCl/cm^3	$[NH_3(CH_3CCl_3)]_{eqm}$ /mol dm^{-3}	

Work out the equilibrium concentrations of ammonia, and put the values that you obtain in the second and fourth columns in the table. These values are obtained as follows.

Ammonia solution reacts with hydrochloric acid according to the equation

$$NH_3(aq) + HCl(aq) \longrightarrow NH_4Cl(aq)$$

one mole of ammonia, NH_3, reacting with one mole of hydrochloric acid, HCl. Suppose x cm^3 of 4.0M HCl is required to neutralize 25 cm^3 of ammonia solution. The amount of HCl required is therefore

$$\frac{x}{1000} \times 4 \text{ moles}$$

The amount of NH_3 in 25 cm^3 of ammonia solution is therefore

$$\frac{x}{1000} \times 4 \text{ moles}$$

The equilibrium concentration of NH_3, $[NH_3(aq)]_{eqm}$, is therefore

$$\frac{x}{1000} \times 4 \times \frac{1000}{25} \text{ mol dm}^{-3}$$

$$\text{that is, } \frac{4}{25} x \text{ mol dm}^{-3}$$

A similar calculation shows that if $y\,cm^3$ of 0.1M HCl is required to neutralize the ammonia in $25\,cm^3$ of 1,1,1-trichloroethane solution, then

$$[NH_3(CH_3CCl_3)]_{eqm} = \frac{0.1}{25}\,y\,mol\,dm^{-3}$$

Finally, for each pair of titrations, divide the equilibrium concentration of ammonia in water by the corresponding value for 1,1,1-trichloroethane, and write the result in the last column in the table.

Questions

1 Why are units given for the figures in the second and fourth columns, but not for those in the last column?

2 What can you say about the ratio

$$\frac{[NH_3(aq)]_{eqm}}{[NH_3(CH_3CCl_3)]_{eqm}}?$$

This ratio is often called the *distribution ratio* (or partition coefficient) for ammonia between water and 1,1,1-trichloroethane.

2 The equilibrium between hydrogen, iodine, and hydrogen iodide, all in the gas phase

The equilibrium

$$H_2(g) + I_2(g) \rightleftharpoons 2HI(g)$$

has been extensively investigated. It is established very slowly indeed at room temperature but is attained in a day or so at higher temperatures. The method of investigation is to seal mixtures of hydrogen and iodine (or pure hydrogen iodide, so that equilibrium can be established from both directions) in glass or silica vessels. The vessels are then heated at a known temperature until equilibrium is attained. (How could this be checked?) After this, the vessels are rapidly cooled to 'freeze' the equilibrium, that is, to preserve the concentrations of reactants and products as they were at high temperature by taking advantage of the fact that any changes at low temperatures will be very slow. The contents of the vessels are then analysed.

One method of doing this is to open the tubes which initially contained hydrogen and iodine, under potassium iodide solution when the iodine and hydrogen iodide dissolve. The solution is then titrated with sodium thiosulphate solution to estimate the iodine. The resulting solution (from which the iodine has now been removed) is titrated with alkali to find the amount of hydrogen iodide present. (Why is it necessary to remove the iodine first?) The hydrogen can be estimated from the fact that conversion of 1 mole of iodine molecules to hydrogen iodide is accompanied by similar conversion of 1 mole of hydrogen molecules. The tubes which contained hydrogen iodide initially, are opened similarly under potassium iodide solution and the iodine estimated by titration with sodium thiosulphate solution. If the initial mass of hydrogen iodide in the tube is known, the composition of the equilibrium mixture can be calculated. From the known volume of the sealed vessel, equilibrium concentrations of hydrogen, iodine, and hydrogen iodide can be calculated.

Concentrations at equilibrium for the system $H_2(g) + I_2(g) \rightleftharpoons 2HI(g)$

Temperature: 698 K (425 °C)

a *Results obtained by heating hydrogen and iodine in sealed vessels*

$[H_2(g)]_{eqm}$ /mol dm^{-3}	$[I_2(g)]_{eqm}$ /mol dm^{-3}	$[HI(g)]_{eqm}$ /mol dm^{-3}
4.56×10^{-3}	0.74×10^{-3}	13.54×10^{-3}
3.56×10^{-3}	1.25×10^{-3}	15.59×10^{-3}
2.25×10^{-3}	2.34×10^{-3}	16.85×10^{-3}

Table 12.1

Note. $[H_2(g)]_{eqm}$ means 'the concentration of hydrogen gas at equilibrium, in moles of hydrogen molecules H_2 per cubic decimetre'; the other symbols have corresponding meanings.
Adapted from Taylor and Crist, Journal of the American Chemical Society, *63, 1381, 1941.*

Can you find a constant numerical relationship between the equilibrium concentrations of the reactants and products? Try

$$\frac{[HI(g)]_{eqm}}{[H_2(g)]_{eqm}[I_2(g)]_{eqm}} \quad \text{and} \quad \frac{[HI(g)]_{eqm}^2}{[H_2(g)]_{eqm}[I_2(g)]_{eqm}}$$

Which gives the better constant? Check your conclusion by using the results given in table 12.2. These were obtained by approaching the equilibrium from the reverse direction, that is starting from hydrogen iodide.

b *Results obtained by heating hydrogen iodide in sealed vessels*

$[H_2(g)]_{eqm}$ /mol dm^{-3}	$[I_2(g)]_{eqm}$ /mol dm^{-3}	$[HI(g)]_{eqm}$ /mol dm^{-3}
0.48×10^{-3}	0.48×10^{-3}	3.53×10^{-3}
0.50×10^{-3}	0.50×10^{-3}	3.66×10^{-3}
1.14×10^{-3}	1.14×10^{-3}	8.41×10^{-3}

Table 12.2

Questions

1 The symbol K_c is used to represent the constant obtained. Hence

$$K_c = \frac{[HI(g)]^2_{eqm}}{[H_2(g)]_{eqm}[I_2(g)]_{eqm}}$$

when this equilibrium is represented by

$$H_2(g) + I_2(g) \rightleftharpoons 2HI(g)$$

K_c is called the *equilibrium constant*. The subscript c indicates that it is expressed in concentrations. By convention, the concentrations of the substances on the *righthand side* of the equation are always put in the *numerator* (top line of fraction) of the equilibrium constant and those of the substances on the *lefthand side* in the *denominator*.
The equilibrium can also be represented by

$$\tfrac{1}{2}H_2(g) + \tfrac{1}{2}I_2(g) \rightleftharpoons HI(g)$$

What expression must be used to calculate K_c from this equation?
Will the values be the same as those obtained from tables 12.1 and 12.2? If not, how will the two sets of values be related?

2 Find the average value of K_c from the calculations that you have done on the results in tables 12.1 and 12.2, correct to 2 significant figures. If an equilibrium mixture of hydrogen, iodine, and hydrogen iodide at 698 K contains 0.5 mole of hydrogen molecules and 5.43 moles of hydrogen iodide molecules, how much iodine is present? Why do you not need to know the volume of the system in order to obtain an answer?

3 Why are the concentrations of hydrogen and iodine the same in each of the first two columns of table 12.2?

3 The equilibrium between ethyl ethanoate, water, ethanoic acid and ethanol

What is the expression for the equilibrium constant, K_c, for the reaction

$$CH_3CO_2C_2H_5(l) + H_2O(l) \rightleftharpoons CH_3CO_2H(l) + C_2H_5OH(l)?$$

Is your answer confirmed by the experimental results in table 12.3? Why is the total volume of the equilibrium mixture not needed to calculate K_c?

The equilibrium constant can be determined by allowing mixtures of ethyl ethanoate, water, and catalyst ($2\,cm^3$ of hydrochloric acid) to stand at room temperature, and then analysing the contents when equilibrium has been reached by titrating the contents of each flask with alkali. $2\,cm^3$ of concentrated hydrochloric acid are titrated separately and the volume of alkali needed is subtracted from all the other results to allow for acid added as catalyst. The amount of ethanoic acid (d mole) in the equilibrium mixture is calculated. As the original mixture contained b mole ethyl ethanoate and c mole water, there must be d mole ethanoic acid, d mole ethanol, (b–d) mole ethyl ethanoate, and (c–d) mole water at equilibrium.

The results in table 12.3 are adapted from the work of sixth form students of Dauntsey's School, Devizes, Wiltshire. The experiment was carried out at about 293 K (20 °C) and results were obtained by allowing mixtures of ethyl ethanoate and water, of differing proportions (between $29\,cm^3$ ester + $1\,cm^3$ water and $15\,cm^3$ ester + $15\,cm^3$ water) plus a fixed volume ($2\,cm^3$) of concentrated hydrochloric acid, to stand for one week. Hydrogen ions from the hydrochloric acid act catalytically in enabling equilibrium to be attained more rapidly.

The quite remarkable effect of the catalyst in this set of experiments can be appreciated when it is known that a mixture of ethyl ethanoate and water alone would take several years to reach equilibrium at room temperature.

Amount of ethyl ethanoate at equilibrium/mole	Amount of water at equilibrium /mole	Amount of ethanoic acid at equilibrium/mole	Amount of ethanol at equilibrium /mole
0.231	0.079	0.065	0.065
0.204	0.118	0.082	0.082
0.150	0.261	0.105	0.105
0.090	0.531	0.114	0.114

Table 12.3

Taking the value of K_c as 0.27, how many moles of ethyl ethanoate molecules would you expect to obtain at equilibrium from a starting mixture of 3 moles of ethanol molecules and 2 moles of ethanoic acid molecules? What mass of ethyl ethanoate (in grams) is this?

The Equilibrium Law

From many investigations such as those described above, the following general statement emerges.

For any system in equilibrium, there is a simple relationship between the concentrations of the substances present. If the reaction for the equilibrium is represented by the equation

$$mA + nB \rightleftharpoons pC + qD$$

the expression

$$\frac{[C]^p_{eqm}[D]^q_{eqm}}{[A]^m_{eqm}[B]^n_{eqm}} = \text{a constant at a given temperature} = K_c$$

K_c is called the *equilibrium constant*.

This is known as *the Equilibrium Law*.

Two important conclusions arise from this law:

1 If K_c is large the equilibrium mixture will contain a high proportion of products; that is, the reaction has gone nearly to completion. If K_c is small, the reaction does not proceed very far at the temperature concerned and the concentration of products is low.

2 The addition of more reactants to a system in equilibrium will result in the formation of more products; the system will adjust itself so that the concentrations again satisfy the value for K_c. Similarly, on addition of more products, the equilibrium will move in the opposite direction and the concentrations of reactants will increase. Once K_c for a reaction is known, the relative proportions of reactants and products at equilibrium for any mixture of reactants used initially can be calculated.

When stating the value of K_c for a particular reaction, it is important to indicate the equation on which the constant is based. For example, in the reaction between hydrogen and bromine to form hydrogen bromide, if we write the equation in the form

$$H_2(g) + Br_2(g) \rightleftharpoons 2HBr(g)$$

$$K_c = \frac{[HBr(g)]^2_{eqm}}{[H_2(g)]_{eqm}[Br_2(g)]_{eqm}}$$

At $500\,K$, the value of this constant is about 10^{12} (reaction very nearly complete).

The same equilibrium can also be represented by

$$\tfrac{1}{2}H_2(g) + \tfrac{1}{2}Br_2(g) \rightleftharpoons HBr(g)$$

but then there is a different equilibrium constant

$$K_c' = \frac{[HBr(g)]_{eqm}}{[H_2(g)]_{eqm}^{1/2}[Br_2(g)]_{eqm}^{1/2}}$$

It will be seen that $K_c' = \sqrt{K_c}$

thus K_c' at 500 K $\approx \sqrt{10^{12}} \approx 10^6$

For reactions in which the number of particles on each side of the equation is the same, as in the example above, the concentration units cancel, and K_c has no units. For all other reactions this is not the case and units for K_c must be stated. Thus in the equilibrium

$$2SO_2(g) + O_2(g) \rightleftharpoons 2SO_3(g)$$

which is the basis of the contact process for manufacturing sulphuric acid

$$K_c = \frac{[SO_3(g)]_{eqm}^2}{[SO_2(g)]_{eqm}^2[O_2(g)]_{eqm}}$$

If a particular equilibrium mixture contains $x\,\text{mol dm}^{-3}$ SO_3, $y\,\text{mol dm}^{-3}$ SO_2, and $z\,\text{mol dm}^{-3}$ O_2

$$K_c = \frac{(x\,\text{mol dm}^{-3})^2}{(y\,\text{mol dm}^{-3})^2(z\,\text{mol dm}^{-3})}$$

$$= \frac{x^2}{y^2z\,\text{mol dm}^{-3}} = \frac{x^2}{y^2z}\,\text{dm}^3\,\text{mol}^{-1}$$

One final point: it is often convenient to assume that, in principle, all reactions can proceed to an equilibrium state, even if for practical purposes many of them appear either to go completely to products, or not to start at all. The usefulness of this assumption will become evident later in this Topic, and in Topic 15.

12.2
THE EFFECT OF PRESSURE AND TEMPERATURE ON EQUILIBRIUM

The equilibrium constant expressed in terms of partial pressures

When dealing with reactions involving gases it is often found more convenient to use an equilibrium constant expressed in terms of pressure (K_p) rather than in terms of concentration (K_c).

The possibility of using a pressure relationship for an equilibrium constant when gases are involved in a reaction arises from the fact that the total pressure of a gas mixture is the sum of separate pressures (the *partial pressures*) exerted by each of the gases in the mixture. The partial pressure of each gas in the mixture is the pressure that it would exert if it alone occupied the volume occupied by the mixture. This was first pointed out by John Dalton, in 1801, and is known as the law of partial pressures. Strictly speaking, it applies to 'ideal' gases only, but under conditions of not too high pressures it is a sufficiently accurate statement to serve for most purposes.

From the general gas equation (discussed in Topic 3),

$$pV = nLkT \qquad (n = \text{moles of gas})$$

$$\therefore \quad \frac{p}{LkT} = \frac{n}{V}$$

but $\dfrac{n}{V}$ = concentration of gas = [gas]

$$\therefore \quad p = LkT[\text{gas}]$$

This means that, for a fixed temperature, the concentration of a gas is proportional to its pressure.

$$[\text{gas}] \propto p$$

Also, since $\quad p = \dfrac{n}{V}LkT$

the partial pressure of each gas in a mixture of gases is proportional to the moles of each gas in the mixture. For example, if a mixture of ammonia, hydrogen, and nitrogen contains 1 mole of ammonia molecules, 3.6 moles of hydrogen molecules and 13.5 moles of nitrogen molecules (total 18.1 moles), and the total pressure is 2 atmospheres partial pressure of ammonia, $p_{NH_3} = \dfrac{1}{18.1} \times 2 = 0.11 \, \text{atm}$

partial pressure of hydrogen, $p_{H_2} = \dfrac{3.6}{18.1} \times 2 = 0.40 \, \text{atm}$

partial pressure of nitrogen, $p_{N_2} = \dfrac{13.5}{18.1} \times 2 = 1.49 \, \text{atm}$

Since $[\text{gas}] \propto p$ (the partial pressure of a gas), if we represent an equilibrium situation involving gases only as

$$aX(g) + bY(g) \rightleftharpoons cZ(g)$$

we can write

$$\frac{p_{Zeqm}^c}{p_{Xeqm}^a p_{Yeqm}^b} = K_p$$

Thus, if the proportions of ammonia, hydrogen, and nitrogen given above are in equilibrium at a temperature T kelvin, from the equation

$$N_2(g) + 3H_2(g) \rightleftharpoons 2NH_3(g)$$

$$K_p = \frac{p_{NH_3eqm}^2}{p_{N_2eqm} p_{H_2eqm}^3}$$

$$\therefore \quad \text{at } T\,\text{K}, K_p = \frac{(0.11 \, \text{atm})^2}{(1.49 \, \text{atm})(0.40 \, \text{atm})^3}$$

$$= \frac{1.3 \times 10^{-1}}{\text{atm}^2}$$

$$= 1.3 \times 10^{-1} \, \text{atm}^{-2}$$

We can, of course, also use the expression

$$\frac{[Z(g)]_{eqm}^c}{[X(g)]_{eqm}^a [Y(g)]_{eqm}^b} = K_c$$

if this is more convenient. However, the values obtained for the equilibrium constant will not necessarily be the same, and the units will always be different except, of course, for a gas reaction in which the number of gas molecules is the same on each side of the stoicheiometric equation; in this case neither K_c nor K_p has units. K_c and K_p are, however, related to each other by the expression

$$K_p = K_c(LkT)^n$$

where $n =$ number of gas molecules on the righthand side of the equilibrium equation minus the number of gas molecules on the lefthand side of the equilibrium equation.

The effect of pressure on equilibrium

For an equilibrium involving gases, if the partial pressure of one of the gases is changed, the system will move in such a way that the partial pressures of reactants and products again give the same value for K_p. If the *total* pressure of the system is increased, the various partial pressures also increase.

Consider the equilibrium

$$2SO_2(g) + O_2(g) \rightleftharpoons 2SO_3(g)$$

for which

$$K_p = \frac{p^2_{SO_3 eqm}}{p^2_{SO_2 eqm} p_{O_2 eqm}}$$

If at a given temperature,

$$p_{SO_3 eqm} = a \, atm$$

$$p_{SO_2 eqm} = b \, atm$$

$$p_{O_2 eqm} = c \, atm$$

then $\quad K_p = \dfrac{a^2}{b^2 c} \, atm^{-1}$

Now let the pressure be doubled so that the three partial pressures are doubled also. This will mean that momentarily,

$$p_{SO_3} \text{ becomes } 2a \, atm$$

$$p_{SO_2} \text{ becomes } 2b \, atm$$

$$p_{O_2} \text{ becomes } 2c \, atm$$

so that

$$\frac{p_{SO_3}^2}{p_{SO_2}^2 p_{O_2}} = \frac{(2a)^2}{(2b)^2 2c} = \frac{a^2}{b^2 c} \times \frac{1}{2}\,\text{atm}^{-1}$$

The system then adjusts itself so that the partial pressures again satisfy the value for K_p, that is more SO_2 and O_2 must react to form SO_3.

This change results in a decrease in volume, with 3 moles of gas molecules changing to 2 moles.

In general, *increasing the pressure on a system in equilibrium produces a change which tends to a decrease in volume.* Conversely, reducing the pressure favours the change which results in an increase in volume. Thus, for the sulphur trioxide equilibrium

$$\xrightarrow{\text{pressure increase}}$$

$$2SO_2(g) + O_2(g) \;\rightleftharpoons\; 2SO_3(g)$$

$$\xleftarrow{\text{pressure decrease}}$$

Questions

1 What would be the result of an increase of pressure on the following equilibria?

$$N_2O_4(g) \rightleftharpoons 2NO_2(g)$$

$$CO(g) + 2H_2(g) \rightleftharpoons CH_3OH(g)$$

$$H_2(g) + Br_2(g) \rightleftharpoons 2HBr(g)$$

$$CO_2(g) + NO(g) \rightleftharpoons CO(g) + NO_2(g)$$

2 How would you alter the pressure in order to increase the yield of ethene from the equilibrium given below?

$$C_2H_6(g) \rightleftharpoons C_2H_4(g) + H_2(g)$$

The effect of temperature change on the value of the equilibrium constant for a reaction

Changes in K_p value with temperature for an equilibrium state

Table 12.4 shows some values of K_p for a number of reactions at different temperatures, together with the enthalpy changes involved when the reactions go to completion.

The unit of these enthalpy changes in each case is $kJ\,mol^{-1}$. You will remember that when moles are mentioned, the particles referred to must always be specified; for example, 'a mole of hydrogen atoms' or '2 moles of oxygen molecules'. In these cases we are referring to a mole of all the particles that appear in the equation. For example, in the reaction

$$N_2(g) + 3H_2(g) \rightleftharpoons 2NH_3(g)$$

the mole referred to consists of one mole of nitrogen molecules and 3 moles of hydrogen molecules reacting to give 2 moles of ammonia molecules.

$N_2(g) + 3H_2(g) \rightleftharpoons 2NH_3(g)$
$\Delta H^{\ominus}_{298} = -92\,kJ\,mol^{-1}$

$N_2(g) + O_2(g) \rightleftharpoons 2NO(g)$
$\Delta H^{\ominus}_{298} = +180\,kJ\,mol^{-1}$

T/K	K_p/atm^{-2}
500	3.55×10^{-2}
700	7.76×10^{-5}
1100	5×10^{-8}

T/K	K_p
700	5×10^{-13}
1100	4×10^{-8}
1500	1×10^{-5}

$2SO_2(g) + O_2(g) \rightleftharpoons 2SO_3(g)$
$\Delta H^{\ominus}_{298} = -197\,kJ\,mol^{-1}$

$H_2(g) + I_2(g) \rightleftharpoons 2HI(g)$
$\Delta H^{\ominus}_{298} = -9.6\,kJ\,mol^{-1}$

T/K	K_p/atm^{-1}
500	2.5×10^{10}
700	3×10^{4}
1100	1.3×10^{-1}

T/K	K_p
500	160
700	54
1100	25

Table 12.4
Variation of K_p values with temperature, and values of $\Delta H^{\ominus}_{298}$, for a series of reactions

The principle of Le Châtelier

From the information given in table 12.4 and from investigations that you may have done in class, it may be concluded that:

An *increase in temperature* of an equilibrium system results in an *increase in the value of the equilibrium constant* if the reaction involved is *endothermic* (ΔH positive), and a *decrease in the value of the equilibrium constant* if the reaction is *exothermic* (ΔH negative).

An endothermic reaction becomes exothermic when carried out in the reverse direction, for example

$$2SO_2(g) + O_2(g) \longrightarrow 2SO_3(g); \qquad \Delta H^{\ominus}_{298} = -197\,kJ\,mol^{-1}$$

$$2SO_3(g) \longrightarrow 2SO_2(g) + O_2(g); \qquad \Delta H^{\ominus}_{298} = +197\,kJ\,mol^{-1}$$

To increase the temperature of a system in equilibrium, thermal energy must be added. The effect of this addition is to cause the equilibrium to change so that thermal energy is absorbed, that is, in the direction of the endothermic reaction.

The effect of change in conditions on a system in equilibrium can be predicted qualitatively by a general statement put forward by H. L. Le Châtelier in 1888. This applies both to changes of *concentration* or *pressure* in which the system is adjusting itself so that the concentrations or partial pressures again satisfy the expressions for K_c or K_p; or to changes of *temperature* in which the system is adjusting itself to a new value of K_c or K_p.

Le Châtelier's principle of equilibrium
'If a system in equilibrium is subjected to a change, the processes which take place are such as to tend to counteract the effect of the change.'

For chemical systems the changes most frequently involved are those of concentration, pressure, and temperature. The results of such changes on systems in equilibrium are summarized below.

Concentration changes

If the concentration of one of the reactants or products in a system in equilibrium is increased, the effect may be counteracted by establishing a new equilibrium, so that the concentration of the added substance is reduced.

Example If more gaseous iodine is added to the equilibrium

$$H_2(g) + I_2(g) \rightleftharpoons 2HI(g)$$

it causes more hydrogen iodide to be formed, thus reducing the effect of the change. Addition of hydrogen iodide would cause more hydrogen and iodine to be formed. (If the value of the equilibrium constant for the reaction is known, the actual concentration changes can be calculated.)

Pressure changes

If the pressure of the system is increased, the effect of this change may be counteracted by the establishment of a new equilibrium so that the volume is decreased. Conversely, decrease of pressure will cause a new equilibrium to be established so that the volume is increased. Obviously, this is only applicable if there is a change in volume in the reaction.

Example If the pressure of the equilibrium system

$$N_2(g) + 3H_2(g) \rightleftharpoons 2NH_3(g)$$

is increased, its effect is to cause more ammonia to be formed, thus reducing the effect of the change by decreasing the volume of the system and hence the pressure. Decrease of pressure would cause more nitrogen and hydrogen to be formed.

Temperature changes

Changing the temperature of a system in equilibrium results in the establishment of a new equilibrium constant. (This does not happen for changes in concentration or pressure.) Qualitatively, increase in temperature favours the change which takes place with absorption of thermal energy. This may be represented as follows:

temperature increase
$$\xrightarrow{\hspace{3cm}}$$

$$A + B \quad \rightleftharpoons \quad C + D; \qquad \Delta H = +z\,\text{kJ}\,\text{mol}^{-1}$$

$$\xleftarrow{\hspace{3cm}}$$
temperature decrease

Example For the reaction

$$H_2O(g) + C(\text{graphite}) \longrightarrow CO(g) + H_2(g); \qquad \Delta H^{\ominus}_{298} = +130\,\text{kJ}\,\text{mol}^{-1}$$

high temperatures therefore favour the production of carbon monoxide and hydrogen, while at low temperatures the equilibrium lies almost completely on the lefthand side of the equation.

Review of changes

If the temperature, pressure, or concentration of one of the substances present in an equilibrium state is altered, there is a shift of equilibrium which can result in a change in the relative concentrations of the substances involved. Changes in pressure and concentration do not affect the value of the equilibrium constant for the reaction. A change in temperature always results in a new equilibrium being established, for which the equilibrium constant is different from that for the initial conditions.

12.3
HETEROGENEOUS EQUILIBRIA

Most of the equilibrium systems discussed so far have involved one *phase* only. (A phase is any part of a system which is of the same composition throughout – it is *homogeneous* – and is separated from the rest of the system by a distinct boundary.) These are called homogeneous equilibria, and they include equilibria in the gas phase and those in solution.

Equilibria which involve two or more phases are called *heterogeneous*.

A common example of a heterogeneous equilibrium is a solid salt in contact with its ions in solution. Figure 12.1 shows such an example.

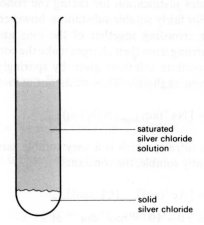

saturated
silver chloride
solution

solid
silver chloride

Figure 12.1

In the figure, a saturated solution of the sparingly soluble salt, silver chloride, appears in contact with solid silver chloride. No more solid can dissolve, but a dynamic equilibrium exists between the silver and chloride ions in the solution and those in the solid

$$AgCl(s) \rightleftharpoons Ag^+(aq) + Cl^-(aq)$$

$$K_p = p_{Br_2 eqm} \qquad (= 0.288 \text{ atm at } 293 \text{ K})$$

Reactions between metals and metal ions are examples of ionic, heterogeneous equilibria, for example

$$Cu(s) + 2Ag^+(aq) \rightleftharpoons Cu^{2+}(aq) + 2Ag(s)$$

For this

$$K_c = \frac{[Cu^{2+}(aq)]_{eqm}}{[Ag^+(aq)]^2_{eqm}}$$

$$= 3 \times 10^{15} \text{ dm}^3 \text{ mol}^{-1} \text{ at } 298 \text{ K}$$

Equilibria of this type are studied in Topic 15.

12.4
ACID-BASE EQUILIBRIA

In this book we shall be giving special consideration to two extremely important types of equilibria. They are acid-base equilibria, to be dealt with in this section, and redox equilibria to be tackled in Topic 15.

You will already have some knowledge of the behaviour of acids and bases from the work that you did before you started on this course, and from earlier Topics. Here we shall be using the Lowry–Brønsted theory of acid-base behaviour. In this theory an *acid* is defined as a substance which can *provide protons* under reaction conditions; that is, an acid is a *proton donor*. A *base* is a substance which can *combine with protons*; that is, it is a *proton acceptor*.

The equilibrium constant for the ionization of water

In practical laboratory situations we generally use acids in aqueous solutions, and we must therefore take into account some of the properties of water.

Water is able to function both as a base, accepting protons

$$H^+ + H_2O \longrightarrow H_3O^+$$

and as an acid, giving protons

$$H_2O \longrightarrow H^+ + OH^-$$

Consequently, an equilibrium exists in water, with some of the molecules acting as an acid and some as a base

$$2H_2O(l) \rightleftharpoons H_3O^+(aq) + OH^-(aq)$$

This equilibrium is very often written as

$$H_2O(l) \rightleftharpoons H^+(aq) + OH^-(aq)$$

and we say that the water is *ionized*. The equilibrium constant for this ionization, K_c, is given by

$$K_c = \frac{[H^+(aq)]_{eqm}[OH^-(aq)]_{eqm}}{[H_2O(l)]_{eqm}}$$

The proportion of water molecules that ionize in this way is so very small that, at a given temperature, the concentration of water molecules, $[H_2O(l)]_{eqm}$, can be treated as constant. Multiplying both sides of the equation by this constant, we have

$$K_c \times [H_2O(l)]_{eqm} = [H^+(aq)]_{eqm}[OH^-(aq)]_{eqm}$$

Since the product of two constants is a constant itself, the lefthand side of this equation is a constant, known as the ionization constant for water, and given the symbol K_w. At 298 K, $K_w = 10^{-14}\,mol^2\,dm^{-6}$. In pure water, and in any absolutely neutral solution,

$$[H^+(aq)]_{eqm} = [OH^-(aq)]_{eqm}$$

and so the value of each is $\sqrt{10^{-14}} = 10^{-7}$.

Acid-base systems and their chemistry

The chemistry of acid-base systems is concerned with equilibria between electrovalently-bonded species and covalently-bonded species. Equal sharing of electrons (a true covalent bond) occurs only between like atoms, as in H_2, Cl_2, etc. Bonding between unlike atoms always results in unequal sharing and polar bonds, *e.g.* $\overset{\delta+}{H}—\overset{\delta-}{Cl}$.

When sharing is carried to its limit, transfer of an electron pair takes place and an electrovalent bond results. The degree of electron sharing in polar molecules is changed when they are dissolved in polar solvents. This often results in the formation of an electrovalent bond. It happens when hydrogen chloride reacts with water

$$HCl(g) + H_2O(l) \longrightarrow H_3O^+ + Cl^-(aq)$$

a process which can be resolved into two stages.

$$HCl(g) \longrightarrow H^+ + Cl^-$$
$$H^+ + Cl^- + H_2O(l) \longrightarrow H_3O^+ + Cl^-(aq)$$

The hydrogen ion (H^+) is a single proton, with no electrons associated with it to produce a comparatively large volume. Thus it is some 50 000 times smaller than the next smallest atom. The possibility of very close approach between the free proton and the oxygen atom in the water molecule results in a strong bond being formed by the lone pair of electrons on the oxygen atom. Many other substances react with water in this way.

In acid-base systems we have a competition for protons. Thus the equilibria involved in the reaction of hydrogen chloride with water are:

$$HCl(g) \rightleftharpoons Cl^-(aq) + H^+$$
$$\quad \text{acid}_1 \qquad \text{base}_1$$

and

$$H_2O(l) + H^+ \rightleftharpoons H_3O^+$$
$$\text{base}_2 \qquad\qquad \text{acid}_2$$

These can be combined to give

$$HCl(g) + H_2O(l) \rightleftharpoons Cl^-(aq) + H_3O^+$$
$$\text{acid}_1 \qquad \text{base}_2 \qquad \text{base}_1 \qquad \text{acid}_2$$

The position of equilibrium is determined by the relative tendencies of the acids and bases to lose and gain protons. In this particular equilibrium the concentrations of $Cl^-(aq)$ and H_3O^+ greatly exceed those of $HCl(g)$. This means that hydrogen chloride loses protons much more readily than H_3O^+, and that $H_2O(l)$ accepts protons more readily than $Cl^-(aq)$.

Towards soluble bases water can act as an acid. In aqueous ammonia, for example, the acid-base equilibrium is

$$NH_3(aq) + H_2O(l) \rightleftharpoons NH_4^+(aq) + OH^-(aq)$$
$$\text{base}_1 \qquad \text{acid}_2 \qquad\quad \text{acid}_1 \qquad\quad \text{base}_2$$

In this solution there is a higher concentration of $NH_3(aq)$ than of $NH_4^+(aq)$. This indicates that $NH_4^+(aq)$ loses protons more readily than $H_2O(l)$, and that $OH^-(aq)$ accepts protons more readily than $NH_3(aq)$. The competing equilibria are

$$NH_4^+(aq) \rightleftharpoons NH_3(aq) + H^+$$

and

$$H_2O(l) \rightleftharpoons OH^-(aq) + H^+$$

If an aqueous solution of hydrogen chloride is added to an aqueous solution of ammonia, the acid H_3O^+ reacts with the base $OH^-(aq)$ to set up the equilibrium

$$H_3O^+ + OH^-(aq) \rightleftharpoons 2H_2O(l)$$

This is always well over to the righthand and leads to removal of H_3O^+ and $OH^-(aq)$ ions from solution, thus altering the equilibrium position of

$$HCl(g) + H_2O(l) \rightleftharpoons H_3O^+ + Cl^-(aq)$$

and

$$NH_3(aq) + H_2O(l) \rightleftharpoons NH_4^+(aq) + OH^-(aq)$$

towards the righthand side also. The final products are mainly ammonium ions and chloride ions in aqueous solution.

Similar equilibria are set up in other solvents. In liquid ammonia, for example, there is the equilibrium

$$2NH_3(l) \rightleftharpoons NH_4^+ + NH_2^-$$

corresponding to $2H_2O(l) \rightleftharpoons H_3O^+ + OH^-(aq)$ in water. Reactions in liquid ammonia are similar to those in water. For example, the compound sodamide, $NaNH_2$, ionizes

$$NaNH_2 \longrightarrow Na^+ + NH_2^-$$

and ammonium chloride also ionizes

$$NH_4Cl \longrightarrow NH_4^+ + Cl^-$$

A 'neutralization' reaction then takes place

$$NH_4^+ + NH_2^- \longrightarrow 2NH_3(l).$$

In this section we shall deal with aqueous systems only, but the wider application of the principles involved should not be forgotten.

The strengths of acids and bases

If a substance (represented by HA) which can function as an acid is dissolved in water, two competing equilibria are established:

$$HA(aq) \rightleftharpoons A^-(aq) + H^+$$

and

$$H^+ + H_2O(l) \rightleftharpoons H_3O^+.$$

If HA loses protons readily, a high concentration of H_3O^+ ions will be produced when the system has reached equilibrium. In this case HA is functioning as a strong acid. On the other hand, if HA is a weak acid, with no pronounced tendency to part with protons, the concentration of H_3O^+ ions will be smaller. In effect, the H_3O^+ ion can be used as a standard against which to compare the relative strengths of acids. The problem now becomes one of measuring $[H_3O^+]_{eqm}$ in aqueous solutions of different acids.

(There is no general agreement about the name to be used for the H_3O^+ ion; it has been called the hydronium ion, the hydroxonium ion, and the oxonium ion. Unless special attention needs to be directed towards the oxonium ion as a combination of a hydrogen ion and a water molecule the usual practice is to use the symbol $H^+(aq)$ and to call this 'the hydrogen ion'.)

One way of comparing hydrogen ion concentrations is to measure pH values, as has been done earlier in the course by using indicators. The indicator method gives rough values only; we now need a more accurate method.

The relationship between the pH value of a solution and the hydrogen ion concentration is

$$pH = -lg[H^+(aq)]$$

where $[H^+(aq)]$ is measured in $mol\,dm^{-3}$. A logarithmic scale is used because the range of possible hydrogen ion concentrations in solution is very large (from about 10 to 10^{-15}). The minus sign is introduced to make pH values positive in almost all cases encountered in practice. A few examples will help to make the relationship clearer.

When

$$[H^+(aq)] = 10^{-3} \, mol \, dm^{-3}; \, pH = 3 \, (-lg[H^+(aq)] = +3)$$
$$[H^+(aq)] = 10^{-8} \, mol \, dm^{-3}; \, pH = 8$$
$$[H^+(aq)] = 5 \times 10^{-6} \, mol \, dm^{-3}; \, pH = 5.3$$

The last example may surprise you; remember that the mantissa of a logarithm (the figures after the decimal point) is *always* positive, whereas the characteristic (before the decimal point) can be positive or negative.

$$lg \, 5 \times 10^{-6} = 6.7 = -6 + 0.7 = -5.3, \, therefore \, pH = 5.3.$$

Instruments known as pH meters can be used to determine the pH of a solution with greater accuracy than can be achieved by using indicators; a glass electrode is immersed in the solution whose pH is to be determined and the value is read off a meter. The way in which these instruments work will become apparent when you have covered the theory explained in Topic 15.

Note that the pH scale is a logarithmic scale using logarithms *to base 10*, symbol lg. In other topics the logarithms that are used are natural logarithms (logarithms to base e, e = 2.718...), symbol ln.

pH values in aqueous solution

As mentioned earlier, in pure water, at 298 K, $[H^+(aq)]_{eqm} = 10^{-7} \, mol \, dm^{-3}$ and so

$$pH = -lg \, 10^{-7} = 7$$

A solution of pH < 7 is acidic; $[H^+(aq)]_{eqm} > [OH^-(aq)]_{eqm}$

A solution of pH > 7 is basic (or alkaline); $[H^+(aq)]_{eqm} < [OH^-(aq)]_{eqm}$. For example, in 0.1M HCl,

$$[H^+(aq)]_{eqm} = 10^{-1} \, mol \, dm^{-3}$$
$$\therefore pH = 1, \, and$$
$$[OH^-(aq)]_{eqm} = 10^{-13} \, mol \, dm^{-3}$$

In 0.01M NaOH (assume it has completely dissociated into ions)

$$[OH^-(aq)]_{eqm} = 10^{-2} \, mol \, dm^{-3}$$
$$\therefore [H^+(aq)]_{eqm} = 10^{-12} \, mol \, dm^{-3}$$
$$\therefore pH = 12$$

Theoretically, assuming there has been complete ionization, pure sulphuric acid (which is 18M) should have a pH of about -1.5. In practice, however, most of the acid at this concentration is in the form of covalent molecules, ionization is low, and the pH value is correspondingly high. The range of pH values actually encountered in aqueous systems varies between just less than zero (negative value) and a little over 14 (concentrated alkaline solutions).

Change of pH with dilution and its relationship to the strength of an acid

If we use a pH meter to measure pH values for progressively diluted hydrochloric acid, we find that there is an increase of about one pH unit per tenfold dilution. For example:

pH of   M HCl(aq) is approximately 0
pH of   0.1M HCl(aq) is approximately 1
pH of 0.01M HCl(aq) is approximately 2
pH of 0.001M HCl(aq) is approximately 3

These observations can be accounted for if we assume that the acid is almost completely ionized at all concentrations, so that the equilibrium

$$HCl(aq) + H_2O(l) \rightleftharpoons H_3O^+ + Cl^-(aq)$$

lies almost completely over to the right. A tenfold dilution will then reduce the value of $[H^+(aq)]_{eqm}$ by 10 and the pH will increase by $\lg 10 = 1$. A few other acid solutions behave in this way. For most acidic solutions, however, the increase in pH for a dilution factor of ten is less than one unit, and the pH values for comparable concentrations are always greater than for hydrochloric acid solutions. This could mean that dissociation into ions is incomplete and a considerable proportion of reactants remains when the equilibrium

$$HA(aq) + H_2O(l) \rightleftharpoons H_3O^+ + A^-(aq)$$

is attained. (HA is used to stand for any acid which behaves in this way.) Thus the hydrogen ion concentration will be smaller than would be expected for complete dissociation and the pH value higher. Further dilution (increase of $H_2O(l)$ concentration) would then move the equilibrium more towards the right until, at high dilution, conversion to products (ions) is virtually complete.

Acids which ionize nearly completely at moderate dilutions (0.1M or 0.01M) are called *strong acids*. Those which ionize slightly, or exist mainly as the covalently bonded form under these conditions, are called *weak acids*. As with electrovalent and covalent bonds, there is no sharp dividing line between strong and weak acids but rather a kind of spectrum of acidic properties.

The arguments outlined above can be pursued quantitatively. We shall consider a relatively weak acid and write the general equilibrium equation as

$$HA(aq) \rightleftharpoons H^+(aq) + A^-(aq)$$

so that we can neglect the $[H_2O(l)]_{eqm}$ term.

For this equilibrium

$$K_c = \frac{[H^+(aq)]_{eqm}[A^-(aq)]_{eqm}}{[HA(aq)]_{eqm}}$$

When dealing with acid-base equilibria the symbol K_a is often used instead of K_c.

$$\frac{[H^+(aq)]_{eqm}[A^-(aq)]_{eqm}}{[HA(aq)]_{eqm}} = K_a \text{ (the \textit{dissociation constant} of the acid)}$$

Using this expression we can calculate the value of K_a for an acid solution if we know the concentration of the solution and its pH. A specific example will show the method.

The pH of a 0.01M solution of methanoic acid (HCO_2H) is 2.90 at 25 °C. Calculate the value of K_a at this concentration and temperature.

$$HCO_2H(aq) \rightleftharpoons HCO_2^-(aq) + H^+(aq)$$

$$K_a = \frac{[HCO_2^-(aq)]_{eqm}[H^+(aq)]_{eqm}}{[HCO_2H(aq)]_{eqm}}$$

Neglecting the hydrogen ions which arise from ionization of the water, since the concentration of these will be very small compared with the concentration of those from the acid, we can say that

$$[H^+(aq)]_{eqm} = [HCO_2^-(aq)]_{eqm}$$

and $[HCO_2H(aq)]_{eqm} = 0.01 - [H^+(aq)] \text{ mol dm}^{-3}$

$$pH = -lg[H^+(aq)]_{eqm} = 2.90$$

$$\therefore lg[H^+(aq)]_{eqm} = -2.90$$

and $[H^+(aq)]_{eqm} = \text{antilg}(-2.90) = \text{antilg}\,\overline{3}.10$
$$= 1.26 \times 10^{-3}\,\text{mol dm}^{-3}$$
$$= [HCO_2^-(aq)]_{eqm}$$

and $[HCO_2H(aq)]_{eqm} = 0.01 - 1.26 \times 10^{-3}$
$$= 10 \times 10^{-3} - 1.26 \times 10^{-3}$$
$$= 8.74 \times 10^{-3}\,\text{mol dm}^{-3}$$

$$K_a = \frac{(1.26 \times 10^{-3})^2}{8.74 \times 10^{-3}} = 1.82 \times 10^{-4}\,\text{mol dm}^{-3}$$

If we measure the pH of a weak acid at different concentrations, calculate values of K_a from these, and find K_a to be fairly constant, we would see that our assumptions about the increasing dissociation of a weak acid on dilution are not unreasonable.

By reversing the calculation, the pH of a solution of a weak acid for any molarity can be found if the value of K_a for the acid is known. Can you do this for the following example?

Calculate the pH of a 0.001M solution of aminoethanoic acid (glycine), $NH_2CH_2CO_2H$ ($K_a = 1.7 \times 10^{-10}\,\text{mol dm}^{-3}$).

Change of pH during acid-base titrations

When a base is added to an acid the reaction

$$H^+(aq) + OH^-(aq) \rightleftharpoons H_2O(l)$$

takes place. As we have already seen, this equilibrium lies far to the right. As a base is added to an acid the hydrogen ion concentration grows progressively less, that is, the pH value of the resulting solution grows progressively greater. In the following experiment you will investigate, using a pH meter, how the pH changes on addition of a base.

There are four different combinations of acid and base possible, namely

strong acid and strong base
strong acid and weak base
weak acid and strong base
weak acid and weak base

You will probably be asked to investigate one of these, and other members of your class will investigate the others.

EXPERIMENT 12.4
An investigation of the change of pH during acid-base titration

Procedure

Using a pipette and pipette filler, or a burette, put $25.0\,cm^3$ of the 1.00M acid that you are using, in a 100-cm^3 beaker. If a magnetic stirrer is available, stand the beaker on it, and place the stirrer bar in the beaker.

Fill a burette with 1.00M alkali, and clamp it so that the alkali can be run into the acid in the beaker.

Connect the electrode to the pH meter, and put it in the acid in the beaker. Clamp it gently in position; if you are using a magnetic stirrer, make sure that the electrode is in such a position that it cannot be struck by the stirrer bar when the stirrer is switched on. Note the pH of the acid solution.

Next, switch on the magnetic stirrer and run the alkali into the acid from the burette, $1.0\,cm^3$ at a time, noting the pH after each addition. Record your results in a table in your notebook as shown below. You should add a total of $35\,cm^3$ of alkali in this way.

Volume of alkali added/cm^3	0	1	2	3	etc.
pH of solution					

If you do not have a magnetic stirrer, then stir the mixture gently with the electrode before recording each pH.

Is there a time during the course of the addition when very little alkali seems to have a great effect on the pH? If so, repeat the experiment but this time add the alkali a drop at a time over this critical stage.

Plot your results on a graph, putting 'pH' on the vertical axis, and 'Volume of 1.00M alkali added to $25\,cm^3$ of 1.00M acid' on the horizontal axis. Collect the results from the other members of your class who have investigated other combinations of acid and base, plotting them all on the same axes, and stick the graph in your notebook.

Are the shapes of the graphs as you expected?

You will be discussing how to interpret the results in class. When you have done this, write a summary of the interpretation in your notebook.

12.5
BUFFER SOLUTIONS AND INDICATORS
Buffer solutions

Ethanoic acid is a weak acid, being only slightly ionized in solution $(K_a = 1.7 \times 10^{-5} \text{mol dm}^{-3})$. In an aqueous solution of ethanoic acid the equilibrium

$$CH_3CO_2H(aq) \rightleftharpoons CH_3CO_2^-(aq) + H^+(aq) \qquad (1)$$
$$\text{acid} \qquad \quad \text{corresponding} $$
$$\text{base}$$

lies well over to the left, and the hydrogen ion concentration is relatively small. There is a second equilibrium involving hydrogen ions in this system

$$H^+(aq) + OH^-(aq) \rightleftharpoons H_2O(l) \qquad (2)$$

What happens when extra ethanoate ions are added to the solution? They can be introduced by adding a soluble salt of ethanoic acid, such as sodium ethanoate, which is highly ionized in solution. Using Le Châtelier's principle, equilibrium (1) moves to the left and the hydrogen ion concentration decreases. This we should expect anyway since we are adding a base (the ethanoate ion) to an acid. The mixture now contains a relatively high concentration of un-ionized ethanoic acid and a relatively high concentration of ethanoate ion. It therefore contains both an acid and a base.

If more hydrogen ions are added to this system, by adding a small volume of a solution of a strong acid, these will combine with ethanoate ions to form more un-ionized ethanoic acid. Equilibrium (1) moves to the left, removing nearly all the added hydrogen ions. The concentration of hydrogen ions, and thus the pH of the solution, will alter a little, but not very much.

Adding a strong base, for example sodium hydroxide, to the system disturbs equilibrium (2) so that $OH^-(aq)$ ions combine with $H^+(aq)$ ions to form $H_2O(l)$. This reduces the hydrogen ion concentration, and more CH_3CO_2H molecules ionize to restore it to near its original value. The two equilibria adjust themselves in this way until nearly all the added hydroxide ions are removed. The pH value of the system will rise a little in consequence, but not very much. The changes in pH resulting from additions of acid or base are much smaller than they would be if the mixture of weak acid and salt were not present.

Solutions of this kind, containing a weak acid and its corresponding base (sometimes called the conjugate base), thus provide a 'buffer' against the effects of adding strong acid or strong base. They are therefore known as *buffer solutions*. Essentially they are solutions possessing readily available reserve supplies of both an acid and a base.

Another example of a buffer solution contains a mixture of ammonium chloride (highly ionized) and ammonia (present mainly as NH_3 molecules). The equilibria present are

$$NH_4^+(aq) \rightleftharpoons NH_3(aq) + H^+(aq)$$

(acid)   (corresponding
base)

and $H^+(aq) + OH^-(aq) \rightleftharpoons H_2O(l)$

The addition of more $H^+(aq)$ ions results in their reacting with the base $NH_3(aq)$ to form $NH_4^+(aq)$ ions. When more $OH^-(aq)$ ions are added the following changes occur

$$H^+(aq) + OH^-(aq) \longrightarrow H_2O(l)$$
$$NH_4^+(aq) \longrightarrow NH_3(aq) + H^+(aq)$$

until the two equilibria are again restored. Again, the pH value remains nearly constant.

Calculations involving buffer solutions

If the relative concentrations of acid and base in a buffer solution are known, the pH of the mixture can be calculated. Alternatively, the composition of the mixture needed to make a buffer solution of a given pH value can be found.

An equation that makes it possible to do these calculations can be derived quite easily. We will consider the equilibrium

$$CH_3CO_2H(aq) \rightleftharpoons CH_3CO_2^-(aq) + H^+(aq)$$

for which

$$\frac{[CH_3CO_2^-(aq)]_{eqm}[H^+(aq)]_{eqm}}{[CH_3CO_2H(aq)]_{eqm}} = K_a$$

re-arranging this, we have

$$[H^+(aq)]_{eqm} = K_a\left(\frac{[CH_3CO_2H(aq)]_{eqm}}{[CH_3CO_2^-(aq)]_{eqm}}\right)$$

taking logarithms of both sides

$$\lg[H^+(aq)]_{eqm} = \lg K_a + \lg\left(\frac{[CH_3CO_2H(aq)]_{eqm}}{[CH_3CO_2^-(aq)]_{eqm}}\right)$$

change the signs all through

$$-\lg[H^+(aq)]_{eqm} = -\lg K_a - \lg\left(\frac{[CH_3CO_2H(aq)]_{eqm}}{[CH_3CO_2^-(aq)]_{eqm}}\right)$$

now

$$-\lg[H^+(aq)]_{eqm} = pH$$

$$\therefore pH = -\lg K_a - \lg\left(\frac{[CH_3CO_2H(aq)]_{eqm}}{[CH_3CO_2^-(aq)]_{eqm}}\right)$$

A similar argument applied to the equilibrium

$$NH_4^+(aq) \rightleftharpoons NH_3(aq) + H^+(aq)$$

leads to the equation

$$pH = -\lg K_a - \lg\left(\frac{[NH_4^+(aq)]_{eqm}}{[NH_3(aq)]_{eqm}}\right)$$

The general equation is

$$pH = -\lg K_a - \lg\left(\frac{[acid]_{eqm}}{[base]_{eqm}}\right)$$

Two points are worth noting from this equation.
  1 The pH of a buffer solution depends on the *ratio* of the concentrations of acid and base, not on the actual values of these concentrations.
  2 When $[acid]_{eqm} = [base]_{eqm}$, $pH = -\lg K_a$. This means that the K_a value of an acid can be found by measuring the pH value of a solution of the acid which has been half neutralized by a strong base (experiment 12.5).
  Two examples will show how the equation can be used.

Example 1

Calculate the pH of a solution which is 0.05M with respect to ethanoic acid and 0.20M with respect to sodium ethanoate (for ethanoic acid, $K_a = 1.7 \times 10^{-5}$ mol dm^{-3}).

In this, and all similar calculations, we are dealing with systems for which K_a is small. We can therefore simplify the calculations by assuming that the ethanoic acid is un-ionized and that all the ethanoate ions come from the sodium ethanoate.

By doing this we can write

$$[acid]_{eqm} = [CH_3CO_2H(aq)]_{eqm} = 0.05 \, mol \, dm^{-3}$$

and $[base]_{eqm} = [CH_3CO_2^-(aq)]_{eqm} = 0.20 \, mol \, dm^{-3}$

$$\therefore pH = -\lg 1.7 \times 10^{-5} - \lg \frac{0.05}{0.20}$$

$$= 4.8 - \lg \frac{0.05}{0.20}$$

$$= 4.8 - \bar{1}.4 = 4.8 + 0.6$$
$$= 5.4.$$

Example 2

In what proportions must 0.1M solutions of ammonia and ammonium chloride be mixed to obtain a buffer solution of pH 9.8? (K_a for the ammonium ion is 5.6×10^{-10} mol dm^{-3}). Here the acid is $NH_4^+(aq)$ and the base $NH_3(aq)$.

$$\therefore 9.8 = -\lg 5.6 \times 10^{-10} - \lg \left(\frac{[NH_4^+(aq)]_{eqm}}{[NH_3(aq)]_{eqm}} \right)$$

$$9.8 = 9.3 - \lg \left(\frac{[NH_4^+(aq)]_{eqm}}{[NH_3(aq)]_{eqm}} \right)$$

$$\therefore 0.5 = -\lg \left(\frac{[NH_4^+(aq)]_{eqm}}{[NH_3(aq)]_{eqm}} \right)$$

$$or \lg \left(\frac{[NH_4^+(aq)]_{eqm}}{[NH_3(aq)]_{eqm}} \right) = -0.5 = \bar{1}.5$$

$$\therefore \frac{[NH_4^+(aq)]_{eqm}}{[NH_3(aq)]_{eqm}} = 0.32$$

The solutions must therefore be mixed in the proportions, 0.32 volume 0.1M ammonium chloride to 1 volume 0.1M ammonia solution.

The effect of adding $H^+(aq)$ or $OH^-(aq)$ ions to a buffer solution can be seen by considering what will happen when these are added to a buffer and a non-buffer solution. For the buffer solution we can take a solution that is 0.1M with respect to both ammonium chloride and ammonia. For this $[NH_4^+(aq)]_{eqm} = [NH_3(aq)]_{eqm}$ and $pH = -\lg K_a = 9.3$. For the non-buffer solution we will take a 10^{-5}M solution of sodium hydroxide, which has $pH = 9$.

What happens if we add $10\,cm^3$ of a 1M solution of a strong acid, $[H^+(aq)] = 1$, to one cubic decimetre of each solution? The actual amount of $H^+(aq)$ that is being added $= \dfrac{10}{1000} = 0.01\,mol$. If the *buffer solution* works perfectly, all this additional hydrogen ion will be removed by the reaction

$$NH_3(aq) + H^+(aq) \longrightarrow NH_4^+(aq)$$

At the same time $0.01\,mol\,NH_3(aq)$ will be removed and $0.01\,mol\,NH_4^+(aq)$ formed. We now have

$[NH_4^+(aq)]_{eqm} = 0.1 + 0.01 = 0.11\,mol\,dm^{-3}$
and
$[NH_3(aq)]_{eqm} = 0.1 - 0.01 = 0.09\,mol\,dm^{-3}$
and

$$pH = 9.3 - \lg\frac{0.11}{0.09}$$

$$= 9.3 - 0.09$$
$$= 9.21,\ \textit{a change of less than 0.1 pH unit}$$

For the non-buffer solution, the value of $[H^+(aq)]_{eqm} = 10^{-9}\,mol\,dm^{-3}$, before the acid is added. It becomes $10^{-9} + 0.01\,mol\,dm^{-3}$ after the acid is added. Clearly the new value of $[H^+(aq)]$ is very nearly 0.01 or $10^{-2}\,mol\,dm^{-3}$. For the new solution $pH = -\lg 10^{-2} = 2$. There is, therefore, *a change of 7 pH units*.

Now see if you can work out the pH change resulting from the addition of $10\,cm^3$ of M sodium hydroxide solution, $[OH^-(aq)] = 1\,mol\,dm^{-3}$, to the buffer and non-buffer solutions.

Many chemical processes, especially those that take place in living organisms, are sensitive to pH changes and either cease or proceed in undesired directions if the pH of the system changes markedly. In such systems buffer solutions provide a means of establishing conditions of fairly constant pH value. Some acid–base chemistry in the human body in which buffer solutions play an important part is discussed in the Background reading which can be found at the end of this section.

Indicators

You will have used acid–base indicators such as methyl orange, phenolphthalein, and litmus, in previous work. They are used to test for alkalinity and acidity, and for detecting the end-point in acid–base titrations. A given indicator cannot be used in all circumstances; some are more suitable for use with weak acids and others with weak bases. From a study of the titration curves that you obtained in experiment 12.4, you will see that there is a rapid change of pH with addition of alkali, when about $25.0 \, cm^3$ of 1.00M alkali have been added to $25 \, cm^3$ of 1.00M acid. At this stage the reaction has reached its 'end-point', as the correct reacting quantities of acid and alkali are present. However, this rapid change of pH occurs at a comparatively low pH during a titration of a strong acid against a weak base, and at a comparatively high pH during a titration of a weak acid against a strong base.

Now, different indicators change colour at different values of pH. Phenolphthalein, for example, changes colour over the range pH 8–10, and so is suitable for a titration involving a weak acid and a strong base, whose end-point occurs within this range. Methyl orange, however, changes colour over the range pH 4–7, and so can be used to find the end-point in a titration involving a strong acid and a weak base. Phenolphthalein would be unsuitable for this type of titration, as it would only change colour when an excess of alkali was present, and then only gradually, instead of sharply on the addition of one drop of extra alkali.

An indicator may be considered as a weak acid, for which either the acid or the corresponding base, or both, are coloured. We can represent this in a general way, using HIn for the acid form, as

$$HIn(aq) \rightleftharpoons H^+(aq) + In^-(aq)$$
colour A   colour B

Addition of acid, $H^+(aq)$ ions for example, displaces the equilibrium to the left and increases the intensity of colour A. Addition of base, $OH^-(aq)$ or $NH_3(aq)$, removes hydrogen ions,

$$H^+(aq) + OH^-(aq) \longrightarrow H_2O(l)$$
or
$$NH_3(aq) + H^+(aq) \longrightarrow NH_4^+(aq),$$

with the result that the equilibrium moves to the right to restore the value of K_a and increase the intensity of colour B. The colour of the system will depend on the relative concentrations, $[\text{HIn(aq)}]_{\text{eqm}}$ and $[\text{In}^-(\text{aq})]_{\text{eqm}}$.

Regarding indicators as weak acids we can apply the general equation

$$\text{pH} = -\lg K_a - \lg \frac{[\text{acid}]}{[\text{base}]}$$

to them. That is, for an indicator HIn,

$$\text{pH} = -\lg K_a - \lg \frac{[\text{HIn(aq)}]_{\text{eqm}}}{[\text{In}^-(\text{aq})]_{\text{eqm}}}$$

If the ratio of $[\text{HIn(aq)}]$ to $[\text{In}^-(\text{aq})]$ at a given point in the colour range of an indicator is known, together with the pH value for this mixture, the value of K_a for an indicator can be calculated (see experiment 12.5).

Taking phenolphthalein as an example, we can explore the possible pH range over which the colour change can be perceived. For this indicator K_a is approximately 10^{-9}. Representing phenolphthalein by HPh in the acid form, the equilibrium is

$$\text{HPh(aq)} \rightleftharpoons \text{H}^+(\text{aq}) + \text{Ph}^-(\text{aq})$$
colourless   red

When the indicator has developed half the full red colour,

$$[\text{HPh(aq)}]_{\text{eqm}} = [\text{Ph}^-(\text{aq})]_{\text{eqm}}$$
and
$$\text{pH} = -\lg K_a = -\lg 10^{-9} = 9$$

In a solution (or very pure water) of pH 7, the ratio of red (base) to colourless (acid) form can be calculated as follows.

$$7 = -\lg K_a - \lg \frac{[\text{HPh(aq)}]_{\text{eqm}}}{[\text{Ph}^-(\text{aq})]_{\text{eqm}}}$$

$$\therefore 7 = 9 - \lg \frac{[\text{HPh(aq)}]_{\text{eqm}}}{[\text{Ph}^-(\text{aq})]_{\text{eqm}}}$$

$$\therefore \lg \frac{[\text{HPh(aq)}]_{\text{eqm}}}{[\text{Ph}^-(\text{aq})]_{\text{eqm}}} = 2$$

and $$\frac{[\text{HPh(aq)}]_{\text{eqm}}}{[\text{Ph}^-(\text{aq})]_{\text{eqm}}} = \frac{100}{1}$$

Out of 101 molecules, 100 will be in the colourless form and one only in the red form. The eye cannot detect this small proportion of colour so the mixture appears colourless.

At pH 8,

$$\lg \frac{[HPh(aq)]_{eqm}}{[Ph^-(aq)]_{eqm}} = 9 - 8 = 1$$

and $\dfrac{[HPh(aq)]_{eqm}}{[Ph^-(aq)]_{eqm}} = \dfrac{10}{1}$

This ratio of colourless to coloured form is just detectable.

At pH 10,

$$\lg \frac{[HPh(aq)]_{eqm}}{[Ph^-(aq)]_{eqm}} = 9 - 10 = -1$$

and $\dfrac{[HPh(aq)]_{eqm}}{[Ph^-(aq)]_{eqm}} = \dfrac{1}{10}$

Out of 11 molecules of indicator, 10 will be in the coloured form. Any increase in the proportion of coloured molecules will not be noticed as an increase in colour.

Most indicators have a detectable colour change over a range of about 2 pH units. This is a consequence of the sensitivity of our eyes to colour changes.

EXPERIMENT 12.5
To measure K_a for (a) an indicator and (b) a weak acid

For an indicator, HIn,

$$pH = -\lg K_a - \lg \frac{[HIn]_{eqm}}{[In^-]_{eqm}}$$

If the ratio $[HIn]/[In^-]$ is found for a solution of known pH, K_a can be calculated.

Similarly, for a weak acid, HA,

$$pH = -\lg K_a - \lg \frac{[HA]_{eqm}}{[A^-]_{eqm}}$$

In a buffer solution containing equimolar concentrations of HA and A^-,

$$pH = -\lg K_a,$$

and K_a can be calculated if the pH is known.

In this experiment K_a for bromophenol blue and K_a for benzoic acid are determined.

Procedure

Make up the following solutions:

Solution X Add one drop of concentrated hydrochloric acid to $5\,cm^3$ of bromophenol blue solution.

Solution Y Add one drop of 4M sodium hydroxide solution to $5\,cm^3$ of bromophenol blue solution.

(Solution X contains the indicator wholly in the HIn form and solution Y has the same concentration of indicator, wholly in the In^- form.)

a *To measure K_a for bromophenol blue* – Arrange 18 test-tubes in nine pairs, one behind the other, in a double test-tube rack so that when you look through a pair of tubes the colour you see will be due to the solutions in both tubes.

Put $10\,cm^3$ distilled water into each of the 18 test-tubes and add drops of solutions X and Y, as follows:

Tube	1	2	3	4	5	6	7	8	9
Drops of X	1	2	3	4	5	6	7	8	9
Tube	10	11	12	13	14	15	16	17	18
Drops of Y	9	8	7	6	5	4	3	2	1

Add $5\,cm^3$ of 0.02M methanoic acid to $5\,cm^3$ 0.02M sodium methanoate in a test-tube of the same size as those in the rack (K_a for methanoic acid is $1.6 \times 10^{-4}\,mol\,dm^{-3}$). Add 10 drops of bromophenol blue solution to the test-tube and shake to mix the contents of the tube. Compare the colour of the mixture with the colours seen by looking through the *pairs* of the test-tubes in the rack.

Questions

1 Which pair of tubes matches most closely the colour in the methanoic acid/sodium methanoate mixture?

2 Calculate the pH of the methanoic acid/sodium methanoate mixture.

3 What is the ratio $[HIn]_{eqm}/[In^-]_{eqm}$ at this pH?

4 Calculate K_a for bromophenol blue.

b *To measure K_a for benzoic acid* – Mix $5\,cm^3\,0.02M$ benzoic acid and $5\,cm^3\,0.02M$ sodium benzoate in a test-tube; add 10 drops bromophenol blue solution. Compare the colour of the mixture with that seen through the pairs of test-tubes in the rack.

Questions

1 Which pair of tubes matches most closely the colour in the benzoic acid/sodium benzoate mixture?

2 What is the ratio $[HIn]_{eqm}/[In^-]_{eqm}$ in the mixture?

3 Calculate the pH of the mixture.

4 Calculate K_a for benzoic acid.

pK_a values

You will have noticed that the logarithm of the dissociation constant of an acid, $\lg K_a$, has appeared in a number of the equations earlier in this section, usually preceded by a minus sign. This expression, $-\lg K_a$, is known as the pK_a value for the acid.

For the acid dissociation

$$HA \rightleftharpoons H^+ + A^-$$

$$K_a = \frac{[H^+][A^-]}{[HA]}\ mol\,dm^{-3}$$

and

$$pK_a = -\lg(K_a/mol\,dm^{-3})$$

pK_a values give a convenient measure of the strength of an acid, and a number of them are given in the *Book of data*. The weaker the acid, the bigger the pK_a value. Typical values of pK_a are: nitric acid, -1.4; ethanoic acid, 4.8; chloric(I) acid, 7.4.

BACKGROUND READING
Acid–base chemistry in the human body

The chemical processes which occur in the body are complex and often present challenging problems to the biochemist. Some of the mechanisms are reasonably well understood, as is the mechanism by which the body regulates the pH value of the blood. There are three general mechanisms: buffers in the blood, both intracellular and extracellular, the removal of carbon dioxide by the lungs, and the slower process of control of acid excretion by the kidneys. Only the first two topics will be dealt with here but you might like to look up information on the role of the kidneys.

Although it is red to the naked eye, blood consists of a pale yellow aqueous fluid, the plasma, in which red and white cells are suspended. The plasma contains proteins, albumin, fibrinogens, and globulins, and also ions such as Na^+, Cl^-, HCO_3^-, and H^+ present in aqueous solution. The red cells contain haemoglobin, an enzyme known as carbonic anhydrase, and ions such as K^+, Cl^-, HCO_3^-, and H^+, also in aqueous solution. The chief functions of the blood are to transport oxygen and food materials to the body tissues and to carry carbon dioxide and other waste products of metabolism from the tissues to the lungs and kidneys, where they are excreted.

Air inhaled into the lungs finds its way into small air sacs, the alveoli, where it is brought into close proximity with blood across thin membranes. In Man, there are some 700 million alveoli, each about 2×10^{-4} cm in diameter. The blood arriving in the lungs is known as venous blood and it contains carbon dioxide collected during its passage round the body. Gaseous exchanges occur across the alveolar membranes, with oxygen passing into the blood and carbon dioxide moving from the blood into the alveoli ready for exhalation. The oxygenated or arterial blood is pumped out of the lungs to circulate through the tissues, giving up oxygen to them and collecting carbon dioxide. The process is summarized in figure 12.2.

Let us now examine the chemistry of the events we have been discussing.

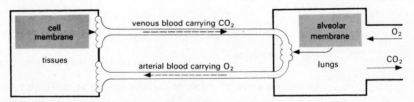

Figure 12.2
The transport of oxygen and carbon dioxide in the blood.

Transport of oxygen in the blood

Oxygen both occurs in the blood as a 'normal' solution, $O_2(aq)$, and is also combined chemically with haemoglobin in the red cells to form a bright red compound, oxyhaemoglobin. Approximately 0.3 % of the oxygen is present as $O_2(aq)$ in solution, the rest being in the oxyhaemoglobin. The concentration of the solution of $O_2(aq)$ follows Henry's Law:

$[O_2(aq)] \propto$ partial pressure of oxygen in alveolar air, p_{O_2}
or
$[O_2(aq)] = Kp_{O_2}$

where K is a solubility constant.

Haemoglobin is a complex organic molecule which contains iron, and normally carries a negative charge: we shall represent it here by the symbol Hb^-. The reaction with oxygen is written

$$Hb^-(aq) + O_2(aq) = \underset{\text{oxyhaemoglobin}}{HbO_2^-(aq)}$$

and one mole of oxygen molecules combines with 16 700 g haemoglobin.

The amount of oxygen entering the blood depends upon the partial pressure of oxygen in the alveolar air, which in turn is dependent upon the partial pressure in the atmosphere. In a healthy adult, p_{O_2} in alveolar air is 100 mmHg. The figures in table 12.5 show partial pressures for oxygen at various altitudes and reveal the necessity for breathing equipment in high altitude flight and mountaineering.

Altitude/m	Partial pressure of oxygen/mmHg
sea level	159
3 000	108
6 000	72
15 000	18

Table 12.5
Partial pressure of oxygen in the air.

Carriage of carbon dioxide in the blood

When carbon dioxide dissolves in water, it can react slowly to form hydrogen carbonate ions:

$$aq + CO_2(g) \longrightarrow CO_2(aq)$$
$$H_2O(l) + CO_2(aq) \longrightarrow H^+(aq) + HCO_3^-(aq)$$

The blood carries carbon dioxide in three ways: about 6 % present in solution in the plasma as $CO_2(aq)$, 86 % in the form of $HCO_3^-(aq)$, and 8 % in the form of a compound which is formed with haemoglobin in the red cells –

$$H^+(aq) + Hb^-(aq) + CO_2(aq) \longrightarrow HHbCO_2(aq)$$

The normally slow reaction between carbon dioxide and water is catalysed, in both directions, by the enzyme carbonic anhydrase. The efficiency of this catalysis can be appreciated when it is realized that, on average, a red cell spends less than half a second in the alveolar membrane during its passage through the lung.

The formation of HCO_3^- ions in the blood is accompanied by the formation of H^+ ions; if the blood were not buffered, its pH would fall when carbon dioxide passed into it. The blood in fact contains a number of buffering systems, and this, in healthy individuals, maintains its pH at about 7.4.

Blood buffers

The plasma proteins and haemoglobin can act as buffers because they contain both acidic ($-CO_2H$) and basic ($-NH_2$) groups in their molecules.

The action of a protein molecule as a base and as an acid is represented in figure 12.3, in the upper diagram as a base and in the lower diagram as an acid.

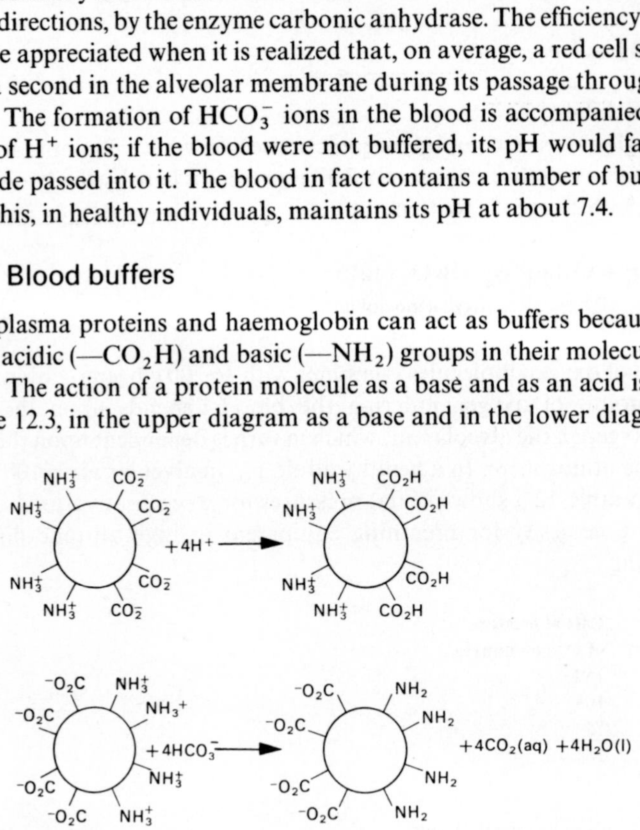

Figure 12.3

The most important buffer, however, is the hydrogen carbonate ion itself, since the equilibrium

$$H_2O(l) + CO_2(aq) \rightleftharpoons H^+(aq) + HCO_3^-(aq)$$

lies well to the left.

The processes which we have been discussing are summarized in figure 12.4. Notice the phenomenon known as the chloride shift. Can you account for it?

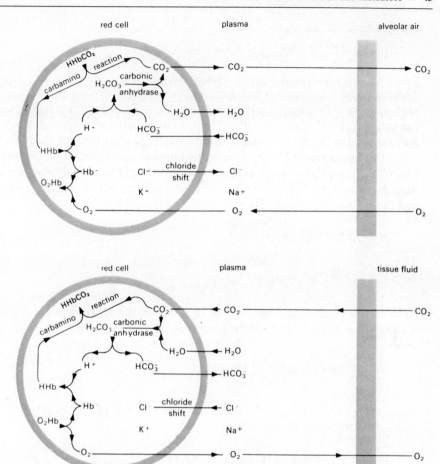

Figure 12.4
A schematic summary of the chemical processes which happen when haemoglobin takes
up oxygen in the lungs (above) and parts with its oxygen in the tissues (below).
The chloride shift is shown as a movement of ions. The shaded part represents the
capillary wall. *(After* ROUGHTON, E. J. W. Physiol. Rev. *15, 293, 1935.)*

Let us now see how the pH of the blood, effectively that of the plasma,
is related to the partial pressure of carbon dioxide in the alveolar air.

The equilibrium constant for the equilibrium

$$CO_2(aq) + H_2O(l) \rightleftharpoons H^+(aq) + HCO_3^-(aq)$$

is $K_a = \dfrac{[H^+(aq)]_{eqm}[HCO_3^-(aq)]_{eqm}}{[CO_2(aq)]_{eqm}}$

and

$$[CO_2(aq)]_{eqm} = bp_{CO_2eqm}$$

where b is a solubility constant for carbon dioxide in blood plasma. It is defined as the amount in millimoles of carbon dioxide dissolved in 1 cubic decimetre of plasma when $p_{CO_2} = 760\,mmHg$ and the temperature is $38\,°C$ (the normal body temperature).

Hence,

$$K_a = \frac{[H^+(aq)]_{eqm}[HCO_3^-(aq)]_{eqm}}{bp_{CO_2eqm}}$$

$$\therefore \; [H^+(aq)]_{eqm} = \frac{K_a bp_{CO_2eqm}}{[HCO_3^-(aq)]_{eqm}}$$

$$\therefore \; pH = -\lg K_a - \lg\frac{bp_{CO_2eqm}}{[HCO_3^-(aq)]_{eqm}}$$

Doctors and biochemists refer to this equation as the *Henderson–Hasselbalch* equation. It is a special form of the general equation relating to buffer solutions:

$$pH = -\lg K_a - \lg\frac{[acid]_{eqm}}{[base]_{eqm}}$$

$bp_{CO_2eqm} = [CO_2(aq)]_{eqm}$ is the acid concentration, and $[HCO_3^-(aq)]_{eqm}$ is the base concentration in the equilibrium

$$H_2O(l) + CO_2(aq) \rightleftharpoons H^+(aq) + HCO_3^-(aq)$$

If the HCO_3^- ion concentration is expressed in millimole dm^{-3} and p_{CO_2} in mmHg, K_a has the value $7.9 \times 10^{-7}\,mol\,dm^{-3}$ and b is 0.03 millimole dm^{-3}. In the plasma, the ratio

$$\frac{[CO_2(aq)]_{eqm}}{[HCO_3^-(aq)]_{eqm}} = \frac{bp_{CO_2eqm}}{[HCO_3^-(aq)]_{eqm}}$$

may be taken to be 0.05. Hence the normal pH value of plasma is 7.4. If the value of $[HCO_3^-(aq)]$ is 25.0 millimole dm^{-3}, verify for yourself that p_{CO_2eqm} in the alveolar air is 42 mmHg. In practice, we may take as standard the values $p_{CO_2eqm} = 40\,mmHg$, $pH = 7.4$, and $[HCO_3^-(aq)]_{eqm} = 25.0$ millimole dm^{-3}, although some variations from these figures are found even in healthy people.

Disturbances in the acid–base chemistry of the body

In a hospital biochemical laboratory the pH, p_{O_2}, and p_{CO_2} of blood samples taken from patients suffering from respiratory diseases are regularly measured. If the pH and p_{CO_2} are known, the $[HCO_3^-(aq)]$ can be calculated from the Henderson–Hasselbalch equation. However, as you will appreciate, repeated calculations of that nature would be time-wasting. Nomograms have therefore been constructed from the Henderson–Hasselbalch equation. Figure 12.5 shows such a nomogram. If pH and p_{CO_2} are measured and a ruler laid across the nomogram so as to join the values found, the total carbon dioxide present in the plasma, the dissolved carbon dioxide, and the $[HCO_3^-(aq)]$ can be read off.

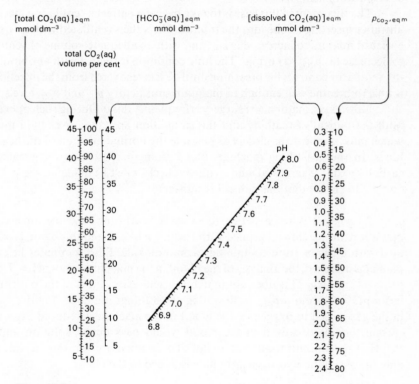

Figure 12.5
A nomogram.
After McLEAN, F. Physiol. Rev. *18, 511, 1938.*

Blood analyses are also carried out in the operating theatre. When patients are anaesthetized or when they are unable to maintain their own ventilation it may be necessary to take over the control of breathing with a mechanical system or ventilator. In such cases, the anaesthetist must ensure that the gas mixture supplied to the patient and the amount of ventilation maintain the p_{CO_2}, p_{O_2}, and pH at suitable values. This will be checked by regular blood analysis.

Example 1 A man suffering from a severe chest infection was admitted to the intensive care unit of a London hospital. He was unconscious and blood samples showed p_{CO_2} = 75 mmHg, p_{O_2} = 25 mmHg, and pH = 7.31. This man's disease prevented him from adequately ventilating his alveoli, so that the partial pressure of carbon dioxide rose to this high value while he failed adequately to oxygenate his blood. The buffer reserve of his blood had been exceeded and the pH had fallen. This is a result of respiratory failure and the condition is known as respiratory acidosis.

The treatment in this case is to take over the patient's ventilation mechanically after inserting a tube into the trachea. He is then ventilated with an oxygen-enriched mixture to increase his p_{O_2}, and with an adequate volume of ventilation to decrease his p_{CO_2} to normal. The lung condition is treated with antibiotics and the ventilator controls his breathing until he has recovered from the infection and is able to breathe well enough to maintain satisfactory p_{O_2} and p_{CO_2} levels.

Nowadays, major heart surgery is frequently done. During such operations, both the patient's breathing and the circulation are taken over by a machine which must duplicate as closely as possible the normal function of the heart and lungs. In the heart-lung machine, blood from the patient is oxygenated and carbon dioxide is removed, and frequent checks must be made on the pH, p_{CO_2}, ¬nd p_{O_2} to ensure that this control is satisfactory.

Example 2 When a patient goes on a heart-lung machine for a cardiac operation, the machine is primed with $10 \, dm^3$ of blood, and this blood is oxygenated with a gas mixture containing carbon dioxide. Fifteen minutes before the patient is attached, the analysis of the blood in the machine shows pH = 7.25 and p_{CO_2} = 42 mmHg. This does not match the acid–base status of the patient, who has a pH = 7.4 and a p_{CO_2} = 30 mmHg. Two things are done. Firstly, the p_{CO_2} in the gases used to oxygenate the blood in the machine is reduced. As a result, carbon dioxide dissolved in the blood would pass out of solution and thus the HCO_3^- concentration would fall. To compensate for this, a calculated amount of sodium hydrogencarbonate is added to the blood.

How many mmol of sodium hydrogencarbonate would you add to the original $10 \, dm^3$ of blood at pH = 7.25, p_{CO_2} = 42 mmHg, in order to stabilize it at pH = 7.45 and p_{CO_2} = 30 mmHg with the new gas mixture?

12.6
EQUILIBRIUM AND ZERO TOTAL ENTROPY CHANGE

Our earlier work in this Topic has shown how it is possible to answer two important and related questions, namely:

In what proportions are substances present at equilibrium?
How can conditions be altered so as to modify those proportions as one wishes?

We shall now try to explain some part of the molecular picture which lies behind the experimental facts.

If the entropy increases when a reaction goes in one direction, the reaction will proceed further in that direction, simply because increasing entropy means an increase in the number of ways of arranging molecules and energy. Of course, as we have seen before, we must consider *all* entropy changes including those in the surroundings. If the total entropy decreases, the reaction will proceed further in the opposite direction, because in that direction, the total entropy increases.

So, at equilibrium, when neither forward nor reverse reactions are preferred, the total entropy change must be exactly zero.

At equilibrium, $\Delta S_{total} = 0$

We must now think systematically about how to calculate the entropy changes in an equilibrium reaction.

Let us start with a very simple example, which we have used before; the freezing of water.

water $\longrightarrow$ ice $-6000\,\mathrm{J\,mol^{-1}}$ (lost from the water to the surroundings)

We know that in fact the equilibrium is at 273 K. Suppose some *small* amount of water freezes; call it x mole.

We lose x mole of water, entropy $63\,\mathrm{J\,K^{-1}\,mol^{-1}}$ at 273 K and 1 atm, so the total entropy changes by $-63x\,\mathrm{J\,K^{-1}}$ on this account (a decrease).

We gain x mole of ice, entropy $41\,\mathrm{J\,K^{-1}\,mol^{-1}}$ at 273 K and 1 atm, so the total entropy changes by $+41x\,\mathrm{J\,K^{-1}}$ on this account (an increase).

The surroundings gain $6000x\,\mathrm{J}$ at 273 K, so the total entropy increases by

$$\frac{6000x}{273} = +22x\,\mathrm{J\,K^{-1}}$$ on this account.

These changes can be set out to give a 'balance sheet' as follows:

Event	Entropy change/J K^{-1}
lose x mole water	$-63x$
gain x mole ice	$+41x$
6000x J to surroundings	$+22x$
x mole, freezes	0

So for this reversible process at the equilibrium temperature, the entropy change is zero when a *small* amount of water freezes. Obviously, if the amount of water involved is not small, the system will no longer be in equilibrium, and our argument will not apply.

If the temperature were 300 K, the last term would be $\dfrac{6000x}{300} = 20x$ J K^{-1};

and although the other terms will change a little, the balance sheet will give a negative value for the entropy change. The water will not freeze.

If the temperature were 250 K, the last term would be $\dfrac{6000x}{250} = 24x$ J K^{-1};

and although the other terms will change a little, the balance sheet will give a positive value for the total entropy change. The water will freeze.

Had we made up a balance sheet for ice melting, the reverse process, it would have been exactly the same but with all the signs reversed. We would have concluded the same about the equilibrium: that at 300 K, with a positive balance, the ice would melt, and that at 250 K, with a negative balance, the ice would not melt.

The point of this example, which tells us nothing that we do not already know, is that the same scheme is needed for cases where we *can* learn something.

Notice, however, one bad thing about the calculation. We needed to know the answer to get the answer. We had to know that the equilibrium was at 273 K to use the entropies of ice and water at 273 K – for we know (Topics 6 and 10) that entropy depends upon temperature. If we had tables of the entropies at a lot of different temperatures we could in principle draw up the balance sheet at each temperature, until we hit upon the one that added up to zero. Sometimes a chemist may have to do just that. But sometimes he or she can do better. Tables give entropies in some standard state. The chemist may know *how* entropy changes with, for example, pressure or concentration. If that is known, then the chemist can write an equation for the entropies in the unknown equilibrium state, and find out what the pressure or concentration is at that state.

So the plan of campaign is this:

write down the entropy balance sheet for the standard state
put in expressions which say how much the entropies change when (say)

concentrations change
say that the new balance sheet must add up to zero, and find the values for
which this is true.

A gas reaction

We know from previous work, in Topic 10, that the entropy of a gas changes with
pressure according to the relation:

$$S = S^\ominus - Lk \ln p/p^\ominus$$

If $p < p^\ominus$, $\ln p/p^\ominus$ is negative and $-Lk \ln p/p^\ominus$ is positive, hence $S > S^\ominus$ and
the entropy of the system increases. For example, if $p = 380 \, \text{mmHg}$, with
$p^\ominus = 760 \, \text{mmHg}$

$$\ln p/p^\ominus = \ln 380/760 = -0.69$$

so that

$$
\begin{aligned}
-Lk \ln p/p^\ominus &= -6 \times 10^{23} \times 1.4 \times 10^{-23} \times (-0.69) \\
&= +5.8 \, \text{J K}^{-1} \, \text{mol}^{-1}
\end{aligned}
$$

and

$$S = S^\ominus + 5.8 \, \text{J K}^{-1} \, \text{mol}^{-1}$$

If $p > p^\ominus$, $\ln p/p^\ominus$ is positive and the entropy of the system decreases.

For $p = 76\,000 \, \text{mmHg}$, $\ln p/p^\ominus = \ln 100 = +4.6$

and

$$
\begin{aligned}
S &= S^\ominus - 6 \times 10^{23} \times 1.4 \times 10^{-23} \times 4.6 \\
&= S^\ominus - 38.6 \, \text{J K}^{-1} \, \text{mol}^{-1}
\end{aligned}
$$

We also know *why* the relation is of this form; it is because the lower the pressure,
the more room the molecules have to move in.

Let us take the example of the synthesis of ammonia

$$N_2(g) + 3H_2(g) \rightleftharpoons 2NH_3(g); \qquad \Delta H^\ominus = -92 \, \text{kJ mol}^{-1}$$

The molar entropies in the standard state, 1 atm, 298 K, are:

Gas	$S_{298}^{\ominus}/\text{J K}^{-1}\text{mol}^{-1}$
N_2	191.6
H_2	130.6
NH_3	192.3

First let us draw up the balance sheet for the standard state, for x mole of nitrogen, N_2, and $3x$ mole of hydrogen, H_2, turning into $2x$ mole of ammonia, NH_3, and giving $-\Delta H = 92x$ kJ to the surroundings.

Event	Entropy change/J K^{-1}
Lose x mole N_2	$-191.6x$
Lose $3x$ mole H_2	$-3(130.6)x$
Gain $2x$ mole NH_3	$+2(192.3)x$
Total so far	$-198.8x$
Give $92x$ kJ to surroundings at 298 K (92 000 x/298)	$+308.7x$
Reaction with reactants at standard state	$+109.9x$

We can conclude that the reaction will proceed (though perhaps rather slowly or even very slowly, for kinetic reasons). Note that the effect of the change in the number of molecules involved – four reacting to give two – is not in favour of the reaction. The number of ways in which molecules can be arranged goes down; the total entropy change on that account is $-198.8\,\text{J K}^{-1}\text{mol}^{-1}$. But the reaction is exothermic, so it spreads energy around in the surroundings. The effect of that at 298 K is bigger than the decrease due to the change in the number of molecules, so the balance sheet shows at ΔS_{total} is greater than zero, that is, $\Delta S_{\text{total}} > 0$.

Would raising the temperature help? Doubling the temperature to 596 K would halve the entropy increase of the surroundings (if ΔH were not affected): from $+308.7x$ to $+154.3x$. The entropies of the gases would all get numerically larger (and by something like the same for each) so the subtotal for the gases would get *even more negative*. Even if it stayed the same, the balance sheet would now be negative ($+154.3$ is not enough to balance -198.8, let alone the larger negative quantity it would be). So just making it hotter would be a bad idea.

So we have a problem – the reaction goes the way we want too slowly to be much use, or the way we do not want when we make it hotter. Fortunately, we can solve the problem by raising the pressure, which is the discovery that made Haber famous.

Let us now draw up the balance sheet for partial pressures p_{N_2}, p_{H_2}, and p_{NH_3}, instead of the partial pressure $p^{\ominus}$ for each. Remember that $-Lk\ln p/p^{\ominus}$ is negative if $p > p^{\ominus}$, so that *losing* molecules makes a gain, and gaining them a loss.

Event	Entropy change at standard state/J K^{-1}	Correction for change in pressure/J K^{-1}
lose x mole N_2	$-191.6x$	$+Lk \ln p_{N_2}/p^\ominus\ x$
lose $3x$ mole H_2	$-3(130.6)x$	$+3Lk \ln p_{H_2}/p^\ominus\ x$
gain $2x$ mole NH_3	$+2(192.3)x$	$-2Lk \ln p_{NH_3}/p^\ominus\ x$
give $92x$ kJ to surroundings at 298 K	$+308.7x$	

The numbers in the entropy change column add up as before, but the values $Lk \ln p/p^\ominus$ now alter the total. We want to know what values of the pressures will make the sheet balance, and the total entropy change zero. Cancelling x, that is, working per mole of nitrogen, we get:

$$+Lk \ln p_{N_2}/p^\ominus + 3Lk \ln p_{H_2}/p^\ominus - 2Lk \ln p_{NH_3}/p^\ominus + 109.9 = 0$$

This is the condition for equilibrium that we have been seeking. The partial pressures p_{N_2}, p_{H_2}, and p_{NH_3} for any mixture of these gases in equilibrium must satisfy this equation. It is helpful to simplify the equation, and one way of doing this is to express the pressures in atmospheres so that $p^\ominus = 1$.

This gives

$$Lk(\ln p_{N_2} + 3 \ln p_{H_2} - 2 \ln p_{NH_3}) + 109.9 = 0$$

or

$$-\ln p_{N_2} - 3 \ln p_{H_2} + 2 \ln p_{NH_3} = 109.9/Lk = 13.1$$

$$\therefore \ln \left[\frac{p_{NH_3}^2}{p_{N_2} \times p_{H_2}^3} \right] = 109.9/Lk$$

$$= \frac{109.9}{6 \times 10^{23} \times 1.4 \times 10^{-23}} = 13.1$$

Since these figures refer to equilibrium conditions

$$\frac{p_{NH_3 eqm}^2}{p_{N_2 eqm} \times p_{H_2 eqm}^3} = K_p$$

and

$$\ln K_p = 109.9/Lk = 13.1$$

$$\therefore\ K_p = \exp(13.1) \approx 5 \times 10^5$$

So the argument says that the quantity K_p, containing the pressures of the components raised to the power of the number of moles in the reaction equation, will be constant. It also tells us what that constant will be.

Now the constant, before we divided by Lk, was just the total entropy change for the reaction with all the reactants in the standard state. Lk times the sum of the logarithm terms was just the modifying effect on the gas entropies of the changes of pressure. The equilibrium lies where the pressures depart far enough from the standard state to balance the total entropy change for reactants in their standard states.

Why do the numbers of reacting moles appear? Because each makes its contribution to change in numbers of ways. Making two moles of NH_3 adds twice as much entropy as one. Three moles of hydrogen removed take away three times as much entropy as one. Why do they appear as *powers* in K_p? Because numbers of ways *multiply*; if adding one mole multiplies W by M, two multiply it by M^2.

The value of the equilibrium constant is large. At equilibrium, there is a lot of NH_3 and not much N_2 or H_2. But there is *some* N_2 and H_2. Why some, and not none? This is because with just a little N_2 and H_2, at *low* pressures, and NH_3 at a higher pressure, making NH_3 reduces the number of ways much more than when the pressures are equal (see the balance sheet). That happens because removing molecules at low pressure makes a big difference to the entropy; adding them at high pressure makes only a little.

Equilibrium constant and free energy

The equilibrium constant was determined by the value $109.9 \, \text{J K}^{-1} \text{mol}^{-1}$. Where did that come from?

It was just the total entropy change, of chemicals and surroundings, at the standard state. It was $\Delta S^{\ominus}_{\text{total}}$.

What was it made up of? Looking at the balance sheet it was

$$\Delta S^{\ominus}_{\text{total}} = 2 \, S^{\ominus}_{NH_3} - S^{\ominus}_{N_2} - 3 \, S^{\ominus}_{H_2} - \Delta H^{\ominus}/T$$

which we could write as

$$\Delta S^{\ominus}_{\text{total}} = \Delta S^{\ominus}_{\text{chemicals}} - \Delta H^{\ominus}/T$$
(and $-\Delta H^{\ominus}/T$ could be written as $+\Delta S^{\ominus}_{\text{surroundings}}$)

So $\Delta S^{\ominus}_{\text{total}}$ is made up of two components

the change of entropy of the chemicals themselves, $\Delta S^{\ominus}_{\text{chemicals}}$, and
the change of entropy of the surroundings, $-\Delta H^{\ominus}/T$.

Multiplying all through by $-T$ we have

$$-T\Delta S^{\ominus}_{total} = \Delta H^{\ominus} - T\Delta S^{\ominus}_{chemicals}$$

The righthand side of this equation is called the *standard free energy change* of the reaction, and is given the symbol $\Delta G^{\ominus}$.

The idea of *free energy* is an extremely useful one for chemists, as will be seen in Topic 15. For the moment we will simply note that

$$\Delta S^{\ominus}_{total} = -\Delta G^{\ominus}/T$$

In the case of the nitrogen, hydrogen, and ammonia equilibrium

$$109.9 = -\Delta G^{\ominus}/T$$

or, with $T = 298$ K

$$\Delta G^{\ominus} = \frac{-109.9\,\mathrm{J\,K^{-1}\,mol^{-1}} \times 298\,\mathrm{K}}{10^3} - 32.75\,\mathrm{kJ\,mol^{-1}}$$

To get $\ln K_p$ we divided 109.9 by Lk, so we can write

$$\ln K_p = -\Delta G^{\ominus}/LkT$$

or $\quad \Delta G^{\ominus} = -LkT \ln K_p$

This important relationship will be used again in Topic 15.

SUMMARY

At the end of this Topic you should:

1 know the characteristics of the equilibrium state, namely that it can be attained only in a closed system, it can be approached from either direction, it has a constancy of intensive properties, and it is a dynamic state;

2 know the Equilibrium Law, and be able to work out expressions for equilibrium constants, both K_c and K_p;

3 know the effect of pressure and temperature on equilibrium;

4 know Le Châtelier's principle, and be able to apply it to changes of concentration, pressure, and temperature;

5 understand what is meant by heterogeneous equilibria;

6 know what is meant by solubility product;

7 understand the definition of acids as proton donors and bases as proton acceptors;
8 know what is meant by the ionization constant for water;
9 understand the term 'strength' as applied to acids and bases;
10 be familiar with the use of pH meters to measure hydrogen ion concentrations;
11 know how the pH changes during various types of acid-base titrations;
12 understand the function of buffer solutions, and be able to calculate the concentrations of specified solutes needed to make buffers of a particular pH;
13 understand how acid-base indicators function;
14 be aware of some aspects of acid-base chemistry in the human body;
15 appreciate how to calculate entropy changes in equilibrium reactions;
16 be aware of free energy, and its relationship with equilibrium constants.

PROBLEMS

* Indicates that the *Book of data* is needed.

1 11 g of ethyl ethanoate were mixed with $18\,cm^3$ of 1.0M hydrochloric acid in a flask and allowed to stand at constant temperature until equilibrium had been reached.

$$CH_3CO_2C_2H_5(l) + H_2O(l) \rightleftharpoons CH_3CO_2H(l) + C_2H_5OH(l)$$

The contents of the flask were titrated with 1.0M sodium hydroxide solution and $106\,cm^3$ of the alkali were required. Calculate the equilibrium constant K_c. (Assume that $18\,cm^3$ of 1.0M hydrochloric acid contain 18 g of water.)

2 Pentene (C_5H_{10}) reacts with ethanoic acid to produce pentyl ethanoate, the equilibrium

$$C_5H_{10} + CH_3CO_2H \rightleftharpoons CH_3CO_2C_5H_{11}$$

being established. When a solution of 0.02 mole of pentene molecules and 0.01 mole of ethanoic acid molecules in $600\,cm^3$ of an inert solvent was allowed to reach equilibrium at $15\,^\circ C$, 0.009 mole of pentyl ethanoate molecules was formed.

a How many moles of **i** pentene molecules and **ii** ethanoic acid molecules were present in the solution at equilibrium?
b Write down the expression for the equilibrium constant, K_c, for the above reaction.

c Complete the following expressions:

$$[C_5H_{10}]_{eqm} = \frac{\times}{} \, mol \, dm^{-3}$$

$$[CH_3CO_2H]_{eqm} = \frac{\times}{} \, mol \, dm^{-3}$$

$$[CH_3CO_2C_5H_{11}]_{eqm} = \frac{\times}{} \, mol \, dm^{-3}$$

d Use your answers to **b** and **c** to calculate the value of K_c at 15 °C.

3 The equilibrium

$$N_2O_4 \rightleftharpoons 2NO_2$$

can be established in trichloromethane solution at temperatures near 0 °C, and the composition of the equilibrium mixture can be calculated from the density of colour of the solution (N_2O_4 is colourless, NO_2 is brown). In a solution of this kind, at 10 °C, the concentration of NO_2 molecules was found to be $0.0014 \, mol \, dm^{-3}$ and the concentration of N_2O_4 molecules, $0.13 \, mol \, dm^{-3}$.
Calculate the value of K_c for the reaction at 10 °C.

4 For the equilibrium

$$PCl_5(g) \rightleftharpoons PCl_3(g) + Cl_2(g)$$

$K_c = 0.19 \, mol \, dm^{-3}$ at 250 °C.

2.085 g phosphorus pentachloride were heated to 250 °C in a sealed vessel of capacity $500 \, cm^3$ and maintained at this temperature until equilibrium was established.
a What would be the concentration of PCl_5, in $mol \, dm^{-3}$, if no change took place? (Assume that the volume of the vessel remains constant.)
b If the concentration of chlorine at equilibrium is $x \, mol \, dm^{-3}$, what are the equilibrium concentrations of **i** PCl_5, **ii** PCl_3, in $mol \, dm^{-3}$?
c What is the expression for the equilibrium constant, K_c, for the above reaction?
d Insert the values obtained in **b** and the value of K_c given above into the equilibrium constant expression, and thus calculate the value of x. (This involves solving a quadratic equation, one root of which is obviously absurd as an answer to this problem.)

e What are the concentrations of PCl_5, PCl_3, and Cl_2 present at equilibrium?
(Relative atomic masses: P = 31; Cl = 35.5.)

5 Propanone and hydrocyanic acid react in ethanol solution to form a product called 2-hydroxy-2-methylpropanenitrile, according to the equilibrium equation

$$CH_3COCH_3 + HCN \rightleftharpoons CH_3 \overset{\displaystyle OH}{\underset{\displaystyle CN}{\overset{|}{\underset{|}{C}}}} CH_3$$

At 20 °C, K_c for this equilibrium is 32.8 $dm^3\,mol^{-1}$. If 100 cm^3 of 0.1M solution of propanone in ethanol are mixed with 100 cm^3 of 0.2M solution of hydrocyanic acid in ethanol, what mass of the product will be formed at equilibrium?
(Relative atomic masses: H = 1; C = 12; N = 14; O = 16.)

6 A sealed vessel containing 0.0023 mole hydrogen iodide gas was heated at 900 K until the equilibrium

$$2HI(g) \rightleftharpoons H_2(g) + I_2(g)$$

was attained. The bulb was then cooled rapidly and opened under potassium iodide solution (in which the iodine dissolves). The resulting solution was titrated with 0.1M sodium thiosulphate, 6.4 cm^3 being required to react completely with the iodine.

$$2S_2O_3^{2-}(aq) + I_2(aq) \longrightarrow S_4O_6^{2-}(aq) + 2I^-(aq)$$

a Calculate the amount of iodine present at equilibrium, in moles.
b How many moles of hydrogen were present at equilibrium?
c How many moles of hydrogen iodide were present at equilibrium?
d Calculate the value of K_c for this equilibrium at 900 K.

7 For the equilibrium

$$C_2H_5OH(l) + C_2H_5CO_2H(l) \rightleftharpoons C_2H_5CO_2C_2H_5(l) + H_2O(l)$$

propanoic acid   ethyl propanoate

K_c = 7.5 at 50 °C. What mass of ethanol must be mixed with 60 g of propanoic acid at 50 °C in order to obtain 80 g of ethyl propanoate in the equilibrium mixture?
(Relative atomic masses: H = 1; C = 12; O = 16.)

8 Sulphur dioxide and oxygen in the ratio 2 moles: 1 mole were mixed at constant temperature and a constant 9 atmospheres pressure, in the presence of a catalyst. At equilibrium, one third of the sulphur dioxide had been converted into sulphur trioxide:

$$2SO_2(g) + O_2(g) \rightleftharpoons 2SO_3(g)$$

Calculate the equilibrium constant (K_p) for this reaction under these conditions.

9 $2NO_2(g) \rightleftharpoons 2NO(g) + O_2(g)$

For this equilibrium, a particular equilibrium mixture has the composition 0.96 mole $NO_2(g)$, 0.04 mole $NO(g)$, 0.02 mole $O_2(g)$ at 700 K and 0.2 atmosphere.

a Calculate the equilibrium constant K_p for this reaction under the stated conditions.

b Calculate the average relative molecular mass of the mixture under the stated conditions.

10 For the following equilibrium K_p at 373 K is 3.8×10^{-5}

$$CO_2(g) + H_2(g) \rightleftharpoons CO(g) + H_2O(g)$$

a Suppose that 1 mole of carbon dioxide molecules and I mole of hydrogen molecules were put into a vessel at 2 atmospheres and 373 K. Calculate how many moles of carbon monoxide molecules would be present when equilibrium was reached.

b Suppose that 1 mole of carbon dioxide molecules and 1 mole of hydrogen molecules were mixed at 4 atmospheres and 373 K. How many moles of carbon monoxide molecules would be present when equilibrium was reached? Comment on the relationship between your answers to **a** and **b**.

c K_p for the reaction at 298 K is 10^{-5}. Would you expect the yield of carbon monoxide at 298 K to be greater than, less than, or the same as, the yield from the same mixture at 373 K and at the same pressure? Give the reasons for your decision.

d Deduce whether the forward reaction is exothermic or endothermic. Give the reasons for your decision.

11 0.2 mole of carbon dioxide was heated with an excess of carbon in a closed vessel until the following equilibrium was attained.

$$CO_2(g) + C(s) \rightleftharpoons 2CO(g)$$

It was found that the average relative molecular mass of the gaseous equilibrium mixture was 36.

a Calculate the mole fraction of carbon monoxide in the gaseous equilibrium mixture.

b The pressure at equilibrium in the vessel was 12 atmospheres. Calculate K_p for the equilibrium at the temperature of the experiment.

c Calculate the mole fraction of carbon monoxide which would be present in the equilibrium mixture if the pressure were reduced to 2 atmospheres at the same temperature.

*12 A saturated solution of strontium carbonate was filtered. When $50\,cm^3$ of the filtrate were added to $50\,cm^3$ of 1M sodium carbonate solution, some strontium carbonate was precipitated. Calculate the concentration of strontium ion (in moles per cubic decimetre) remaining in the solution. All the work was done at $25\,°C$.

*13 Which of the following pairs of 0.001M solutions should form a precipitate when equal volumes are mixed at $25\,°C$? Explain how you arrive at your decision and, if there is a precipitate, name it.

a silver nitrate and sodium chloride
b calcium hydroxide and sodium carbonate
c silver nitrate and potassium bromate(v)
d magnesium sulphate and sodium hydroxide

14 Calculate the pH of the following solutions at $25\,°C$. In parts a to d assume complete ionization.

a 0.2M HCl
b 0.2M KOH
c 0.125M HNO_3
d A mixture of $75\,cm^3$ of 0.1M HCl and $25\,cm^3$ of 0.1M NaOH
e 0.1M bromoethanoic acid (CH_2BrCO_2H; $K_a = 1.35 \times 10^{-3}\,mol\,dm^{-3}$)

*15 What is the concentration of methanoate ion (in $mol\,dm^{-3}$) in 0.01M methanoic acid solution at $25\,°C$?

16 In a 0.1M solution of a certain acid HA,
$[A^-(aq)]_{eqm} = 1.3 \times 10^{-3}\,mol\,dm^{-3}$; calculate K_a for the acid.

17 A 0.1M solution of a certain acid HA has a pH of 5.1; calculate K_a for the acid.

*18 Calculate the pH of a 0.001M solution of phenylammonium chloride at 25 °C. Assume that phenylammonium chloride $(C_6H_5NH_3Cl)$ is fully ionized.

*19 Calculate the pH of a solution which is 0.1M with respect to propanoic acid and 0.05M with respect to sodium propanoate.

*20 In what proportions must 0.1M solutions of ethanoic acid and sodium ethanoate be mixed to obtain buffer solutions of

   a pH 4.7
   b pH 4.4?

*21 Draw up an entropy balance sheet for the equilibrium

$$2SO_2(g) + O_2(g) \rightleftharpoons 2SO_3(g)$$

Use it
   a to work out the total entropy change for the reaction with the compounds at standard state, and
   b to obtain a value for K_p for the equilibrium.

Carbon compounds with acidic and basic properties

This third Topic on the organic chemistry of carbon is concerned with some compounds that have acidic properties, the *carboxylic acids*, and some compounds that have basic properties, the *amines*. We will also consider the chemistry of some derivatives of carboxylic acids, the *esters*, the *acyl chlorides*, and the *nitriles*.

Finally, we will consider the important naturally occurring compounds, the *proteins* which are made from *amino acids*. Amino acids are considered in this Topic because they contain both a carboxylic acid functional group and an amine functional group.

$$CH_3-\overset{\displaystyle O}{\overset{\|}{C}}-OH \qquad \text{ethanoic acid, a carboxylic acid}$$

$$CH_3-CH_2-NH_2 \qquad \text{ethylamine, an amine}$$

$$CH_3-\overset{\displaystyle O}{\overset{\|}{C}}-O-CH_3 \qquad \text{methyl ethanoate, an ester}$$

$$CH_3-\overset{\displaystyle O}{\overset{\|}{C}}-Cl \qquad \text{ethanoyl chloride, an acyl chloride}$$

$$CH_3-\overset{\displaystyle O}{\overset{\|}{C}}-NH_2 \qquad \text{ethanamide, an amide}$$

$$CH_3-C\equiv N \qquad \text{ethanenitrile, a nitrile}$$

$$\underset{\displaystyle NH_2}{CH_2}-\overset{\displaystyle O}{\overset{\|}{C}}-OH \qquad \text{glycine (aminoethanoic acid), an amino acid}$$

13.1
CARBOXYLIC ACIDS

We have already met the hydroxyl group, —OH, and the carbonyl group,

$\diagdown$
$\quad$C$=$O but in the carboxylic acids we meet them on one carbon atom,
$\diagup$

O
$\parallel$
—C—OH, rather than separately. We might expect some significant alterations in reactivity rather than the sum of the separate reactions of the two groups.

The most significant alteration is the ionization of the functional group

$$CH_3-C\overset{\displaystyle O}{\underset{\displaystyle OH}{\diagup}} + H_2O \rightleftharpoons CH_3-C\overset{\displaystyle O^-}{\underset{\displaystyle O}{\diagup}} + H_3O^+$$

When the functional group ionizes, the resulting anion is stabilized by the π electrons from the double bond and an electron pair from the other oxygen atom, forming a delocalized system. This results in a symmetrical structure with both C—O bonds having the same length. Delocalization was introduced in Section 8.4 of Topic 8 in *Students' book I* and you should read this section again to refresh your memory on this subject. You should also compare the structure of the carboxylate anion with the structure of the phenoxide anion which was described in Topic 11 on page 381.

The presence of the C$=$O double bond in the $\overset{\displaystyle O}{\overset{\parallel}{—C—}}$OH group means that the carbon atom of the functional group is electrophilic in nature. Reagents such as ammonia, however, react as bases forming a salt

$$CH_3-CO_2H + NH_3 \longrightarrow CH_3-CO_2^- + NH_4^+$$

rather than acting as nucleophiles and attacking the carbon atom.

Very weak nucleophiles, such as alcohols $CH_3CH_2\ddot{O}H$, will react, attacking the carbon atom, but only when catalysed by H^+ ions.

The infra-red spectrum of a carboxylic acid such as ethanoic acid shows a broad absorption around $3100\,cm^{-1}$ due to the hydrogen bonding of the O—H group, and another characteristic absorption at $1740\,cm^{-1}$ due to the C$=$O group. The C$=$O absorption is in the same position as that produced by carbonyl compounds but is a broader trough (figure 13.1).

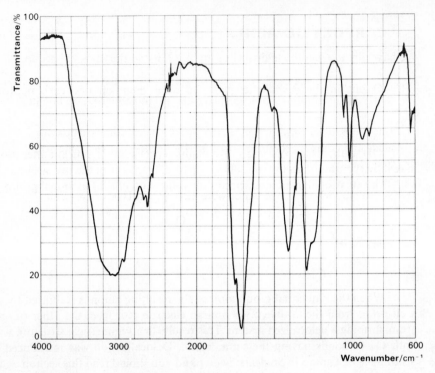

Figure 13.1
The infra-red absorption spectrum of ethanoic acid.

Experiments with ethanoic acid

Pure ethanoic acid is sometimes known as *glacial* ethanoic acid because it freezes at 17 °C and its ice-like crystals were often observed in unheated laboratories. An alternative name for ethanoic acid is acetic acid (from the Latin *acetum* or vinegar, of which it is the active constituent). In these experiments use pure, that is, glacial, ethanoic acid. *TAKE CARE:* it has a pungent odour, and can cause painful blisters if left in contact with the skin. Wash it off immediately with plenty of water.

EXPERIMENT 13.1
An investigation of the reactions of ethanoic acid

Procedure

Safety glasses should be worn throughout this experiment.

1 *Solubility and pH*   To 1 cm³ of ethanoic acid (*TAKE CARE*) in a test-tube, add water in drops. Do they mix in all proportions? Add a few drops of Full-range Indicator. Finally add sodium carbonate solution.

Is ethanoic acid a strong or a weak acid?
Is it a strong enough acid to displace carbon dioxide from carbonates?

Use the *Book of data* to look up the pK_a values of some carboxylic acids and compare the values with the pK_a values for carbonic acid and phenol.

When ethanoic acid and water were mixed what type of molecular interaction was helping them to dissolve?

2 *Formation of salts*   To 10 cm³ of 0.1M ethanoic acid in a beaker add a few drops of Full-range Indicator; then add, while stirring, 3-cm³ portions of 0.1M sodium hydroxide until you have added a total of 15 cm³. Note the pH after each addition. Also measure the pH of a solution of sodium ethanoate.

Explain what happens.

3 *Formation of the acid chloride*   To 1 cm³ of *pure* ethanoic acid in a dry beaker add a *little* solid phosphorus pentachloride. (*TAKE CARE:* this reacts readily with moisture; *do not leave the lid off the container.*)

Is there any sign of reaction? Is a gas evolved?

What bond is broken in the ethanoic acid?

Is this nucleophilic or electrophilic attack by the phosphorus pentachloride?

4 *Laboratory preparation: formation of an ester, methyl benzoate*   If you have time you can carry out the preparation of methyl benzoate, an experiment which uses benzoic acid and not ethanoic acid. (An alternative experiment is the preparation of cholesteryl benzoate on page 74 or methyl 2-hydroxybenzoate on page 300.)

methanol   benzoic acid   methyl benzoate

To a 50-cm^3 pear-shaped flask add 8 g of benzoic acid, 15 cm^3 of methanol, and 2 cm^3 of concentrated sulphuric acid (*TAKE CARE*). Fit the flask with a reflux condenser and boil the mixture for about 45 minutes. Cool the mixture to room temperature and pour it into a separating funnel that contains 30 cm^3 of cold water. Rinse the flask with 15 cm^3 of 1,1,1-trichloroethane and pour this into the separating funnel. Mix the contents of the separating funnel by vigorous shaking, releasing the pressure carefully from time to time, allow them to settle, and run the two layers into separate conical flasks. Return the (lower) 1,1,1-trichloroethane layer to the separating funnel and wash with 15 cm^3 of water and then 15 cm^3 of 0.5M aqueous sodium carbonate solution. Dry the 1,1,1-trichloroethane extract over anhydrous sodium sulphate, filter, and remove the 1,1,1-trichloroethane (T_b 74 °C) by careful distillation. Complete the distillation, collecting the distillate boiling above 190 °C as methyl benzoate. The yield should be about 70%.

What is the percentage yield from the benzoic acid in your preparation?

Which bonds could have broken in the reaction? You should find that two alternative patterns of bond breaking are possible.

The methyl benzoate can be used for the experiment with esters in the next section.

Reactions of carboxylic acids

Acidic properties involving the O—H bond

1 *Solubility and pH*   The carboxylic acids C_1 to C_4 mix with water in all proportions but at C_5 and thereafter solubility rapidly reduces. The molecules will hydrogen bond to each other as well as to water molecules.

$$
\begin{array}{c}
\text{H} \\
| \\
\text{O}\cdots\text{H}\!-\!\text{O} \\
\diagup\!\diagup \\
\text{CH}_3\!-\!\text{C} \\
\diagdown \\
\text{O}\!-\!\text{H}\cdots\text{O}\!-\!\text{H} \\
| \\
\text{H}
\end{array}
$$

The carboxylic acids are only weak acids in water, as can be seen from their pK_a values (table 13.1).

$$CH_3CO_2H(aq) \rightleftharpoons H^+(aq) + CH_3CO_2^-(aq)$$

Compound	pK_a
Methanoic acid	3.8
Ethanoic acid	4.8
Propanoic acid	4.9
Butanoic acid	4.8

Table 13.1
pK_a values for some carboxylic acids.

They are usually less than 1% ionized in water and do not readily evolve hydrogen by reaction with metals.

2 *Formation of salts* The carboxylic acids are strong enough acids to displace carbon dioxide from sodium carbonate, unlike phenol, and will neutralize sodium hydroxide, forming a sodium salt.

$$CH_3CO_2H + Na^+{}_2CO_3^{2-} \longrightarrow CH_3CO_2^-Na^+ + H_2O + CO_2$$
$$CH_3CO_2H + Na^+OH^- \longrightarrow CH_3CO_2^-Na^+ + H_2O$$

Because ethanoic acid is a weak acid, a sodium ethanoate–ethanoic acid mixture can be used as a buffer solution, changing pH only slowly on the addition of a strong acid or a strong base (see Topic 12).

3 *Formation of acyl chlorides* Carboxylic acids can be converted to acyl chlorides by reaction with phosphorus pentachloride. The —OH group of the acid is replaced by a Cl atom. The mechanism is complex but the nucleophile is $Cl^{\delta-}$

benzoic acid   benzoyl chloride

The acyl chlorides are stable volatile liquids which can be useful in the synthesis of other compounds because they are more reactive to nucleophiles than the parent carboxylic acids. Thus benzoyl chloride is used in preference to benzoic acid when preparing benzoate esters.

Nucleophilic reactions involving the C=O bond

4 *Formation of esters*   Alcohols have a lone pair of electrons on their oxygen atom and can therefore react as nucleophiles, forming *esters* with carboxylic acids. The reaction is reversible and is carried out in strongly acidic conditions

$$\langle\bigcirc\rangle\!-\!CO_2H + CH_3OH \underset{}{\overset{H^+}{\rightleftharpoons}} \langle\bigcirc\rangle\!-\!CO_2CH_3 + H_2O$$

benzoic acid   methanol   methyl benzoate

From this equation you can see that the water produced may have derived its oxygen atom from either the acid or the alcohol. Check the structural formulae and notice that both molecules have an O—H group. Two American chemists, Roberts and Urey, found out which compound provided the oxygen for the water in 1938, by using methanol containing a high proportion of the isotope ^{18}O.

either

$$\langle\bigcirc\rangle\!-\!\overset{\overset{\displaystyle O}{\|}}{C}\!-\!{}^{18}OCH_3 + H_2O \quad (1)$$

$$\langle\bigcirc\rangle\!-\!\overset{\overset{\displaystyle O}{\|}}{C}\!-\!OH + CH_3{}^{18}OH$$

or

$$\langle\bigcirc\rangle\!-\!\overset{\overset{\displaystyle O}{\|}}{C}\!-\!OCH_3 + H_2{}^{18}O \quad (2)$$

Using a mass spectrometer to determine the relative molecular masses of the products they were able to establish that ^{18}O isotope appears in the ester (equation 1) and not the water (equation 2). Thus the new bond was formed between the C atom in the acid and the ^{18}O atom in the alcohol, with the loss of the —OH group from the acid and not from the alcohol. Chemists consider that the first step in an acid-catalysed esterification is the addition of a proton to the C=O group of the acid:

$$\langle\bigcirc\rangle\!-\!\overset{\overset{\displaystyle O^{\delta-}}{\|}}{C^{\delta+}}\!-\!OH \underset{\text{protonation}}{\overset{H^+}{\rightleftharpoons}} \left[\langle\bigcirc\rangle\!-\!\overset{\overset{\displaystyle OH}{|}}{\underset{\underset{\displaystyle OH}{|}}{C^+}}\right]$$

$$\underset{\text{addition}}{\overset{CH_3OH}{\rightleftharpoons}} \left[\langle\bigcirc\rangle\!-\!\overset{\overset{\displaystyle OH}{|}}{\underset{\underset{\displaystyle OH\,H}{|}}{C}}\!-\!O\!-\!CH_3\right]^+$$

$$\xrightarrow[\text{elimination}]{-H_2O} \left[\underset{}{\bigcirc}\overset{\displaystyle OH}{\underset{+}{-C}}-O-CH_3 \right]$$

$$\xrightarrow[\text{elimination}]{-H^+} \underset{}{\bigcirc}\overset{\displaystyle O}{\overset{\|}{-C}}-O-CH_3$$

This is an addition-elimination reaction in which the overall process is substitution of the —OH group by the nucleophilic CH_3O-group. You are not expected to learn this set of equations.

5 *Reduction* Carboxylic acids can be reduced to alcohols by lithium tetra-hydridoaluminate, $LiAlH_4$.

$$\underset{\text{benzoic acid}}{\bigcirc-CO_2H} \xrightarrow{LiAlH_4} \underset{\text{phenylmethanol}}{\bigcirc-CH_2OH} + H_2$$

The reaction proceeds by a nucleophilic attack of the carboxylate ion by the hydride ion, $:H^-$. Sodium tetrahydridoborate, $NaBH_4$, is too mild a reducing agent to react with carboxylic acids.

BACKGROUND READING 1
Naturally occurring carboxylic acids

Lipids are a class of cell materials that are insoluble in water. The most common lipids are known as fats or oils. These compounds are esters of the trihydric alcohol propane-1,2,3-triol (glycerol), CH_2OH—$CH(OH)$—CH_2OH, and long chain carboxylic acids such as octadecanoic acid (stearic acid) CH_3—$(CH_2)_{16}$—CO_2H. A typical fat molecule therefore has the structure

$$CH_2-O-CO-(CH_2)_{n_1}-CH_3$$
$$CH-O-CO-(CH_2)_{n_2}-CH_3$$
$$CH_2-O-CO-(CH_2)_{n_3}-CH_3$$

where n_1, n_2, and n_3 may be the same, or different, but are almost always *even* numbers. Other lipids exist with a more complex structure.

An important function of lipids is to act as an energy store, as they have the best energy value of all foods; plant seeds are particularly rich in lipids for this reason. Nutmegs, for example, yield up to 40 per cent of fat. A feature of lipids, as already noted, is that nearly all the carboxylic acids from which they are derived have an even number of carbon atoms. This arises because the acids are synthesized in living organisms by extending the carbon chain two atoms at a time. The acids are also normally metabolized by degrading the chain two carbon atoms at a time. This mechanism was demonstrated by feeding animals with synthetic arene-substituted acids. Animals fed substituted acids with an even number of carbon atoms were found to have derivatives of phenylethanoic acid in their metabolic waste products.

$$\langle\bigcirc\rangle\!-\!(CH_2)_n\!-\!CO_2H \longrightarrow \langle\bigcirc\rangle\!-\!CH_2\!-\!CO_2H$$

n is ODD, so that the total number of carbon atoms including the CO_2H group is EVEN.

Degradation one atom at a time would have led to benzoic acid. On the other hand, animals fed substituted acids with an odd number of carbon atoms did produce benzoic acid. Again, this was to be expected on the basis of a degradation by two carbon atoms at a time.

$$\langle\bigcirc\rangle\!-\!(CH_2)_n\!-\!CO_2H \longrightarrow \langle\bigcirc\rangle\!-\!CO_2H$$

n is EVEN, so that the total number of carbon atoms including the CO_2H group is ODD.

A number of lipids are of economic importance, as they are used to make foodstuffs, such as margarine, and soap. Soap is made by boiling fats with alkali.

$$
\begin{array}{l}
CH_2\!-\!O\!-\!CO\!-\!(CH_2)_n\!-\!CH_3 \\
|\\
CH\!-\!O\!-\!CO\!-\!(CH_2)_n\!-\!CH_3 + 3NaOH \longrightarrow 3CH_3\!-\!(CH_2)_n\!-\!CO_2Na + \\
|\\
CH_2\!-\!O\!-\!CO\!-\!(CH_2)_n\!-\!CH_3
\end{array}
\qquad
\begin{array}{l}
CH_2OH \\
|\\
CHOH \\
|\\
CH_2OH
\end{array}
$$

| fat | + alkali $\longrightarrow$ | soap | + glycerol |

Because of the value of lipids for foodstuffs, synthetic detergents based on petrochemicals were introduced, but the original synthetic detergents caused serious pollution in rivers. These problems did not arise with soap because soap had straight chain alkyl groups and could be degraded by bacteria. The original synthetic detergents had branched-chain alkyl groups and could not be degraded (figure 13.2).

Figure 13.2
One result of the use of detergents with a branched chain alkyl group (Batford Weir on the River Lee). Taken in 1961. *(Photograph, Crown Copyright.)*

Chemists were able to solve this problem by developing biodegradable synthetic detergents which have straight-chain alkyl groups similar to the alkyl chains in natural fats and oils.

a soap

$CO_2^- Na^+$

a synthetic detergent

a biodegradable synthetic detergent

$SO_3^- Na^+$

$SO_3^- Na^+$

Another use for lipids is to provide carboxylic acids as starting materials for synthesis of organic compounds with eight or more carbon atoms (see table 13.2). The synthesis of C_{even} compounds is straightforward but C_{odd} compounds require the introduction of an extra carbon atom (or removal of one atom). So the synthesis of C_{odd} compounds is more costly.

Number of carbon atoms	Naturally occurring compound	Systematic name	Trivial name	Source
8	$CH_3(CH_2)_6CO_2H$	octanoic acid	caprylic acid	coconut oil
10	$CH_3(CH_2)_8CO_2H$	decanoic acid	capric acid	coconut oil
12	$CH_3(CH_2)_{10}CO_2H$	dodecanoic acid	lauric acid	coconut oil
14	$CH_3(CH_2)_{12}CO_2H$	tetradecanoic acid	myristic acid	nutmeg seed fat
16	$CH_3(CH_2)_{14}CO_2H$	hexadecanoic acid	palmitic acid	palm oil
	$CH_3(CH_2)_{14}CH_2OH$	hexadecanol	cetyl alcohol	sperm whale oil
18	$CH_3(CH_2)_{16}CO_2H$	octadecanoic acid	stearic acid	animal fats
	$CH_3(CH_2)_7CH\!\!=\!\!CH(CH_2)_7CO_2H$	octadec-9-enoic acid	oleic acid	olive oil
	$CH_3(CH_2)_7CH\!\!=\!\!CH(CH_2)_7CH_2OH$	octadec-9-en-1-ol	oleyl alcohol	sperm whale oil

Table 13.2
Some naturally occurring carboxylic acids and alcohols.

13.2
CARBOXYLIC ACID DERIVATIVES

The derivatives of the carboxylic acids are considered to be those compounds in which another group appears in the $-CO_2H$ group in the place of the $-OH$. In this section, we shall be concerned with acyl chlorides, esters, and amides.

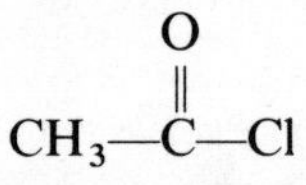

$$CH_3-\overset{\overset{\displaystyle O}{\|}}{C}-Cl$$ ethanoyl chloride

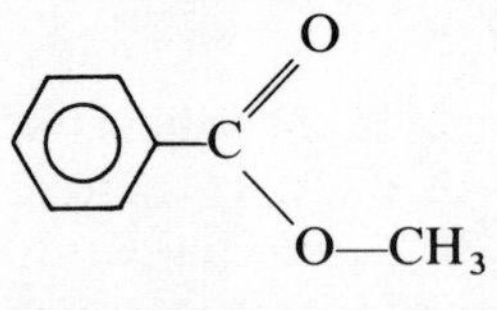

methyl benzoate

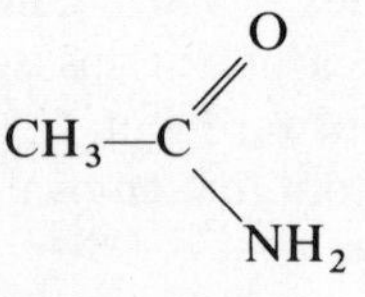

ethanamide

The reactions of nitriles form a common pattern with the carboxylic acid derivatives and will also be dealt with in this section.

Experiments with carboxylic acid derivatives

Some of these experiments will be demonstrated to you; you may be able to carry out others for yourself. No experiments with nitriles are included because of their toxic properties.

EXPERIMENT 13.2
An investigation of some reactions of carboxylic acid derivatives
Procedure

CAUTION: ethanoyl chloride is volatile and forms pungent fumes in moist air. Eye protection should be worn throughout this experiment, even if the teacher is demonstrating.

1 *Teacher demonstration: acyl chlorides*
a Put 5 cm^3 of water in a beaker and *very carefully* add a little ethanoyl chloride, one drop at a time. Note the vigour of the reaction.
b Put 1 cm^3 of ethanol in a beaker and *carefully* add 1 cm^3 of ethanoyl chloride one drop at a time. When the reaction has subsided, cautiously add sodium carbonate solution to neutralize any acids present and then see if you can detect the presence of an ester by its smell.
c Put 1 cm^3 of concentrated ammonia (*TAKE CARE:* there are fumes) in a beaker and add 5 *drops* of ethanoyl chloride. Note the fumes produced, then evaporate off the water; a solid product should be obtained.

> For each of the experiments **a** to **c**, write an equation showing the structures of the compounds, name the nucleophile involved, and identify the bond which breaks in the ethanoyl chloride.

2 *Esters*   Put 2 cm^3 of methyl benzoate and 30 cm^3 of 2M sodium hydroxide in a 50-cm^3 pear-shaped flask and fit the flask with a Liebig condenser. Boil the mixture under reflux for about half an hour. Cool the reaction mixture and acidify with 2M hydrochloric acid. What is the precipitate that forms? If you have time, the identity of the precipitate can be confirmed by collecting it by suction filtration, recrystallizing from water, and determining its melting point.

Write an equation showing the structures of the compounds, name
the nucleophile involved, and identify the bond which breaks in the
ester.

3 *Acid amides* Place 5 cm³ of water in a beaker and add some solid
ethanamide (note the smell). Is there any sign of reaction? Now add 5 cm³ of
2M sodium hydroxide and boil gently. (*TAKE CARE*). Is any gas evolved?
Test by smell and pH.

Write an equation showing the structures of the compounds, name
the nucleophile involved, and identify the bond which breaks in the
ethanamide.

Devise an experiment to see if the same reaction can be catalysed by acid.

4 *Laboratory preparation: cholesteryl benzoate, 'liquid crystals'*

benzoyl
chloride

cholesterol

cholesteryl benzoate

Place 1 g of cholesterol in a 50-cm³ conical flask. *In a fume cupboard* add
3 cm³ of pyridine (*TAKE CARE*: the vapour is harmful; avoid contact with
skin and eyes) and swirl the mixture to dissolve the cholesterol. Then add
0.4 cm³ of benzoyl chloride (*TAKE CARE*: it is a lachrymator – that is, it will
bring tears to your eyes). Heat the resulting mixture on a steam bath for about
10 minutes. At the end of the heating period cool the mixture.

Dilute with 15 cm³ of methanol and collect the solid cholesteryl benzoate
by suction filtration, using a little methanol to rinse the flask and to wash the
crystals. Recrystalize the cholesteryl benzoate by heating it in a conical flask
with the minimum volume of ethyl ethanoate (about 15–20 cm³) which will
completely dissolve the crystals, cooling in an ice bath, and collecting the crystals
by suction filtration. Yield: from 0.6 to 0.8 g.

The formation of the 'liquid crystal' phase of cholesteryl benzoate can be
seen by placing 0.1 g of the compound on the end of a microscope slide and
heating the sample by holding the slide with a pair of tongs above a small burner

flame. The solid will turn first to a cloudy liquid and then, with further heating, to a clear melt. On cooling, the cloudy liquid crystal phase will appear first, and then it will change to a hard, crystalline solid. With good lighting from the side, for example at a window, a band of colour should be seen at the boundary between the clear melt and the cloudy liquid on both heating and cooling. The more cautious the heating, the better you can see the changes. You can repeat the heating and cooling many times with the same sample.

Reactions of carboxylic acid derivatives

In previous sections, you have met most of the reactions suitable for the preparation of these compounds and you should draw up your own list of methods of preparation.

Nucleophilic reactions

The hydrolysis of the carboxylic acid derivatives amounts to a nucleophilic substitution

$$CH_3-\overset{\displaystyle O}{\underset{\displaystyle W:}{C}} \ + :OH^- \longrightarrow CH_3-\overset{\displaystyle O}{\underset{\displaystyle OH}{C}} \ + :W^-$$

where —W: and :W$^-$ can be —Cl and Cl$^-$

or —OCH$_3$ and OCH$_3^-$

or —NH$_2$ and NH$_2^-$

The ease with which the substitution occurs is related to the power of —W: to attract electrons.

If $-$W: is strongly electron-attracting then the acid derivative can be hydrolysed by a weak nucleophilic reagent such as water; if $-$W: is only weakly electron-attracting, then the hydrolysis of the acid derivative will need a strong nucleophilic reagent such as hydroxide ions.

The relative electron-attracting powers can be summarized as

$$Cl^- < OH^- < CH_3O^- < NH_2^-$$

1 *Acyl chlorides* which mix with water react rapidly, even violently, to give the parent acid

$$CH_3—COCl + H_2\ddot{O} \longrightarrow CH_3—CO_2H + HCl$$

Other nucleophiles such as alcohols, ammonia, and amines also react readily and this is an excellent method of preparing esters and amides.

$$CH_3—COCl + CH_3\ddot{O}H \longrightarrow CH_3—CO_2—CH_3 + HCl$$
methyl ethanoate

$$CH_3—COCl + \ddot{N}H_3 \longrightarrow CH_3—CO—NH_2 + HCl$$
ethanamide

$$CH_3—COCl + CH_3\ddot{N}H_2 \longrightarrow CH_3—CONH—CH_3 + HCl$$
N-methyl-
ethanamide

The hydrogen chloride produced in the last two examples will react with any excess of ammonia and amine to form ammonium salts, for example, NH_4Cl.

2 *Esters* react quite slowly with water and an acid or base catalyst is necessary

$$\langle\bigcirc\rangle—CO_2CH_3 + H_2O \underset{}{\overset{H^+}{\rightleftharpoons}} \langle\bigcirc\rangle—CO_2H + CH_3OH$$

If a base is used the product is the anion of the carboxylic acid. The anion is stable to attack by weak nucleophiles such as alcohols and the reaction is therefore not reversible when a base is used.

$$\langle\bigcirc\rangle—CO_2CH_3 + OH^- \longrightarrow \langle\bigcirc\rangle—CO_2^- + CH_3OH$$

3 *Amides* are hydrolysed very slowly. Even with an acid catalyst the reaction is slow but a base is more effective.

$$CH_3CONH_2 + H_2O \xrightarrow{OH^-} CH_3CO_2H + NH_3 \longrightarrow CH_3CO_2^- NH_4^+$$
ammonium
ethanoate

4 *Nitriles* will also react with water if a strong acid or strong base catalyst is used. The reaction can be stopped at the stage when an amide has been formed or carried right through to the parent acid.

$$\langle\bigcirc\rangle—CN \xrightarrow[H^+]{H_2O} \langle\bigcirc\rangle—CONH_2 \xrightarrow[H^+]{H_2O} \langle\bigcirc\rangle—CO_2H$$
benzonitrile   benzamide   benzoic acid

The reactions can also be carried out in reverse and form a suitable method of preparing nitriles. The ammonium salt of a carboxylic acid is first heated, when the amide will be formed with loss of water; then the amide can be dehydrated by heating with phosphorus(v) oxide.

$$CH_3CO_2H \xrightarrow{NH_3} CH_3CO_2^- NH_4^+ \xrightarrow{heat} CH_3CONH_2 \xrightarrow{P_2O_5} CH_3CN$$

ethanoic   ammonium   ethanamide   ethane-
acid   ethanoate   nitrile

Reduction

5 The reduction of the derivatives of carboxylic acid by lithium tetrahydridoaluminate, $LiAlH_4$, proceeds in the same manner as the reduction of the parent acids. The products from acyl chlorides and esters are alcohols.

phenylmethanol

The reagent lithium tetrahydridoaluminate is too expensive to use on a large scale but hydrogen can be made to react by using a high pressure and a suitable catalyst, such as a mixed oxide catalyst of chromium(III) oxide and copper(II) oxide.

6 The reduction of nitriles produces amines

phenylmethylamine

BACKGROUND READING 2
Liquid crystal displays

Electronic watches and calculators have made liquid crystal displays commonplace, whereas before the 1980s liquid crystals were only used in specialized applications. Collaborative research between the Royal Signals and Radar Establishment at Malvern and the Chemistry Department at Hull University has been largely responsible for the rapid advance of this new technology.

An Austrian botanist, Friedrich Reinitzer, made the first observations of liquid crystal behaviour in 1888. When investigating the formula of cholesterol he prepared cholesteryl benzoate and in determining its melting point noticed that it melted first to a turbid liquid, at 147 °C, and then changed to a clear liquid at 180 °C. The phase between 147 °C and 180 °C is a true liquid phase but the molecules have a degree of orientation and the liquid therefore has some of the properties of a crystal.

Compounds which display liquid crystal behaviour usually have relatively rigid rod-like molecules and in the liquid crystal phase the rod-like molecules take up various parallel arrangements. For example, in a nematic liquid crystal phase (figure 13.3a) the molecules lie parallel, without any regular arrangement of the centres of the molecules. That is, no layer arrangement exists, as for example is the case in the smectic type of liquid crystal.

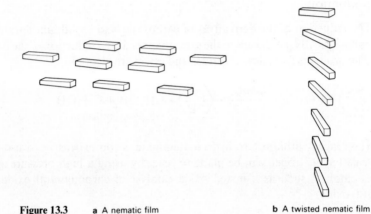

Figure 13.3   **a** A nematic film   **b** A twisted nematic film

Nematic is the ancient Greek for a thread and the name is derived from the thread-like lines which can be seen when a nematic liquid crystal is observed through a polarizing microscope.

If the rod-like molecules are chiral, we obtain, instead of a nematic phase, a spontaneously twisted nematic phase – frequently called a cholesteric liquid crystal phase. The molecules may be thought of as constituting sheets stacked on top of one another. In any one sheet, the molecules are arranged with their long axes parallel, as in a nematic phase, but each sheet is rotated slightly, relative to adjacent sheets. On passing through a stack of sheets, the long axes of the molecules trace out a spiral or helical path. The pitch of the spiral is affected by temperature and this changes the optical properties of the phase. Cholesteric liquid crystals are therefore used in thermochromic temperature sensing devices.

Another form of twist can be imposed on a nematic, this time by the external forces exerted by the upper and lower surfaces of two glass plates confining a thin film of the nematic phase. If the molecules lie flat to each plate, but the directions of the long axes at the two surfaces are at 90°, the orientation of the long axes changes gradually on passing through the film, that is, a quarter helix is formed as shown in figure 13.3b.

An important feature of the mechanically induced twisted nematic liquid crystal film is that the orientation of the molecules can be changed by small electric fields, and this is used as the basis of a display unit for watches and calculators. A typical display unit consists of a thin film of liquid crystal sandwiched between two glass plates which have a transparent conductive coating on their inner surfaces. The inner surfaces are also treated so that the molecules are orientated with their long axes at 90° and form a quarter helix (see above) between the glass plates.

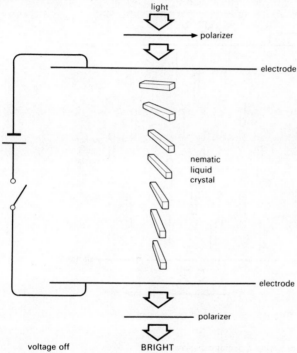

Figure 13.4
The off-state of a twisted nematic display device.

Light which enters the display unit first passes (figure 13.4) through a polarizer. This polarizer is set so that the emergent plane polarized light is vibrating in a plane which contains the direction of orientation of the molecules at that surface. When the plane polarized light passes through the twisted nematic liquid crystal, the plane of polarization is guided or rotated through 90°.

The light then emerges from the display unit with its plane of polarization oriented so that it passes through a second polarizer set at 90° to the first polarizer. The display unit therefore appears bright, either in reflection or transmission. If a small voltage is now applied, using the conductive coatings on the glass plates as electrodes (figure 13.5), the orientation of the molecules is changed. They now stand up at right angles to the surfaces and no longer rotate the plane of polarization of the light. The plane polarized light therefore does not pass through the second polarizer and the display unit appears dark. On switching off the voltage , the molecules relax back to the original quarter helical arrangement of figure 13.4, and the cells becomes bright again.

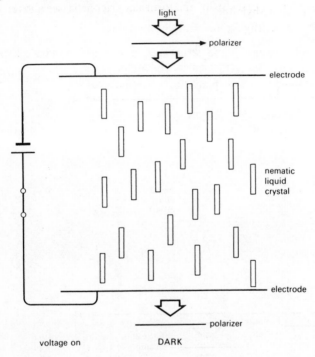

Figure 13.5
The on-state of a twisted nematic display device.

By producing the conductive layer as separate small areas any pattern can be displayed and switched on and off by the application of a small voltage (figure 13.6). As a test of the nature of the light emerging from a liquid crystal display, try the effect of observing the display through a piece of polaroid film or polarizing sunglasses.

Figure 13.6
A watch display using a nematic liquid crystal.
(Photograph, Casio Electronics Company Ltd.)

13.3
AMINES

Amines contain nitrogen. You can think of the amines as derived from ammonia, with one or more of the hydrogen atoms replaced by an alkyl group.

CH_3NH_2 methylamine,   a primary amine
$(CH_3)_2NH$ dimethylamine,   a secondary amine
$(CH_3)_3N$ trimethylamine,   a tertiary amine

You should compare these structures with the structures of primary, secondary, and tertiary halogenoalkanes and alcohols.

Phenyl groups may also replace the hydrogen atoms

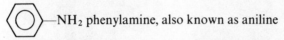—NH_2 phenylamine, also known as aniline

If you consider the properties of ammonia you should expect amines to be basic, to be good nucleophiles, and to form complex ions with metal cations. In the next section we shall be testing these expectations.

The N—H bond absorbs in the infra-red in the region 3500–3300 cm^{-1}; the trough is broadened by hydrogen bonding if it occurs (figure 13.7).

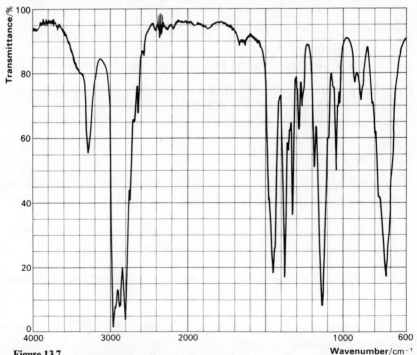

Figure 13.7
The infra-red absorption spectrum of diethylamine.

Experiments with amines

In these experiments, you will be looking at the reactions of ammonia, butyl-amine, and phenylamine (aniline).

Before starting the experiments, make sure that you are familiar with the formulae of the three compounds, and the shapes of their molecules, if possible by building models such as those shown in figure 13.8.

Figure 13.8
Space-filling models representing molecules of butylamine and phenylamine.

Experiment 13.3
An investigation of the reactions of amines

CAUTION: phenylamine is harmful by skin absorption; use it in a fume cup-board. Butylamine is extremely flammable and an irritant to the skin, eyes, and respiratory system. Eye protection should be worn throughout this experiment.

Procedure

1 *Solubility and pH*   Prepare or obtain dilute aqueous solutions of the three compounds and add drops of Full-range Indicator.

Are the compounds readily soluble or only sparingly soluble? What type of molecular interaction will be helping them dissolve?

Are the compounds acidic or basic, strong or weak? Use the *Book of data* to look up the pK_a values of their conjugate acid.

2 *Formation of salts*   Add drops of the three compounds to separate por-tions of 2M hydrochloric acid.

Does the odour disappear?
Are the amines more soluble in hydrochloric acid than water?
Could they be reacting with the hydrochloric acid?

Write down equations representing any reactions which you consider to be taking place.

How would you recover the butylamine or phenylamine from their mixture with 2M hydrochloric acid?

Test your suggestion experimentally.

3 *Reaction with metal ions*   Add drops of the three compounds to separate portions of 0.1M copper(II) sulphate solution until present in excess.

Are somewhat similar results obtained?
What type of interaction do you think is taking place?

4 *Ethanoylation*   To 0.2 cm³ of ethanoyl chloride add, *very carefully, one drop at a time,* 0.2 cm³ of butylamine or phenylamine. A solid derivative should be obtained whose melting point is characteristic of the original amine. The product from phenylamine can be recrystallized by adding 1 cm³ of water, warming until the solid dissolves, and then allowing the solution to cool slowly.

In this reaction, which reagent is the electrophile and which the nucleophile? Write an equation for the reaction.

5 *Laboratory preparations*   If the time is available, you could tackle one of the following preparations: 2-ethanoylaminobenzoic acid (an unusual ethanoyl compound) or nylon.

Preparation of 2-ethanoylaminobenzoic acid

2-aminobenzoic acid   (intermediate)   2-ethanoylaminobenzoic acid

Place 3.5 g of 2-aminobenzoic acid in a 50-cm³ flask fitted with a reflux condenser. Add 10 cm³ of ethanoic anhydride, $(CH_3CO)_2O$ (handle with care; it irritates skin and respiratory system). Heat slowly to boiling and reflux for a quarter of an hour. Allow the solution to cool, add 5 cm³ of water, bring the mixture slowly back to the boil, and then allow it to cool *slowly*. Collect the crystals of 2-ethanoylaminobenzoic acid by suction filtration, having the apparatus behind a safety screen, and wash them with a small quantity of cold methanol. The acid (melting point 183–5 °C) may be recrystallized from a mixture of ethanoic acid and water. Allow the crystals to dry thoroughly.

Place several crystals of the product between two watch-glasses. In a darkened room, grind the two glasses together, when flashes of light should be seen. The emission of light has been explained in terms of an electric discharge between the surfaces of the fractured crystal which causes excitation of the molecule with subsequent fluorescence. For the phenomenon to be observed, the crystals must be well formed and free from solvent.

This phenomenon was named *triboluminescence* in 1895 (from the ancient Greek *tribein*, to rub) although the effect had been described by Francis Bacon as long ago as 1605. Francis Bacon noticed that flashes of light are emitted when crystals of sugar are scraped or crushed. Some sugar products also emit light when crushed as well as other compounds such as benzene-1,2-dicarboxylic anhydride and uranyl(VI) nitrate hexahydrate.

The 'nylon rope trick'
This experiment was originated by P. W. Morgan and S. L. Kwolek of E. I. du Pont de Nemours and Co. Inc.

Prepare a solution of 0.5 cm³ of decanedioyl dichloride in 15 cm³ 1,1,2,2,-tetrachloroethane in a 100 cm³ beaker and, separately, a solution of 0.7 g of hexane-1,6-diamine (*CAUTION*: this is caustic) and 2 g of sodium carbonate in 15 cm³ of water.

Clamp the beaker, and alongside it clamp a pair of glass rods as shown in figure 13.9. If possible allow a drop of about 2 metres from the roller to the receiver.

Now pour the aqueous solution carefully onto the tetrachloroethane solution and, using crucible tongs, pull the interfacial film out, over the rods, and down towards the receiver. When a long enough rope has formed, the process should go on of its own accord until the reagents are used up, but the rope may need to be pulled out gently, using the crucible tongs. Take care not to get either solvent or reagent on your fingers.

To obtain a dry specimen of nylon polymer, wash it thoroughly in 50% aqueous ethanol and then in water until litmus is not turned blue by the washings. Note that, because of the way in which it is formed, the nylon 'rope' is likely to be a hollow tube, containing solvent and possibly reagent. You

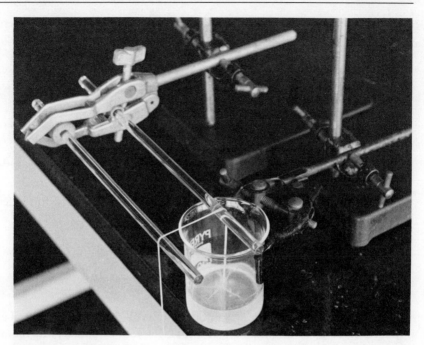

Figure 13.9
The 'nylon rope trick'.
Photograph, University of Bristol, Faculty of Arts Photographic Unit.

should therefore take care when handling it, even after washing in this way.

Write an equation for this polymerization reaction.

Reactions of amines

1 *Basic properties*   Amines are readily soluble in water. Solubility is assisted by hydrogen bonding through a lone pair of electrons on the nitrogen atom

$$CH_3-\overset{\displaystyle H}{\underset{\displaystyle H}{N}}\cdots H-O-H$$

Amines are strong bases, as can be seen from the pK_a values of their conjugate acids (table 13.3). Phenylamine is an exception; electron pair interaction with π bonds in the molecule of this compound much reduces the acceptance of protons by the NH_2 group. You should compare this interaction with the interaction that occurs in phenol.

Compound	pK_a
Phenylammonium ion	4.6
Ammonium ion	9.3
Ethane-1,2-diammonium ion	9.9
Butylammonium ion	10.8

Table 13.3

pK_a values of the conjugate acids of some amines; 'K_a values are defined on page 41. They refer to K_a for the equilibrium $RNH_3^+(aq) \rightleftharpoons H^+(aq) + RNH_2(aq)$.

In the presence of strong acids the equilibrium is shifted to the left and the amines form soluble salts

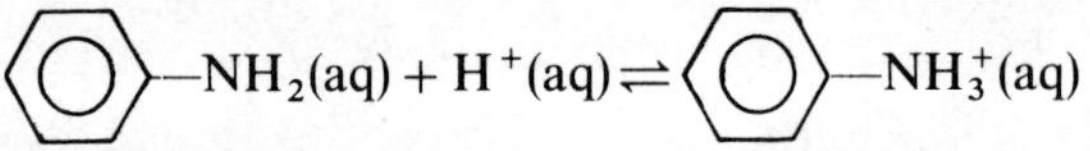

phenylamine   phenylammonium ion

2 *Complex ions*   The lone pair of electrons on the nitrogen atom of amines makes the compounds suitable ligands for metal cations, parallel with the familiar deep blue-coloured complex ion formed between ammonia and copper(II) cations

$$4NH_3 + Cu(H_2O)_4^{2+} \rightleftharpoons Cu(NH_3)_4^{2+} + 4H_2O$$

but the stoicheiometry of the reaction may be different

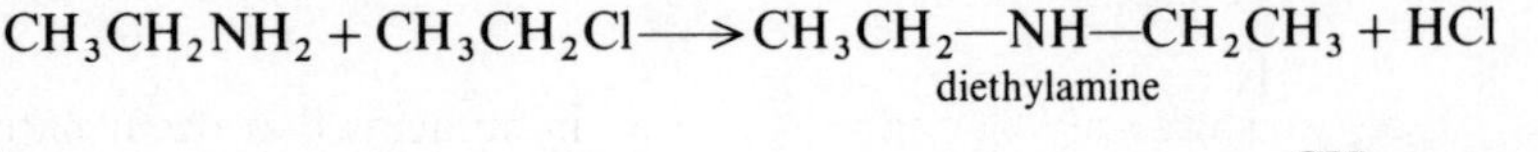

3 *Reaction with electrophiles*

a Amines, being themselves nucleophiles, will react with the electrophilic carbon atom of halogenoalkanes to give secondary and tertiary amines.

$$CH_3CH_2NH_2 + CH_3CH_2Cl \longrightarrow CH_3CH_2{-}NH{-}CH_2CH_3 + HCl$$
diethylamine

and then

$$CH_3CH_2{-}NH{-}CH_2CH_3 + CH_3CH_2Cl \longrightarrow CH_3CH_2{-}\underset{|}{\overset{|}{N}}{-}CH_2CH_3 + HCl$$

$$or\ (CH_3CH_2)_3N\ \text{triethylamine}$$

When chloroethane and ammonia react, ethylamine is produced (see Topic 9, section 9.3), but a mixture of products is obtained because the ethylamine will react with more chloroethane, according to the above sequence of reactions.

b Amines also react as nucleophiles with acyl chlorides forming amides

$$CH_3COCl + \langle\bigcirc\rangle\!-\!NH_2 \longrightarrow \langle\bigcirc\rangle\!-\!NH\!-\!CO\!-\!CH_3 + HCl$$

ethanoyl   phenylamine   *N*-phenylethanamide
chloride

A difficulty with the reaction is that the amine can react with the hydrogen chloride produced, forming a salt (see reaction **1** on page 85), and this prevents reaction with the acid chloride. For this reason, the 'nylon rope trick' was carried out in the presence of sodium carbonate.

$$nH_2N(CH_2)_6NH_2 + nClOC(CH_2)_8COCl$$

hexane-1,6-diamine   decanedioyl dichloride

$$\longrightarrow \left[\!NH(CH_2)_6NH\!-\!\overset{\overset{O}{\|}}{C}(CH_2)_8\overset{\overset{O}{\|}}{C}\!\right]_n + 2nHCl$$

nylon 6–10

This very important polymer will be discussed further in Topic 17.

13.4
AMINO ACIDS AND PROTEINS

Amino acids possess the two functional groups that you have just studied, namely the amino group, $-NH_2$, and the carboxylic acid group, $-CO_2H$. The simplest amino acid is aminoethanoic acid (glycine)

$$NH_2\!-\!CH_2\!-\!CO_2H$$

glycine (gly)

As we shall only be concerned with amino acids that occur naturally we shall be using the non-systematic names that are favoured by biochemists. These names are often abbreviated to a 3-lettered 'code', which usually consists of the first three letters of the name. The amino acids that occur naturally are all 2-amino acids with the amino group on the carbon atom adjacent to the acid group.

$$NH_2-CH-CO_2H$$
$$| $$
$$CH-CH_3$$
$$| $$
$$CH_3$$

valine (val)

$$NH_2-CH-CO_2H$$
$$| $$
$$CH_2OH$$

serine (ser)

The significance of the 2-amino acid structure is that the naturally occurring amino acids commonly have chiral molecules. Nature appears to be stereospecific because almost without exception the naturally occurring amino acids in living material have an L configuration. After death a slow conversion from the L to the D form occurs until eventually an equilibrium mixture results. The configuration of amino acids is shown below.

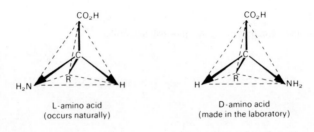

L-amino acid
(occurs naturally)

D-amino acid
(made in the laboratory)

Chemical tests reveal the presence of nitrogen compounds in most animal tissues and, more specifically, the acidic hydrolysis of hair, blood, and muscle tissue shows that they consist almost entirely of amino acids. Naturally occurring compounds made from amino acids are known as proteins.

$$NH_2-CH-CO_2H$$ an amino acid
$$| $$
$$R$$

$$-NH-CH-CO-NH-CH-CO-NH-CH-CO-$$ part of a
$$|\qquad\qquad|\qquad\qquad| $$ protein
$$R\qquad\qquad R\qquad\qquad R$$

The term *peptide group* is used to describe the amide group —CO—NH— when it links together amino acid residues.

When a number of amino acid residues are connected by peptide groups the molecules are known as *polypeptides*.

The biological importance of proteins is apparent from their widespread occurrence and variety of function (table 13.4 and figure 13.10).

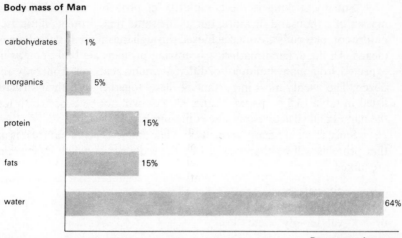

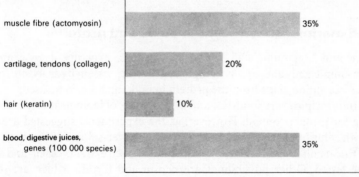

Figure 13.10
The importance of proteins in Man.

Protein	Occurrence (examples)	Function	Relative molecular mass	Approximate number of amino acid units
Insulin	animal pancreas	governs sugar metabolism	5 700	51
Myoglobin	muscle	oxygen carrier	17 000	153
Trypsin	animal pancreas	digests food proteins	23 800	180
Haemoglobin	blood	oxygen carrier	66 000	574
Urease	soya beans	converts urea to ammonia	480 000	4 500

Table 13.4
The occurrence and function of some proteins.

Equally striking is the complexity of proteins, with relative molecular masses of a thousand or more; but at the same time proteins illustrate how in nature complex ends are often achieved through the infinitely varied use of simple means. All the different naturally occurring proteins are built, not, as might be expected, from many hundreds of different amino acids, but from only about two dozen. The twenty most important of these, together with their structures, are listed in table 13.5 on pages 92 and 93. You will not be expected to remember the names and structures of these compounds.

Since there are many proteins and few amino acids in nature it is apparent that proteins will be characterized by the sequence in which their amino acids are linked:

         –gly–val–ser–
or   –val–gly–ser–
or   –val–ser–gly–   etc.

The determination of the amino acid sequence of a protein is considered in the next section.

Experiments with amino acids and proteins

Glycine and L-glutamic acid can be used in these experiments as examples of simple amino acids and, as an example of a mixture, casein hydrolysate (containing the free amino acids from the protein in milk) might also be used.

In principle every foodstuff and all materials of biological origin are worth testing for protein content. However, as the experiments suggested are carried out with solutions it would be as well to select water-soluble substances.

Fresh milk, egg (white and yolk), or the derived extracts casein and albumin can be tested. Other possibilities are pepsin and trypsin, which are digestive enzymes; gelatin, from the hydrolysis of the connective tissue of animals; and commercial meat extracts.

EXPERIMENT 13.4a
An investigation of protein materials
Procedure

Make solutions in water of your samples of protein materials, warming if necessary, and use them for the following experiment.

1 *Acidity and basicity*   To 2 cm^3 of 0.01M hydrochloric acid add a few drops of Full-range Indicator and note the effect on the pH of adding 0.01M sodium hydroxide in $0.5 \, cm^3$ portions. Now repeat the experiment, using a solution of 0.01M glycine or L-glutamic acid in place of the hydrochloric acid.

Does the pH change gradually or sharply?
What type of acid–base behaviour is occurring?

2 *Biuret test*   Add an equal volume of 2M sodium hydroxide to one of your protein samples in solution followed by 1 *drop* of 0.1M copper(II) sulphate. A mauve colour will develop if proteins are present. This is known as the *biuret test* and detects peptide groups:

$$
\begin{matrix}
& O & & & & O & \\
& \parallel & & & & \parallel & \\
-C & -NH & -CH & -C & -NH- \\
& & & | & & & \\
& & & R & & &
\end{matrix}
$$

Carry out a blank test using water as your sample.

What type of reaction and bonding would you expect between copper(II) ions and peptide groups?

3 *Ninhydrin test*   On a piece of chromatography paper place small drops of your solutions and allow to dry. Spray lightly with 0.02M ninhydrin solution in propanone *in a fume cupboard* and again allow to dry. Avoid getting the spray on your fingers. Heat for 10 minutes in an oven at 110 °C or heat cautiously over a Bunsen flame. Red to blue-coloured spots will develop if proteins or amino acids are present. Make a note of any unexpected coloured areas.

Ninhydrin is a reagent used as a specific colour test for amino acids. You do not need to be concerned with the formula of the compound (which is complicated) nor with the details of the chemical reaction which produces the colours.

4 *Chirality*   If there is time investigate the ability of the amino acid solutions to affect polarized light by examining their solutions in a polarimeter.

Formula	Name	Abbreviation	Nature of side chain	R_f value in butan-1-ol/ ethanoic acid/ water
H_2NCHCO_2H | H	glycine	gly	non-polar	0.26
H_2NCHCO_2H | CH_3	alanine	ala	non-polar	0.38
H_2NCHCO_2H | $CHCH_3$ | CH_3	valine	val	non-polar	0.60
H_2NCHCO_2H | CH_2 | $CH(CH_3)_2$	leucine	leu	non-polar	0.73
H_2NCHCO_2H | CHC_2H_5 | CH_3	isoleucine	ile	non-polar	0.72
$HN-CH-CO_2H$ $CH_2 \quad CH_2$ CH_2	proline	pro	non-polar	0.43
$H_2N-CH-CO_2H$ | CH_2 | C indole ring (CH, NH)	tryptophan	try	non-polar	0.50
H_2NCHCO_2H | CH_2 | CH_2SCH_3	methionine	met	non-polar	0.55
H_2NCHCO_2H | $CH_2-C_6H_5$	phenyl-alanine	phe	non-polar	0.68
H_2NCHCO_2H | CH_2OH	serine	ser	polar	0.27

Structure	Name	Abbr.	Class	Value
H_2NCHCO_2H $\mid$ $CHOH$ $\mid$ CH_3	threonine	thr	polar	0.35
H_2NCHCO_2H $\mid$ CH_2SH	cysteine	cys	polar	0.08
H_2NCHCO_2H $\mid$ CH_2 $\mid$ $CONH_2$	asparagine	asn	polar	0.19
H_2NCHCO_2H $\mid$ CH_2 $\mid$ CH_2CONH_2	glutamine	glu	polar	—
H_2NCHCO_2H $\mid$ $CH_2-\bigcirc-OH$	tyrosine	tyr	polar	0.50
H_2NCHCO_2H $\mid$ CH_2 $\mid$ $C{=}CH$ $\mid \quad \mid$ $HN \quad N$ $\diagdown \diagup$ CH	histidine	his	basic	0.20
H_2NCHCO_2H $\mid$ $(CH_2)_3$ $\mid$ NH $\mid$ $HN{=}C{-}NH_2$	arginine	arg	basic	0.16
H_2NCHCO_2H $\mid$ $(CH_2)_3$ $\mid$ CH_2NH_2	lysine	lys	basic	0.14
H_2NCHCO_2H $\mid$ CH_2CO_2H	aspartic acidic	asp	acidic	0.24
H_2NCHCO_2H $\mid$ CH_2 $\mid$ CH_2CO_2H	glutamic acid	glu	acidic	0.30

Table 13.5
The twenty 'standard' amino acids.

EXPERIMENT 13.4b
The chromatographic separation of amino acids

The experiment is designed to give you experience and understanding of an important method. To separate and identify naturally occurring amino acids by paper chromatography would require an effort spread over about three days, so this brief experiment can only suggest the potentialities of the method.

To obtain satisfying results you will have to work with care and keep the experimental materials scrupulously clean. Touch the chromatography papers only on their top corners and never lay them down except on a clean sheet of paper.

Procedure

Put spots of 0.01M amino acids in aqueous solution 1.5 cm from the bottom edge of the chromatography paper cut to the dimensions shown in the diagram (see figure 13.11). To do this, dip a *clean* capillary tube in the stock solution and apply a small drop to the chromatography paper, using a quick delicate touch. Practise on a piece of ordinary filter paper until you can produce spots not more than 0.5 cm in diameter. Apply spots of individual amino acids and also mixtures, making identification marks in *pencil* at the top of the paper. Allow the spots to dry.

Meanwhile prepare a fresh solvent mixture of butan-1-ol ($12 \, cm^3$), glacial ethanoic acid ($3 \, cm^3$), and water ($6 \, cm^3$) in a covered 1-dm^3 beaker. Cover the beaker to produce a saturated atmosphere.

Now roll the chromatography paper into a cylinder and secure it with a paper clip. Stand the cylinder in the covered solvent beaker and leave it for the solvent to ascend to nearly the top of the paper. As 20 minutes are needed to complete the experiment after removing the paper from the beaker, there may not be time to allow the solvent to rise the full distance.

Remove the chromatography paper from the beaker and mark the solvent level. Dry the paper (without unfastening it), in an oven if possible, but *not* over a Bunsen flame, because the solvent is both pungent and flammable.

Detect the amino acids by spraying the paper sparingly with 0.02M ninhydrin solution in a fume cupboard and then heating in an oven at 110 °C for 10 minutes. Purple spots should appear at the positions occupied by the amino acids.

Preserve the spots by spraying with a mixture made up of methanol ($19 \, cm^3$), M aqueous copper(II) nitrate ($1 \, cm^3$), and 2M nitric acid (a drop), and then expose *in a fume cupboard* to the fumes from 0.880 ammonia. Determine the R_f value of amino acid samples.

R_f values are obtained using the expression

$$R_f \text{ value} = \frac{\text{distance moved by amino acid}}{\text{distance moved by solvent}}$$

Have the mixtures separated?

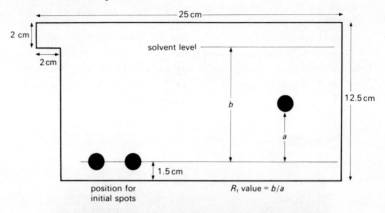

Figure 13.11
Apparatus for simplified chromatography of amino acids.
Photograph, University of Bristol, Faculty of Arts Photographic Unit.

The result of a more complicated procedure is illustrated in figure 13.12.

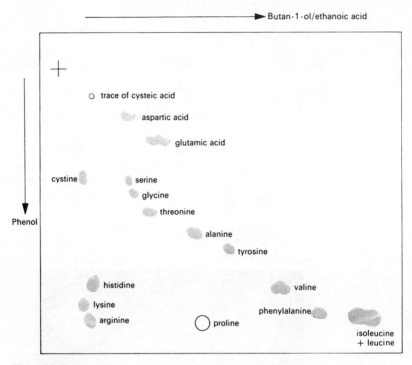

Figure 13.12

Paper chromatography separates the 17 amino acids of insulin. In the chromatogram represented by this diagram, insulin was broken down by hydrolysis and a sample of the mixture placed on the upper left corner of the sheet of paper. The sheet was hung from a trough filled with the solvent which carried each amino acid a characteristic distance down the paper. The sheet was then turned through 90 degrees and the process repeated. The amino acids, with the exception of proline, appear as purple spots when sprayed with ninhydrin.

After THOMPSON, E. O. P. (1955) 'The insulin molecule' Scientific American 192, 5. Copyright © 1955, by Scientific American Inc. All rights reserved.

BACKGROUND READING 3
The chemical and structural investigation of proteins

If the correct molecular formula of a substance is to be established it must be available pure, and the necessary experimental techniques must be available. F. Sanger, working at Cambridge, was the first chemist to establish the molecular formula of a protein. When he started work in 1944, he had to develop new experimental techniques because the problems of protein composition were unsolved at that time, and the only simple protein available pure was insulin. Ten year's work was necessary to establish the correct amino acid sequence for insulin.

The amino acid composition of proteins

The relative molecular mass of insulin is about 5700 and its formula is $C_{254}H_{377}N_{65}S_6$! As is the case with any protein, the first stage in the investigation of insulin was to discover the nature, and number, of the amino acid residues present.

Hydrolysis of a protein by heating in a sealed tube with 6M hydrochloric acid for 24 hours produces the free amino acids. Quantitative separation of the hydrolysate will determine which amino acids are present and their relative amounts. Sanger found that the insulin molecule contains 17 different amino acids ranging from six molecules of cysteine and leucine to one molecule of lysine, the molecular formula being accounted for by a total of 51 amino acid units.

Mixtures of amino acids are usually separated by ion exchange chromatography on a column of a cation exchange resin in the sodium form. The amino acid mixture is added to the column in an acidic solution at pH 3. Under these conditions the basic amino acids are readily exchanged for sodium ions and become absorbed onto the resin. The most acidic amino acids are only slightly absorbed, and if the column is eluted by a solvent at pH 3, they will move down the column faster than the basic amino acids. Amino acids can be eluted from the column in sequence from the most acidic first to the most basic last by progressively changing the pH of the eluting solvent. The amino acids are detected and their concentration measured by their colour reaction with ninhydrin. Automated instruments are available which will carry out a full analysis, using less than a milligram of material, in about two hours.

After the chromatographic identification of the amino acids present in a protein the next stage in the investigation is to determine which amino acids are on the ends of the protein chain. The commonly used technique to identify the amino acid at the end of the chain with a free amino group ($—NH_2$) was devised by Edman.

The technique involves the addition of phenylisothiocyanate, ⟨O⟩—NCS, which reacts with free amino groups. The resulting derivative of the terminal amino acid is then hydrolysed off the chain and identified by gas chromatography (you are not expected to learn the formulae of these compounds).

$$\langle\bigcirc\rangle\!\!-\!\!NCS + NH_2\!\!-\!\!CH\!\!-\!\!CO\!\!-\!\!NH\!\!-\!\!CH\!\!-\!\!CO\!\!-\!\!NH\!\!-\!\!CH\!\!-\!\!CO\!\!-\quad etc.$$

$$R_1R_2R_3$$

$$\downarrow$$

$$\overset{\displaystyle S}{\underset{\displaystyle \parallel}{}}$$

$$\langle\bigcirc\rangle\!\!-\!\!NH\overset{S}{\overset{\parallel}{C}}\!\!-\!\!NH\!\!-\!\!CH\!\!-\!\!CO\!\!-\!\!NH\!\!-\!\!CH\!\!-\!\!CO\!\!-\!\!NH\!\!-\!\!CH\!\!-\!\!CO\!\!-\quad etc.$$

$$R_1\qquad\qquad R_2\qquad\qquad R_3$$

$$\downarrow$$

$$\langle\bigcirc\rangle\!\!-\!\!\underset{\underset{\displaystyle O}{\overset{\displaystyle |}{C}\!-\!CH}}{N}\!\!-\!\!\overset{S}{\overset{\parallel}{C}}\!\!-\!\!NH + NH_2\!\!-\!\!CH\!\!-\!\!CO\!\!-\!\!NH\!\!-\!\!CH\!\!-\!\!CO\!\!-\quad etc.$$

$$R_1\qquad\qquad R_2\qquad\qquad R_3$$

The next amino acid residue in the shortened protein chain can be identified by a repetition of the procedure. The procedure has been so well standardized that automated instruments are available that will rapidly determine the amino acid sequence in polypeptides of up to 20 residues.

The amino acid at the other end of the chain has a free carboxylic acid group ($-CO_2H$) and this amino acid can be reduced with lithium tetrahydridoborate to the corresponding amino alcohol. After complete hydrolysis of the protein chain the amino alcohol can be identified by chromatographic analysis.

$$etc.\!\!-\!\!NH\!\!-\!\!CH\!\!-\!\!CO\!\!-\!\!NH\!\!-\!\!CH\!\!-\!\!CO\!\!-\!\!NH\!\!-\!\!CH\!\!-\!\!CO_2H + LiBH_4$$

$$R_3\qquad\qquad R_2\qquad\qquad R_1$$

$$\downarrow$$

$$-NH\!\!-\!\!CH\!\!-\!\!CO\!\!-\!\!NH\!\!-\!\!CH\!\!-\!\!CO\!\!-\!\!NH\!\!-\!\!CH\!\!-\!\!CH_2OH$$

$$R_3\qquad\qquad R_2\qquad\qquad R_1$$

$$\downarrow$$

$$NH_2\!\!-\!\!CH\!\!-\!\!CO_2H + NH_2\!\!-\!\!CH\!\!-\!\!CO_2H + NH_2\!\!-\!\!CH\!\!-\!\!CH_2OH$$

$$R_3\qquad\qquad\qquad R_2\qquad\qquad\qquad R_1$$

To determine the amino acid sequence along the whole length of a protein chain, the next step is the partial hydrolysis of the chain into relatively short polypeptides. The polypeptides are separated by ion exchange chromatography, their terminal amino acids identified, and their complete amino acid sequence determined if the chain is short enough. The procedures are the same as those already described. If the partial hydrolysis of the protein is carried out two or three times by different reagents so that polypeptides of overlapping composition are obtained, then the amino acid sequence in the original protein can be deduced.

For example the enzyme *trypsin* is very specific in its hydrolysis of proteins and will only attack peptide groups derived from the carboxyl groups of *lysine* (lys) and *arginine* (arg). The use of the enzyme *thermolysin* produces different polypeptides because the protein chain will be broken at the peptide groups derived from the amino groups of *leucine* (leu), *isoleucine* (ile), and *valine* (val).

Sanger used less refined techniques in his work on insulin and the result that he obtained is shown in figure 13.13.

Using today's procedures, it is possible to determine much longer amino acid sequences and the use of automated instruments has greatly speeded up the rate at which the work can be carried out.

In the future, mass spectrometry is likely to become increasingly important in the determination of the amino acid sequence of proteins; dihydrofolate reductase in 1979 was the first example. The advantages of the mass spectrometer are speed, the small amount of sample needed (only $5–30 \times 10^{-9}$ mol are required) and the polypeptides produced by partial hydrolysis need not be completely separated from each other because analysis is possible on a mixture of two to five peptides.

The shapes of protein molecules

There are countless possibilities for the shape of a large molecule of a protein such as myoglobin. These possibilities arise because of the rotation that can take place at the single bonds in the polypeptide chain. A given protein molecule may actually adopt several of these possible shapes, or *conformations*, as they are called. However, if a protein can be obtained in a crystalline state it is very likely that all the molecules making up the crystal have the same conformation.

The first requirement in investigating the possible conformations of a protein is to know as much as possible about the shape and properties of the peptide group. To this end X-ray diffraction studies (figure 13.14) have been made of simple compounds such as ethanoylglycine:

$CH_3CONHCH_2CO_2H$

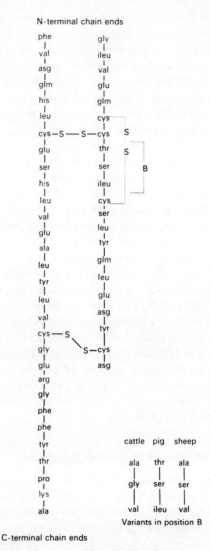

Figure 13.13
The arrangement of the fifty-one amino acids in the molecule of human insulin.
Insulins from other sources differ slightly in the short section marked B.

By giving due weight to all the experimental evidence from the study of a number of dipeptides, an average peptide group can be described. The data are given in figure 13.15.

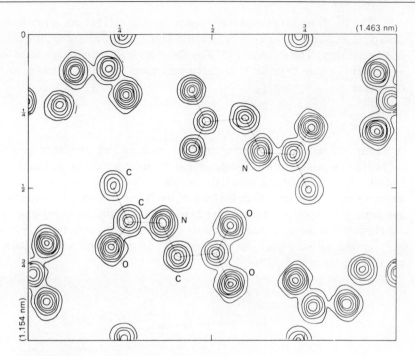

Figure 13.14
An electron density map of ethanoylglycine.
Reprinted with permission. After CARPENTER, G. B. *and* DONOHUE, *J. A. C. S.*
72, *2315, 1950. Copyright 1950 American Chemical Society.*

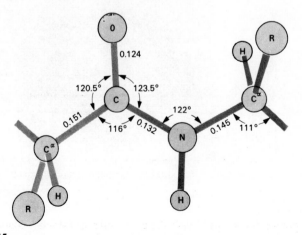

Figure 13.15
The average dimensions of the peptide group (bond lengths in nm).
From Advances in protein chemistry, **22**, *Academic Press, 1967, 249.*

The C—N bond in the average peptide group, at 0.132 nm, is much shorter than the average value for a C—N bond (0.147 nm) and so must have some double bond character. This accounts for the observation that the four atoms of the —CONH— group, together with the carbon atoms on either side of it, all lie in one plane.

The α-helix

A number of naturally occurring proteins such as hair and wool give X-ray diffraction photographs which suggest that the arrangement of their molecules is rod-like. However, natural materials contain a wide variety of amino acids and are not usually well crystallized. Study of the X-ray diffraction photographs of synthetic polypeptides such as polybenzyl glutamate (figure 13.16) leads to the conclusion that the amino acid residues are arranged as rods in the crystalline polypeptide, with a regular repeat every 0.15 nm along the peptide chain, and another repeat every 0.54 nm.

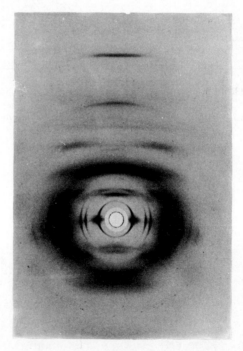

Figure 13.16
An X-ray diffraction photograph of polybenzyl glutamate fibres. The outermost reflection is the 0.15 nm reflection and the large diffuse reflection on the meridian is in the 0.52 nm region.
Photograph, Courtaulds plc.

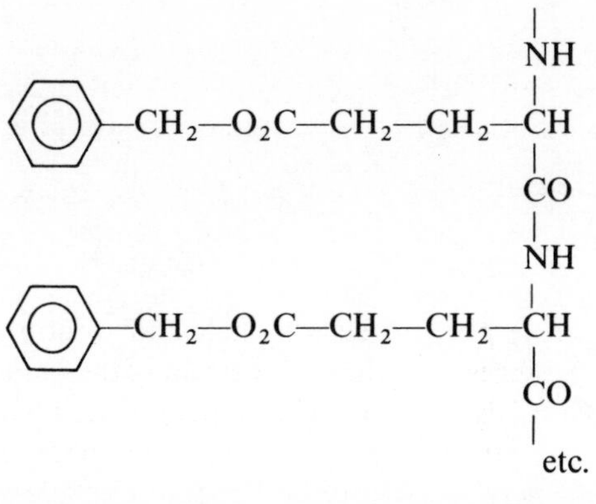

$$\langle\!\langle\bigcirc\rangle\!\rangle\!-\!CH_2\!-\!O_2C\!-\!CH_2\!-\!CH_2\!-\!\overset{\displaystyle\overset{|}{NH}}{\underset{\displaystyle\underset{|}{CO}}{CH}}$$

$$\langle\!\langle\bigcirc\rangle\!\rangle\!-\!CH_2\!-\!O_2C\!-\!CH_2\!-\!CH_2\!-\!\overset{\displaystyle\overset{|}{NH}}{\underset{\displaystyle\underset{|}{CO}}{CH}}$$ polybenzyl-glutamate

etc.

The problem now is to build a scale model of the polypeptide chain which will have regular repeats every 0.54 nm and also every 0.15 nm, as well as having peptide groups and hydrogen bonds of an appropriate size. Look again at Topic 10 for a discussion of hydrogen bonds. After a plausible scale model has been built, it can be analysed and its diffraction pattern calculated. Comparison with actual diffraction patterns for reflection positions and intensities will help to decide if it is a successful model.

The most successful model to be built of the polypeptide chain in rod-like protein molecules was proposed by Pauling and Corey.

They suggested a helical structure of 18 amino acid residues in 5 turns of the helix. The structure is known as the α-helix (figure 13.17) because it is an important component of the structure of the α-keratins which are proteins such as wool, hair, and nails. The β-keratins such as silk, feathers, claws, and beaks have a different structure.

The structure of myoglobin

Myoglobin was the first protein to have its three-dimensional structure worked out. Although myoglobin is a globular protein much of its polypeptide chain has the α-helix conformation (figure 13.18).

The myoglobin molecule structure shows several straight sections of α-helix linked by short lengths of non-helical chain. The haem group, with an iron atom at its centre, is held in place between the chains.

The function of myoglobin in the body is to act as a temporary storehouse in cells for oxygen brought to the cell by the haemoglobin in blood.

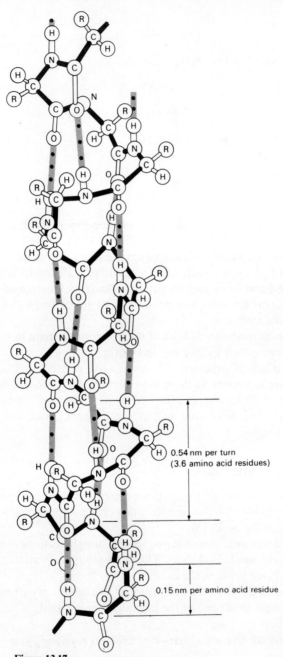

0.54 nm per turn
(3.6 amino acid residues)

0.15 nm per amino acid residue

Figure 13.17
A two-dimensional representation of the α-helix.

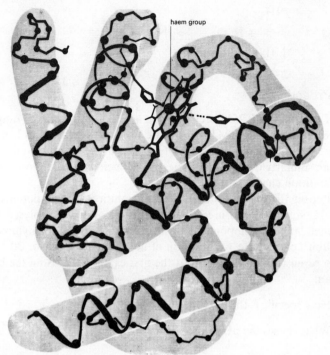

Figure 13.18
The structure of the myoglobin molecule shows the several straight sections of α-helix linked by short lengths of non-helical chain. The haem group, with an iron atom at its centre, is held in place between the chains. *After* DICKERSON, R. E. Ed. NEVRATH, H. The proteins, **2**, *Academic Press, 1964.*

13.5
ENZYMES

Enzymes are proteins whose function in a living organism is to help to bring about necessary biochemical reactions: thus, enzymes can be said to act as a type of catalyst. Any compound whose reaction occurs through the intervention of an enzyme is known as a *substrate* of that enzyme.

The following experiment is designed to illustrate qualitatively the catalytic properties of an enzyme. In later Topics you will be able to compare your results with the characteristics of inorganic catalysts.

EXPERIMENT 13.5
An investigation of the enzyme-catalysed hydrolysis of urea

Urease is an enzyme found in plants; jack beans or water melon seeds are convenient sources. It converts urea to ammonia by a hydrolysis reaction:

$$O{=}C{\overset{\displaystyle NH_2}{\underset{\displaystyle NH_2}{\big\langle}}} + H_2O \xrightarrow{\text{enzyme}} 2NH_3 + CO_2$$

Procedure

1 To $5\,cm^3$ of a 0.25M urea solution add five drops of Full-range Indicator, followed by drops of 0.01M hydrochloric acid until the indicator has just changed to a distinct red colour. Add $1\,cm^3$ of a $1\,\%$ solution of urease active meal (which has been similarly treated with indicator and acid) and note how quickly the pH of the solution changes.

2 Repeat the experiment with 0.25M solutions of compounds which have structural similarities to urea and might therefore be hydrolysed by urease to ammonia. Suitable compounds include ethanamide, and methylurea.

3 Boil $1\,cm^3$ of the $1\,\%$ solution of urease active meal for 30 seconds, then cool to room temperature. Repeat the first experiment, using the boiled urease solution.

How specific is the enzyme activity of urease?

What causes the pH of the solution to change?

The properties of enzymes

As a result of your experimental work you should have some ideas about the properties of enzymes. Unlike an inorganic catalyst such as platinum which catalyses a variety of reactions, enzymes are highly specialized and often for a particular enzyme there is only one reaction of one substrate which it can catalyse. That is, enzymes are *specific* catalysts.

As an example of the specific activity of enzymes, fumarase catalyses only the addition of water to *trans*-butenedioic acid and, furthermore, of the two possible chiral products only one isomer is produced:

trans-butenedioic acid
(fumaric acid)

$+ H_2O \longrightarrow$

L($-$)-2-hydroxybutanedioic
acid (malic acid)

Fumarase is also highly efficient, with a conversion rate at 25 °C of 10^3 molecules of substrate every second per molecule of enzyme! In fact the reaction is so fast, it is controlled by the rate at which the molecules meet by diffusion.

The influence of temperature on catalyst activity is considerable. If the temperature is too high, or too low, enzymes cease to function. So for each enzyme there is an optimum temperature at which, for a given set of conditions, the enzyme will catalyse the greatest amount of chemical change.

Thus a digestive enzyme from a sea squirt was found to have an optimum temperature of 50 °C for a two hour reaction period. Yet being in a sea animal the enzyme will be required to function at a temperature of 15 °C. This does not seem a very efficient digestive situation for the sea squirt until it is realized that it takes up to 60 hours to digest its food. When the optimum temperature over a 60-hour period was investigated it was found to be 20 °C. This illustrates the care needed for the proper investigation of enzyme properties.

What reason can you suggest for enzymes being deactivated by high temperatures?

Enzymes also function best at a particular pH which is not usually very different from neutrality, being mostly in the range pH 5–7. By the use of buffer solutions the pH dependence of α-amylase from saliva can be studied (figure 13.19).

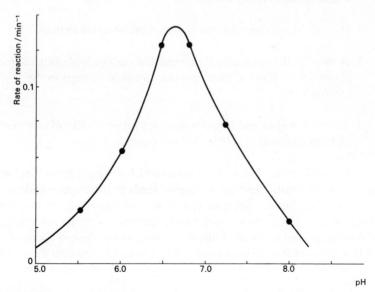

Figure 13.19
The hydrolysis of starch by saliva (a student result).

How might an enzyme be influenced by pH? Remember that enzymes are proteins, so amino acids such as lysine and glutamic acid (see table 13.5) may be present.

An initial understanding of the mechanism of enzyme activity was obtained by kinetic studies. The results suggest that enzymes act catalytically on their substrates and that their reaction, for example for addition of water to the substrate, might be represented as follows:

enzyme + substrate ⟶ enzyme–substrate complex

complex + water ⟶ enzyme + hydrated substrate

The study of enzymes is taken further in the Special Study *Biochemistry*.

Biochemists study the chemical processes that take place in living cells. Most biochemical reactions are catalysed by enzymes and enzyme studies are therefore an integral part of biochemistry. Enzymes are easily damaged and the biochemist cannot employ the extremes of temperature and pressure often required in organic chemistry. Nevertheless the variety of biochemical reactions is quite remarkable. Biochemists attempt to find answers to the following types of question.

1 What is the chemical composition and the function of the compounds which make up living tissues?

2 How are these chemicals made and interconverted in the cell?

3 How is the flow of material through the multitude of chemical pathways controlled so that the organism may adapt to changes in the environment?

4 What is the chemical basis for complex processes like muscle contraction and cell division?

It is perhaps the last two questions which distinguish the biochemist from the organic chemist. The organic chemist tends to investigate single reactions in a closed system but no biochemical reaction can be considered in isolation. The living cell is an open system, continually exchanging matter and energy with its surroundings, and within its walls a multitude of interlinked chemical processes are in a constant state of flux. It is the unravelling of these complex reactions and their inter-relationships which provides much of the challenge of biochemistry.

BACKGROUND READING 4
The industrial uses of enzymes

Enzymes are recognized as being the cornerstones of industrial biochemistry and play an increasingly important role in modern as well as traditional industries.

The introduction of enzymes has revolutionized many industrial processes and their resulting products. The benefits brought to industry by the application of enzymes are increased productivity, lower costs, less wastage, and greater commercial viability.

Enzymes are working proteins manufactured by living cells. Their role is to catalyse the chemical reactions involved in the organic processes necessary for life. If an enzyme is isolated from its cell it will continue to catalyse the same reaction.

Enzymes, like all catalysts, increase the rate of a chemical reaction by providing a pathway with a lower activation energy. At the end of the reaction the enzyme is totally unchanged by the reaction. The rate of an enzyme-catalysed reaction is measured in terms of amount of products, per unit time at stated temperature.

For industrial purpose enzymes are needed in large quantities, in a more or less pure state depending on the process, and in such a state that they can be re-used again and again.

The most effective way of obtaining enzymes in large quantities is by fermentation, using micro-organisms. This process is similar to that used in the brewing industry in which yeasts convert sugars to ethanol. Micro-organisms can be selected (and improved by mutation) to produce larger amounts of the required enzyme. Many microbial enzymes are secreted directly into the culture medium, making their isolation easier.

There are many separation techniques available for the isolation of particular enzymes in a pure state. The degree of purity needed depends on the use of the enzyme; for example, an enzyme which is used to treat cowhide in the leather industry can be used in a much less pure condition than an enzyme in a medical preparation. One common separation technique is selective precipitation by using a solution of ammonium sulphate of a suitable concentration. The resulting enzyme can be purified further by a variety of methods. These methods often involve separation from contaminating enzymes and other proteins by the technique of chromatography, using a column of suitable absorbent or resin. Affinity chromatography, a related technique, relies on the specificity of enzymes to achieve purification, often by preferential binding to selected materials first. The purified enzyme is collected as a particular fraction from the column.

In order to go on using enzymes again and again, certain conditions need to be observed. Enzymes are sensitive to heat and the temperature at which an enzyme works best is called its optimum temperature. This is generally between 25 °C and 50 °C. Enzymes are also sensitive to acidity and alkalinity and have an optimum range of pH. Excess of heat, or extremes of acidity or alkalinity, destroy enzyme activity, and the enzyme is said to be 'denatured'. There are also several compounds which inactivate enzymes, including salts of heavy metals such as mercury or copper. Some enzymes therefore need to be handled and stored carefully under refrigeration, and the temperature and pH conditions within the batch or continuous process have to be maintained close to the optimum in order to prevent denaturing of the enzyme.

For prolonged use in a continuous process or repeated use in a batch process, an enzyme needs to be immobilized, that is, attached to a solid surface so that it is separable from the reaction mixture. Enzymes can be immobilized by attaching them to solid supports (such as glass beads), often by chemically linking the molecules involved. Other methods of enzyme immobilization include micro-encapsulation, absorption, and entrapment in a polymer lattice. Each method has its advantages and disadvantages, depending upon the enzyme and its applications. Enzymes which have not been immobilized may have protective agents added to them to prolong their working life.

Enzymes are extremely valuable to industry because they are very efficient in converting chemicals. Specificity is a most useful characteristic of an enzyme, and usually means that each enzyme produced by the cell can catalyse one reaction only, the conversion of its own individual substrate, which is the substance it changes. The specificity of enzymes is important to their industrial usage. This is because in processes where there are several substances present, an enzyme will be able to recognize its substance from among the diverse mixture present.

The enzyme–substrate relationship is such a close one that many enzymes were named simply by adding -ase to part of the chemical name of the substrate: for example, proteases, which degrade proteins, breaking them down to their constituent amino-acids which are the basic units of proteins. Lipase breaks down lipids (fats) to their constituent molecules of fatty acids and glycerol, and maltase acts on maltose, breaking this disaccharide into two molecules of its component sugar, glucose.

The industry which first publicized the incorporation of enzymes into its merchandise is the washing powder industry. Enzyme washing powders are now well known. They contain a powerful proteolytic enzyme. This is the protease called subtilisin, which is obtained from the growth medium of a bacterium (*Bacillus subtilis*). Subtilisin is relatively stable to heat and is not rapidly destroyed by the detergent also present in the washing powders. Enzyme washing powders remove blood and grease stains by digesting them, biologically. The idea is not a new one, for enzymes from the pancreas of domestic animals had been used in washing powders during the First World War, to remove blood stains from clothing.

Many of the applications of enzymes are in the food industry. This is because foods are made up of proteins, fats, sugars, etc., the very chemicals that nature's catalysts have evolved to deal with. Many of the conversions involve the use of hydrolase enzymes (see table 13.6). Hydrolase is the name given to the group of enzymes which catalyse reactions involving the breakdown of the substrate by hydrolysis. Hydrolytic reactions are ones in which chemical bonds are broken with the addition of water molecules. For example, starch is a polymer consisting of long chain-like molecules of glucose, linked by chemical bonds. The enzyme that acts on starch adds water and splits up the macromolecule in a hydrolytic conversion.

Enzymes are used in the food industry for two main reasons. The first is to improve manufacturing processes and make them cheaper. The second is to improve the foodstuff itself, by using an enzyme on the nearly finished foodstuff. In such an improvement process, the aim is to change the chemical composition of only a portion of the materials involved. The enzyme used for this is chosen because it is known to be specific for the conversion, and will do nothing else.

Enzymes are often used as food additives. However, even though they are natural extracts they have to be tested for toxicity. If an enzyme is to be used as an additive it must first be established that the micro-organism from which it was derived was not pathogenic, nor a likely producer of toxins. Enzymes can be used to change the taste, colour, texture, smell, or even the digestibility of a food stuff. If a food is found to have a toxic effect it may be possible to remove the toxins by using an enzyme.

A fungal enzyme, alpha-amylase, is widely used in the baking industry. Its job is to increase the production of carbon dioxide, needed to make the bread rise. It does this by speeding up the process of fermentation, in which carbon dioxide is a by-product, by degrading starch to glucose and maltose. The advantages of this process are that bread has a better crust and stays fresh longer.

Brewing also depends on sugars being converted by yeast to carbon dioxide and ethanol. Expensive barley can be largely replaced by wheat starch which, when degraded by amylase, provides maltose and glucose. These are needed by the yeast so it can continue to grow and ferment the sugars to alcohol.

In the confectionery industry a yeast enzyme, invertase, is used to convert sucrose to invert sugar, which is an equimolar mixture of glucose and fructose. Soft centred chocolates are produced by surrounding a hard sugar centre mixed with invertase by a chocolate casing. The hard sugar is slowly broken down inside the sealed chocolate to form soft liquid centres.

Other applications in the food industry include the use of specially developed microbial enzymes to convert milk to cheese, replacing scarce natural rennet which is taken from the stomach of unweaned calves.

The textile industry also uses enzymes. The amylases are used to remove starch that has been added to strengthen fibres during processing. Enzymes are also used to remove wool from sheepskin pieces. The enzyme used for this is an unusual protease that can dissolve wool by attacking the disulphide cross links formed by the amino acid cystine present in large proportion in wool keratin.

The photographic industry uses the enzyme ficin, a protease, to dissolve gelatin off scrap film, allowing recovery of valuable silver present.

In toilet preparations, proteases are used in tooth powders to remove protein deposited on teeth, and may be included in some skin creams.

The pharmaceutical industry uses bacterial enzymes in making new penicillins and other antibiotics. New penicillins are made by using an enzyme from the common gut bacterium *Escherichia coli*, to remove the side chain of the penicillin G molecule commonly produced by fermentation techniques using the fungal organism, *Penicillium chrysogenum*. This enzyme is called penicillin acylase, and it has been successfully immobilized to achieve a most important success in processing.

Enzymes also have important uses in medicine; an account of this is given in the next section of Background reading.

Throughout the World, research is being carried out in universities and other research establishments to discover new enzymes – and better uses for existing ones!

Enzyme	Industry	Use
Proteases		
Papain	Meat	'Tenderization'
	Brewing	Haze prevention
Bacterial proteases	Milk	To make milk protein hydrolysate
Pronase	Fish	To help the skinning process
Cellulases	Wood	To produce glucose (ligninase also needed)
Amylases		
Bacterial alpha amylase	Brewing	Degrades unmalted cereals used in the mashing process
Fungal alpha amylase	Baking	Yeast produces more carbon dioxide
Bacterial amylases	Syrups	To hydrolyse starch to produce glucose syrups
Bacterial alpha amylase	Textile	To remove starch in the desizing process
Lipases	Cheese and chocolate	To break down fats
Invertase	Confectionery	In production of soft-centred chocolates and invert sugar
Pectinases	Fruit juices	Clarification, also eases filtration and concentration processes
Lactase	Milk	To convert lactose in milk products to glucose, to sweeten dairy products

Table 13.6
Typical uses of hydrolase enzymes in industry

BACKGROUND READING 5
Enzymes in medicine

Enzymes in the diagnosis of heart and pancreatic disease

Enzymes are found in the blood serum of healthy individuals but much greater concentrations can be detected in serum from patients suffering from a number of illnesses. Disease often increases the permeability of a cell's outer membrane and intracellular material, including enzymes, leaks into the bloodstream. Indeed, even slight damage to an organ rich in enzymes, like the liver, results in increased serum enzyme activity. Since certain enzymes are associated with particular organs the identity of the enzymes with increased serum concentrations can be used to determine which organ is damaged, while the magnitude of the increase reflects the severity of the damage.

Most hospital biochemistry laboratories can, as a matter of routine, investigate the activity of about twelve enzymes in blood samples. The rate of a particular reaction, catalysed by the enzyme in question, is determined in the presence of a known volume of blood serum; this figure is compared with the normal range of values, measured under identical conditions and using blood samples from healthy individuals. The reaction can be followed by studying the absorbence change at a particular wavelength, using a spectrometer.

The determination of enzyme activities in blood serum is of particular use in confirming a diagnosis; this is illustrated by acute pancreatitis and myocardial infarction.

The pancreas produces digestive enzymes which are secreted into the intestine and which catalyse the hydrolysis of food into smaller, readily absorbed molecules. An amylase, similar to the enzyme present in the saliva, brings about the hydrolysis of the polysaccharide starch into the disaccharide maltose and other simple saccharides. Acute pancreatitis, or inflammation of the pancreas, is a serious illness with symptoms which may be confused with those of inflammation of the gall bladder, other acute abdominal emergencies, and certain heart conditions. Normal blood serum contains only small amounts of amylase but the cell damage which occurs during acute pancreatitis results in the extensive leakage of the intracellular enzymes into the bloodstream. Thus, greatly increased amylase activity in the plasma is an easily measured and unambiguous feature of this disease.

Serum enzyme activities have also proved useful in investigating heart disease. It is often difficult to distinguish between severe attacks of angina pectoris and myocardial infarctions. During an anginal attack the blood supply to the heart muscle is reduced but although severe chest pain is experienced there is no actual damage to the heart muscle. In myocardial infarction, a condition commonly described as a heart attack or coronary, the blood supply to a part of the heart muscle, large or small, is blocked, with the result that muscle tissue is seriously damaged. It is of great clinical importance to differentiate between these two conditions and tests which can assist the doctor in clarifying the diagnosis are extremely helpful. The two conditions can often be distinguished by using an electrocardiogram but this may sometimes be inconclusive. However, infarction always results in cell damage with resulting extensive enzyme leakage, whereas an anginal attack does not affect serum enzyme levels. If blood samples display elevated enzyme activities, a heart attack can be diagnosed even if there is no electrocardiographic evidence. Furthermore the magnitude of the increase gives some measure of the extent of the damage.

Enzymes in the detection of diabetes

For good health the concentration of glucose in the blood must be maintained within narrow limits. After a meal blood sugar levels rise but the hormone insulin is released into the bloodstream. This stimulates the uptake of glucose into the cells of the body. As described in the Background reading at the end of Topic 11, diabetics are unable to produce sufficient insulin. The failure to control glucose uptake results in abnormally high blood sugar levels and usually leads to the excretion of glucose in the urine. Normally, urine does not contain glucose although traces may be present after heavy meals. Diabetes may therefore be detected during a routine examination by testing the urine for glucose. The presence of glucose is readily demonstrated by using a test stick; this is impregnated with two enzymes, glucose oxidase and peroxidase, and a dye.

Glucose oxidase shows marked specificity for glucose, catalysing the oxidation of any glucose present in the urine to gluconic acid and hydrogen peroxide. The hydrogen peroxide is rapidly decomposed in the presence of the peroxidase and the oxygen produced oxidizes the dye. During oxidation the dye changes colour, indicating the presence of glucose in the urine.

$$\text{glucose} \quad + \text{oxygen} + \text{water} \xrightarrow{\text{glucose oxidase}} \text{gluconic acid} + \text{hydrogen peroxide}$$

$$\begin{array}{l}\text{hydrogen} + \text{reduced form} \xrightarrow{\text{peroxidase}} \text{water} \quad + \text{oxidized form} \\ \text{peroxide} \quad \text{of dye} \qquad\qquad\qquad\qquad\qquad\qquad\quad \text{of dye}\end{array}$$

A positive result, even in an apparently healthy patient during a routine medical, would indicate the need for a more detailed clinical examination and further tests.

Asparaginase and chymotrypsin in medical treatment

The human body rapidly rejects foreign protein introduced into the bloodstream. Thus, the administration of enzyme preparations has been of limited therapeutic use. However, asparaginase isolated from bacteria has been used in the treatment of certain types of leukaemia. Healthy human cells can make the amino acid asparagine from readily available intermediates but the tumour cells responsible for acute lymphocytic leukaemia are unable to synthesize asparagine; they therefore depend on the blood to supply it. The intravenous injection of asparaginase results in the hydrolysis of the free asparagine present in the bloodstream and the tumour cells are deprived of their supply.

$$
\begin{array}{cc}
\underset{\text{asparagine}}{\overset{\displaystyle NH_2CHCO_2H}{\underset{\displaystyle CONH_2}{\overset{|}{\underset{|}{CH_2}}}}} \;+\; H_2O \;\xrightarrow{\text{asparaginase}}\; \underset{\text{aspartic acid}}{\overset{\displaystyle NH_2CHCO_2H}{\underset{\displaystyle CO_2H}{\overset{|}{\underset{|}{CH_2}}}}} \;+\; NH_3
\end{array}
$$

In the absence of an essential amino acid, protein synthesis is disrupted and further development of the tumour cells is prevented, although healthy cells which can synthesize asparagine are unaffected. Unfortunately, the enzyme is rapidly inactivated and therefore asparaginase treatment alone is of little use. However, a complicated regime involving the administration of drugs and asparaginase injections has proved effective in treating acute lymphocytic leukaemia.

The proteolytic enzyme chymotrypsin has been used with great success during the surgical treatment of cataracts. The lens, which focuses light rays on the retina, is a biconvex, transparent structure, attached to the walls of the eye by tough fibres known as zonules. Many people in later life develop cataracts; the lens becomes opaque and vision is blurred. The condition cannot be treated but sight can be restored by the removal of the lens and the subsequent use of glasses or contact lenses. The operation involves cutting the zonules to release the lens, which can then be removed, but the zonules, which may be particularly tough in the younger patients, are difficult to sever without permanently damaging other regions of the eye. This problem has been overcome by injecting a dilute solution of chymotrypsin into the eye and leaving it in contact with the zonules for two or three minutes before thoroughly washing the eye. The enzyme hydrolyses certain peptide bonds and results in the partial hydrolysis of the protein within the zonules:

$$\underset{\underset{\text{H}}{\underset{|}{|}}\overset{\overset{R^1}{|}}{C}-\underset{\underset{O}{\parallel}}{C}-\underset{\underset{R^2}{|}}{N}-\underset{H}{C}- + H_2O \xrightarrow{\text{chymotrypsin}} \underset{\underset{H}{|}}{\overset{\overset{R^1}{|}}{C}}-\underset{\underset{O}{\parallel}}{C}-OH + \overset{H}{\underset{H}{N}}-\underset{\underset{R^2}{|}}{\overset{\overset{H}{|}}{C}}-}$$

Provided the eye is thoroughly washed, there is no significant damage to other tissues and the partially digested zonules are more easily severed. Since chymotrypsin treatment was introduced the success rate for the operation has increased dramatically.

13.6
SURVEY OF REACTIONS IN TOPIC 13

The chart surveys the main reactions that you have met in this Topic. Follow the same procedure as suggested in the summaries to Topics 9 and 11 and make copies of this chart on which to record the reagents, their chemical nature, and the types of reaction involved. You should also draw up a chart on which to summarize the main chemistry of amino acids and proteins.

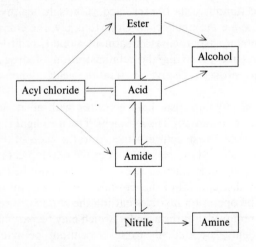

SUMMARY

At the end of this Topic you should:

1 understand, and be able to use, the systematic methods of nomenclature used for the carbon compounds that have been described in the Topic;

2 be familiar with some further organic practical procedures, including the chromatographic separation of amino acids;

3 know the chemical reactions of the various classes of compounds that are given in the parts of the Topic headed 'Reactions of ....' and summarized in the preceding section (page 116);

4 understand something of the nature and function of enzymes;

5 from your study of the Background reading –

be aware of the chemical nature of fats and oils (lipids) and of their uses in the manufacture of margarine and that of soap, and as sources of carboxylic acids;

be aware of the nature and function of liquid crystals;

be aware of the means available for the chemical and structural investigation of proteins, and of some of the results of such investigations;

be aware of the industrial and medical applications of enzymes.

PROBLEMS

Carboxylic acids

1 The boiling points of a series of compounds with similar relative molecular masses are given below.

Structure	Relative molecular mass	T_b/K
$CH_3CH_2CH_2CH_3$	58	273
CH_3CH_2CHO	58	322
$CH_3CH_2CH_2OH$	60	370
CH_3CO_2H	60	391

Name the compounds and suggest reasons for the difference between the boiling points, stating clearly the main types of molecular interaction involved in each of the compounds.

2 Arrange the following groups of acids in order of their strength as acids, putting the weakest acid first.

a (A)

(B) CH_3CHCH_3 with OH below

(C) $CH_3CH-\overset{O}{\overset{\|}{C}}-OH$ with Cl below

(D) $CH_3CH_2-\overset{O}{\overset{\|}{C}}-OH$

b (A) $CH_3-\overset{O}{\overset{\|}{C}}-OH$ (B) $ClCH_2-\overset{O}{\overset{\|}{C}}-OH$ (C) $Cl_2CH-\overset{O}{\overset{\|}{C}}-OH$ (D) $Cl_3C-\overset{O}{\overset{\|}{C}}-OH$

c (A) CH_3—$\overset{\overset{\displaystyle O}{\|}}{C}$—OH   (B) FCH_2—$\overset{\overset{\displaystyle O}{\|}}{C}$—OH   (C) $ClCH_2$—$\overset{\overset{\displaystyle O}{\|}}{C}$—OH   (D) $BrCH_2$—$\overset{\overset{\displaystyle O}{\|}}{C}$—OH

(E) ICH_2—$\overset{\overset{\displaystyle O}{\|}}{C}$—OH

Explain, briefly, why the acidic strength of these compounds differs.

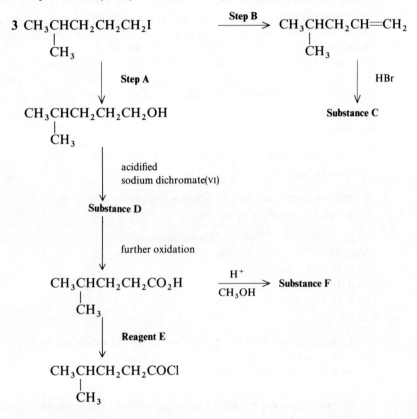

3
$\underset{\underset{\displaystyle CH_3}{|}}{CH_3CHCH_2CH_2CH_2I}$
  $\xrightarrow{\textbf{Step B}}$ $\underset{\underset{\displaystyle CH_3}{|}}{CH_3CHCH_2CH{=}CH_2}$

Step A

HBr

$\underset{\underset{\displaystyle CH_3}{|}}{CH_3CHCH_2CH_2CH_2OH}$

Substance C

acidified
sodium dichromate(VI)

Substance D

further oxidation

$\underset{\underset{\displaystyle CH_3}{|}}{CH_3CHCH_2CH_2CO_2H}$ $\xrightarrow[CH_3OH]{H^+}$ **Substance F**

Reagent E

$\underset{\underset{\displaystyle CH_3}{|}}{CH_3CHCH_2CH_2COCl}$

a Name the starting substance.
b Name the reagents and conditions required for step A. Classify this reaction in terms of reaction and reagent type.
c Name the reagents and conditions necessary for step B. What type of reaction is this?
d Give the structure and name of substance C. Classify the reaction for its formation in terms of reaction and reagent type.
e Write the structural formula of substance D.
f Name a suitable reagent E.

g Write the structural formula of substance F.

h Which substance in the scheme would be obtained if the carboxylic acid $CH_3CH(CH_3)CH_2CH_2CO_2H$ were treated with $LiAlH_4$?

i Which substance in the scheme is chiral?

j Which substance in the scheme would probably fume in moist air?

4 The labels are missing from four bottles known to contain the solids phenol, hexanoic acid, dodecanol, and paraffin wax (a long chain alkane). Devise a simple series of tests which would enable you to distinguish between them.

5 The carboxyl group $-\overset{\overset{\textstyle O}{\|}}{C}-OH$ contains both the carbonyl group $-\overset{\overset{\textstyle O}{\|}}{C}-$ and the hydroxyl group $-OH$. Discuss the points of similarity and difference between the behaviour of the groups when present alone in molecules and when combined in the carboxyl group.

Amines

6 Arrange the following compounds of similar relative molecular mass in the order in which you would expect their boiling points to increase, giving reasons for your choice.

$CH_3CH_2CH_2CH_3$, $CH_3CH_2CH_2OH$, $CH_3CH_2CH_2NH_2$, CH_3CONH_2.

7 Arrange the following compounds in the order in which you would expect their basic strength to increase, putting the weakest base first.

NH_3

Give reasons for your choice.

8 Study the following list of compounds containing nitrogen:

$CH_3CH_2CH_2NH_2$

A

H_3C
$\diagdown$
$\quad N—H$
$\diagup$
CH_3CH_2

B

NO_2 attached to benzene ring

C

CH_3 and CH_3 attached to N on benzene ring

D

$$CH_3—\underset{\underset{CN}{|}}{\overset{\overset{H}{|}}{C}}—C_2H_5$$

E

$CH_3CH_2CONH_2$

F

a Which substance is **i** a primary amine, **ii** a secondary amine, **iii** a tertiary amine, **iv** an acid amide, **v** a nitrile?
b Which substances react readily with acids to form salts?
c Which amine would be the weakest base?
d Which substances could be reduced to form primary amines?
e Which substances could be hydrolysed to form carboxylic acids?
f Which substances might form complexes with metal cations?
g Which amines could react with ethanoyl chloride?
h Which amine would form a quaternary ammonium salt with iodomethane?
i Which substance has a carbon atom which is chiral?

9 A substance, **A**, of molecular formula $C_5H_{13}N$, exists as two optical isomers. On treatment with hydrochloric acid, **A** forms **B**, $C_5H_{14}NCl$. **A** also reacts with ethanoyl chloride to form **C**, $C_7H_{15}NO$.

i Give structural formulae which could apply to **A**, **B**, and **C**.
ii Give the structural formulae of two isomers of **A** which are not optically active and name them.
iii Suggest a method by which one of your suggestions for part **ii** could be synthesized starting
a from the appropriate alcohol containing 5 carbon atoms.
b from the appropriate alcohol containing 4 carbon atoms.

10 Write an account of the reactions of amines, comparing their properties with those of ammonia.

Acid derivatives

11 Study the following group of compounds and answer the questions below:

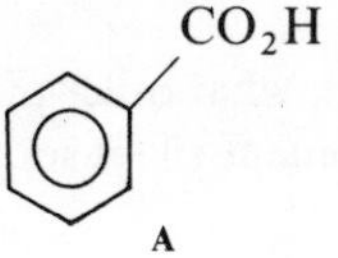

a Which substance should fume in moist air?
b Which substance should have a pleasant smell?
c Which substance is a dibasic acid?
d Write the structural formula of the substance which would be obtained if substance A were dissolved in sodium hydroxide solution.
e With what reagent would you treat substance A to form substance D?
f With what reagent would you treat substance D to form substance C?
g With what reagent would you treat substance D to form substance E?
h What products would be obtained if substance C were refluxed with sodium hydroxide solution?
i Which substances might be dehydrated by heating or chemical means?
j Write structural formulae for the substances you have suggested in **i**.

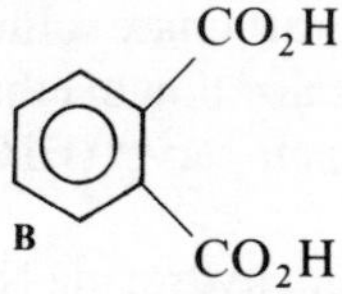

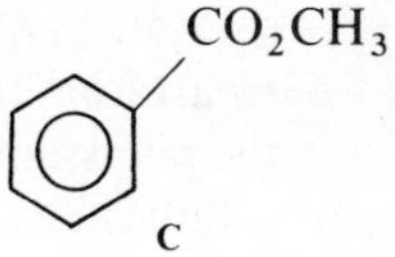

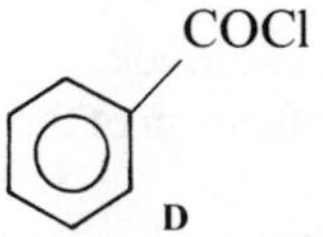

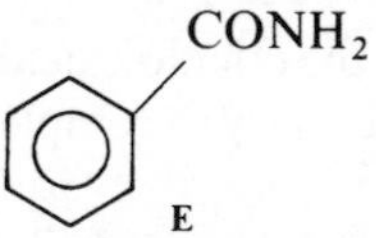

12 In a preparation of ethanenitrile (methyl cyanide), using a reaction which can be represented by

$$CH_3CONH_2 \longrightarrow CH_3CN + H_2O$$

4 g of dry ethanamide (acetamide) were placed in a clean, dry flask. 8 g of phosphorus(v) oxide were weighed and transferred as quickly as possible to the flask, which was shaken to mix the solids. The flask was fitted for distillation using a water condenser and a dry receiver.
The flask was heated gently so that the ethanenitrile distilled slowly. The contents of the flask became molten and frothing occurred during the reaction.
The crude ethanenitrile was transferred to a separating funnel and 1 cm^3 of water was added, followed by sufficient potassium carbonate to saturate the lower aqueous layer. The separating funnel was stoppered, shaken, then left for about 20 minutes. Ethanoic acid (acetic acid), present as an impurity, was neutralized by the potassium carbonate. The aqueous layer was removed and discarded.

a Why must the phosphorus(v) oxide be dealt with 'as quickly as possible'?
b Why is distillation carried out *slowly*?
c What precaution should be taken when shaking the crude product with saturated potassium carbonate solution?
d Apart from the neutralization of ethanoic acid, what other purpose does the potassium carbonate serve? (Hint: the ethanenitrile is somewhat soluble in water.)
e Why should potassium hydroxide NOT be used instead of potassium carbonate?
f By what reaction is the ethanoic acid present produced?
g Outline the steps necessary to complete the purification of the ethanenitrile.

13 Compound A is a colourless acidic liquid with the molecular formula $C_2H_4O_2$. When heated with ethanol and sulphuric acid it gives a colourless neutral liquid B, $C_4H_8O_2$. When A is treated with ammonia solution, and the resulting solution is evaporated, a solid is formed, which on heating gives colourless crystals C, C_2H_5NO. Heating with phosphorus pentoxide converts C into D, C_2H_3N, and treatment of D with a powerful reducing agent converts it into E, C_2H_7N. This last compound, when added to copper sulphate solution, causes a deepening in its colour. Write structural formulae for A to E.

14 Suggest simple chemical tests for distinguishing one compound from the other in the following pairs:

a CH_3CO_2H and CH_3CHO

b
and

c $CH_3CH_2COCH_2CH_3$ and $CH_3CH_2CO_2CH_2CH_3$
d CH_3COCl and CH_3CH_2Cl

e
and $CH_3CH_2CH_2CH_2NH_2$

Amino acids and proteins

15 Aspartic acid has the formula

$$HO_2C—CH_2—CH—CO_2H$$
$$\underset{NH_2}{|}$$

a State how the formula
i resembles
ii differs from that of glycine
b Deduce how the properties of aspartic acid
i resemble
ii differ from those of glycine.

16 The technique of electrophoresis enables a solution of amino acids to be separated and identified by observing their relative movement on chromatography paper under the influence of an applied electric field. At pH 7 little movement of glycine (aminoethanoic acid) is observed; at pH 2 glycine migrates to the cathode, but at pH 12 glycine migrates to the anode. Suggest reasons for these results.

17 The following questions concern the optical isomers of alanine,

$$CH_3$$
$$|$$
$$CHNH_2$$
$$|$$
$$CO_2H$$

What differences would there be in:
a the melting points of the two isomers?
b the rates of their reaction with ethanoyl chloride?
c their effect on copper(II) sulphate solution?
d their effect on plane polarized light?
e their occurrence in natural protein?

18 The formula below represents a molecule of alanine:

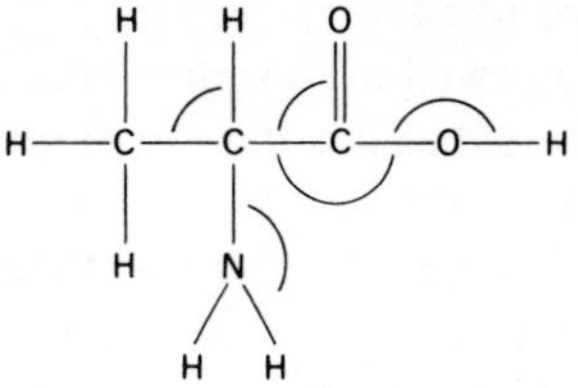

a Which atoms in this molecule could be readily detected by X-ray crystallography?
b The infra-red absorption spectrum of alanine has certain features which are very similar to features in the infra-red spectra of aminoethane, $CH_3CH_2NH_2$, and chloromethane, CH_3Cl.
Which is the smallest part of the molecule which could be said to be

responsible for these features of the infra-red spectrum?

c Draw a 'dot and cross' diagram to show the distribution of electrons in the outer shells of the atoms in the molecule of alanine.

d Write down the approximate bond angles marked, for example, $-\overset{\diagup}{\underset{}{C}}$ as they would be in an actual alanine molecule.

e Alanine can behave both as an acid and a base.

i Draw the structural formula for the anion formed from an alanine molecule on adding a strong base.

ii Draw the structural formula for the cation formed from an alanine molecule on adding a strong acid.

19

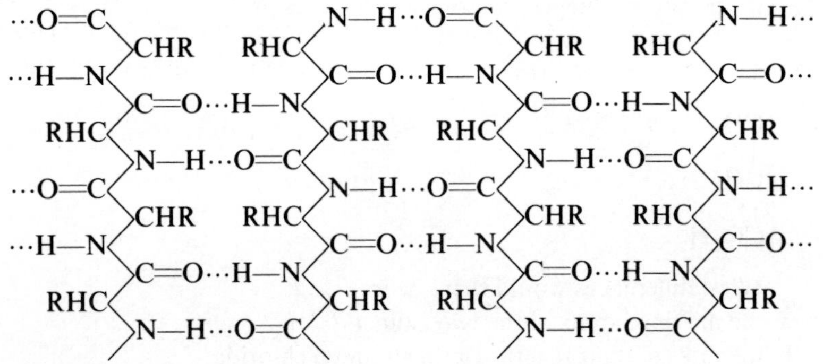

The diagram above is a representation in two dimensions of a very small part of the structure of a naturally occurring material. The letter R represents not just one group but a variety of relatively small structures.

a To what class of naturally occurring material does this substance belong?

b State TWO physical properties which a substance of this structure would be expected to have.

c Assuming that you swallowed some of this substance and that it could be digested, indicate the chemical reactions by which the process of digestion would probably begin.

d By what practical process would you attempt in the laboratory to break down the substance in order to investigate its structure?

e Having broken down the structure, outline the practical procedure you would use next in your investigation.

f After the chemical structure had been worked out, what physical method would be used to help to elucidate the three-dimensional structure of the molecule?

Reaction rates – an introduction to chemical kinetics

14.1
INTRODUCTION

What actually takes place during a chemical reaction? Why do some reactions take longer than others to go to completion? What factors influence the rate of a reaction? What principles govern the interactions that take place on the molecular level during a chemical reaction? Chemical kinetics, the study of the rates of chemical reactions, is concerned with all of these questions.

From your previous work in chemistry you may remember that the rate of a chemical reaction can depend on any or all of the following general factors:

the concentration of the reactants
the temperature
the presence of catalysts

In particular cases the rate may also depend on the state of subdivision of a solid reactant or the influence of electromagnetic radiation such as visible or ultra-violet light. This Topic is concerned with the study of the three general factors mentioned above. We shall not be concerned with the other influences, not because they are unimportant in specific instances (see, for example, the account of the effect of ultra-violet radiation on the rate of halogenation of alkanes in section 9.2), but because there are definite reasons for studying the three general factors. We shall start by seeing what the reasons are, and what we hope to gain by involving ourselves in the detail which follows.

Why we study rates of reactions

Rates of reaction are studied for three main reasons. Firstly, you will realize from the work that you have done on enthalpy and entropy changes that it is possible to predict whether or not a particular chemical reaction will take place when you try it. In Topic 15 you will learn that there are further ways of deciding in a wide variety of cases whether a reaction is likely to proceed or not. What is not immediately obvious is that it is not possible to predict how quickly a feasible reaction will take place. We would like to understand why reaction rates are not predictable, and how to change them if we want to.

A second, more specific reason why a study of reaction kinetics is worth while is that a knowledge of the effect of concentration on the rate of a reaction provides important evidence about the mechanism of the reaction – the individual steps by which a reaction takes place. The mechanisms of some organic reactions have been described in Topics 9, 11, and 13 and others will be described in Topic 17. Mechanisms are suggestions which are backed up by experimental evidence, some of which is derived from a study of rate of reaction. It must not be thought that kinetic evidence can enable us to deduce what the mechanism of a reaction is, but at least we can sometimes decide between a number of rival suggestions.

Thirdly, the study of the effect of temperature on the rate of a reaction leads to an understanding of what happens at the molecular level during the reaction, an obvious follow-on from a study of its mechanism. This in turn leads to an understanding of the role of catalysts in changing the rates of reactions.

14.2
THE EFFECT OF REACTANT CONCENTRATION ON THE RATE OF A REACTION

Firstly we must decide what we mean by the *rate of a reaction*. In a reaction which takes place between two substances A and B, we might follow the reaction by observing how quickly substance A is used up. For a reaction in solution, we would probably measure the concentration of A, symbol [A], at various times, to see how it changes. In such a case, our measure of the rate of the reaction would be the rate of change of concentration of A, symbol r_A.

We might well have chosen to do our measurements on substance B, and obtained a value of r_B, the rate of change of concentration of B. This might well have a different numerical value, so it is obviously essential to specify the substance to which we are referring.

In a reaction of the type

$$xA + yB \longrightarrow \text{products}$$

the rate of change of concentration of substance A will be found to follow an empirical mathematical expression of the form

$$r_A = k[A]^a [B]^b [C]^c$$

An expression of this kind is called a *rate equation*. Substance C is included in this general expression as well as A and B, because rate equations sometimes include concentrations of substances that do not appear in the stoicheiometric equation for the reaction. An example of this appears in experiment 14.2b. It is important to understand the significance of each part of such an equation, so let us consider each part in turn.

1 r_A. The meaning of this has already been explained. It may have units of $mol\,dm^{-3}\,s^{-1}$ but, as we shall see, it is not always necessary to know what it is in these units; it is often enough to know it in some other units. Minutes might be used instead of seconds, or some property which is proportional to the concentration might be used instead of the concentration itself. We shall meet an example of this in experiment 14.2a, the first experiment.

2 *Concentrations* [A], [B], *etc.* As in previous Topics, the use of square brackets in this context denotes 'concentration in moles per cubic decimetre'. Again, in particular cases it may be more convenient to use a quantity which is proportional to the concentration instead of the concentration itself.

3 *The indices a, b, etc.* The index '*a*' is called the order of the reaction with respect to the reactant A, '*b*' is the order of the reaction with respect to the reactant B, and so on. The sum of all the indices is called the overall order of the reaction. In most work at this level, the order of a reaction with respect to a particular reactant is a whole number having one of the values 0, 1, or 2, though it is possible for individual orders to be non-integral or higher than 2.

When we speak of orders of reaction we may say that a reaction is 'first order with respect to A' or 'zero order with respect to B' or perhaps 'second order overall'.

Note that the orders of a reaction are experimental quantities; they cannot be deduced from the chemical equation for the reaction.

4 *The constant k. k* is a constant of proportionality called the rate constant. The units of k depend on the order of the reaction, and can be worked out from the rest of the rate equation. For example, if the rate equation for the change of concentration of a substance A in a particular reaction is

$$r_A = k[A]^2[B]^1$$

then, substituting the units of the various quantities;

$$mol\,dm^{-3}\,s^{-1} = k(mol\,dm^{-3})^2(mol\,dm^{-3})$$

from which the units of k are $dm^6\,mol^{-2}\,s^{-1}$.

A knowledge of the rate constant has a practical use. It is often used to compare rates of reaction at different temperatures or for comparing the rates of different reactions with similar rate equations. The justification for this is that, whatever the orders of reaction, when the reactant concentrations are constant the rate of the reaction is proportional to the rate constant (but note that it has different units).

When studying the dependence of rate on reactant concentration the problem is usually to identify the orders of reaction and perhaps the rate constant.

In designing an experiment or series of experiments to investigate the rate of a reaction there are several problems which have to be overcome. One problem is that usually all the reactant concentrations vary at the same time and it is difficult to discover the effect of a particular reactant. This problem can be overcome by making the concentrations of all the reactants which we do not wish to study much larger than the one concentration we are interested in. The effect of this is that only this one concentration varies significantly and we can then see what the effect of this variation is on the rate of the reaction. In experiment 14.2b, method 1, for example, propanone will be reacting with iodine in the presence of an acid. There are three reactants but we shall want to find the order of the reaction with respect to only one of them, the iodine. It is arranged, therefore, that the propanone and acid concentrations are much higher than the iodine concentration so that effectively the iodine concentration is the only one of the three concentrations which varies. The other two concentrations do change, of course, but only very slightly.

A second problem with experimental design is that there is usually no easy way of finding directly the rate of a reaction at a particular time from the start of the reaction, since the rate may be changing as reactant concentration changes. In practice what we often do is to measure reactant concentration at various times from the start of the reaction, and from the gradient of a graph of reactant concentration against time it is usually possible to deduce what the order is. The order may then be checked by graphical or mathematical means.

A third problem is the very wide variation in the rates themselves. The time needed for a simple electron transfer between atoms to take place is about 5×10^{-16} second, whereas the time needed to bring about the change of half of a sample of L-aspartic acid to D-aspartic acid in fossil bones at ordinary temperatures is about 3×10^{12} seconds (which is about 100 000 years), probably the slowest known reaction. Such very fast or very slow reactions can only be studied by using special techniques. The reactions that we shall study must be chosen with care, so that they have rates that are measurable by techniques that are available in a school laboratory.

EXPERIMENT 14.2a
The kinetics of the reaction between calcium carbonate and hydrochloric acid

$$CaCO_3(s) + 2HCl(aq) \longrightarrow CaCl_2(aq) + CO_2(g) + H_2O(l)$$

In this experiment the problem is to find the order of the reaction with respect to hydrochloric acid. The calcium carbonate is in the form of marble and fairly large pieces are used so that the surface area does not change appreciably during the reaction. On the other hand the hydrochloric acid is arranged to be in such quantity and concentration that it is almost all used up during the reaction.

Two methods are described; one of them may be chosen according to the apparatus available.

Procedure (method 1)

Put all the following items on the pan of a direct-reading, top-loading balance:

a small conical flask containing about 10 g of marble in 6 or 7 lumps
a measuring cylinder containing 20 cm^3 of 1 M hydrochloric acid
a plug of cottonwool for the top of the conical flask

Adjust the balance so that it is ready to weigh all these items. Pour the acid into the conical flask, plug the top with the cottonwool, and replace the measuring cylinder on the balance pan. Allow a few seconds to pass so that the solution is saturated with carbon dioxide; then start timing and taking mass readings. Record the mass of the whole reaction mixture and apparatus at intervals of 30 s until the reaction is over and the mass no longer changes. Record your results in the form of a table and work out the total loss of mass at each time interval. This loss of mass is due to the carbon dioxide escaping into the atmosphere.

Time t/s	Total mass $/g$	Mass of CO_2 m_t/g	$m_{final} - m_t$ $/g$

The fourth column in the table needs a little thought. When the reaction is over, the total mass of carbon dioxide evolved (m_{final}) is proportional to the concentration of the hydrochloric acid at the moment when timing started. Thus ($m_{final} - m_t$) is proportional to the concentration of hydrochloric acid at each time t.

Procedure (method 2)

Set up the apparatus as in figure 14.1.

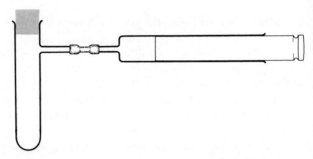

Figure 14.1

Place about 10 g of marble in six or seven lumps in the test-tube and have ready 10 cm³ of 1M hydrochloric acid. Put the acid into the test-tube and allow a few seconds for the solution to become saturated with carbon dioxide. Put the stopper in place and start timing. Taking readings of volume every 30 s until the reaction is over and the volume no longer changes. Record your results in the form of a table:

Time t/s	Volume of CO_2 V/cm^3	$V_{final} - V_t$ $/cm^3$

The third column in the table needs a little thought. When the reaction is over the total volume of carbon dioxide collected (V_{final}) is proportional to the concentration of hydrochloric acid at the moment when timing started, so ($V_{final} - V_t$) is proportional to the concentration of hydrochloric acid at each time, t.

Plot a graph of ($m_{final} - m_t$) or ($V_{final} - V_t$) against t, putting t on the horizontal axis.

Discussion of results

Consider the rate equation in the form

$$r_{HCl} = k[HCl]^a$$

(where r_{HCl} is the rate of change of concentration of hydrochloric acid)
The problem is: 'What is the order of reaction, a?'.

Let us consider the various possibilities:

i If $a = 0$ (zero order) the graph will be a straight line since

$r_{HCl} = k[HCl]^a$ is the same as $r_{HCl} = k$ (see figure 14.2a).

ii If $a = 1$ (first order) the graph will be a curve such that the time it takes for the concentration of the reactant to be halved (the 'half life') is constant whatever value of the concentration you start from (see figure 14.2b).

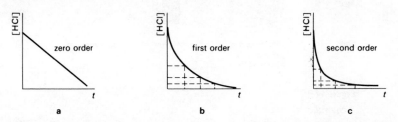

Figure 14.2

iii If $a = 2$ (second order) the graph will again be a curve but as a second order curve is much 'deeper' than a first order one, the half life is not constant but will increase dramatically as the reaction proceeds (see figure 14.2c).

Compare your results with the descriptions given for the various orders and identify the value of a.

Notice that this method assumes that the order is one of the three possibilities 0, 1, or 2. A more sophisticated method of checking this is given in Appendix I and the justification for the remarks made about half life is also given in this appendix. If you wish, you can plot the graphs suggested in the appendix and confirm the order of the reaction you studied in experiment 14.2a.

The method of using half life times to find out about order of reaction can be quite a sophisticated one and an account is given of its scope in Appendix II.

Now that you have used one method of 'following' a reaction so as to investigate its order you are in a better position to appreciate other possible methods. The account which follows mentions several methods and you will be using one of these in the next experiment.

Methods of 'following' a reaction

1 *By titration.* A reaction mixture is made up and samples are withdrawn from it, using a pipette. Some means is then found of 'quenching' the reaction – slowing it abruptly at a measured time from the start of the reaction, perhaps by rapid cooling in ice or by removing the catalyst. The samples can then be titrated in some way which depends on what is in the reaction mixture. This is the principle of the method which is going to be used in experiment 14.2b (method 1) and experiment 14.4.

2 *By colorimetry.* If one of the reacting substances or products has a colour, the intensity of this colour will change during the reaction. The intensity could be followed by a photoelectric device in a colorimeter, as shown in figure 14.3.

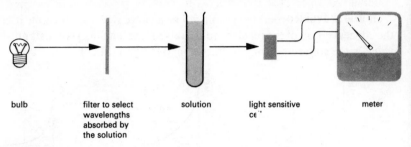

bulb   filter to select   solution   light sensitive   meter
                wavelengths   ce ˙
                absorbed by
                the solution

Figure 14.3

A colorimeter

3 *By dilatometry.* In some reactions, in the liquid phase the volume of the whole mixture changes slightly during the reaction. If it does the change of volume could be followed by using an enclosed apparatus fitted with a capillary tube. This type of apparatus is called a dilatometer. It is illustrated diagrammatically in figure 14.4.

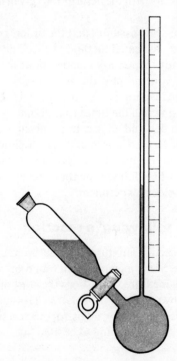

Figure 14.4

A dilatometer.

4 *By measurements of electrical conductivity.* If the total number of ions in solution changes during a reaction it might be possible to follow the reaction by measuring the changes in electrical conductivity of the solution, using a 'conductance bridge'. This uses alternating current so that actual electrolysis of the solution is avoided.

Figure 14.5
A conductance bridge.

5 *By measuring the volume of a gaseous product.* If a gas is given off it can be collected or its mass measured. This was the principle behind the method used in experiment 14.2a.

6 *By measurements of any other physical property which shows significant change.* Possible physical properties not already discussed include chirality and refractive index.

Now you can try another determination of reaction order by experiment.

EXPERIMENT 14.2b
The kinetics of the reaction between iodine and propanone in acid solution

Method 1

The reaction between iodine and propanone follows the overall equation

$$CH_3COCH_3(aq) + I_2(aq) \longrightarrow CH_3COCH_2I(aq) + H^+(aq) + I^-(aq)$$

The reaction is acid-catalysed and is first order with respect to propanone and also first order with respect to hydrogen ions. This experiment is to investigate the order of the reaction with respect to iodine.

For convenience the solutions required are given identifying letters:

A is a 0.02M solution of iodine in potassium iodide
B is an aqueous solution of propanone (1.0M)
C is sulphuric acid (1.0M)
D is a solution of sodium hydrogencarbonate (0.5M)
E is a solution of sodium thiosulphate (0.01M)

Procedure

1 In a titration flask labelled number *1* place 50 cm³ of A and in another, labelled number *2*, place 25 cm³ of B and 25 cm³ of C. Use measuring cylinders for these solutions.

2 Into each of several titration flasks labelled numbers *3*, *4* etc. put 10-cm³ portions of D, again using a measuring cylinder.

3 Noting the time, pour the contents of flask *2* into flask *1* and shake well for about 1 minute. Then, noting the time, withdraw 10 cm³ of the reaction mixture, using a pipette and safety filler, and run this liquid into flask number *3*. Shake until bubbling ceases. The sodium hydrogencarbonate solution neutralizes the acid catalyst and quenches the reaction.

4 Titrate the contents of flask number *3* with E.

5 At measured and noted intervals from 5 to 10 minutes, repeat the extraction of a 10-cm³ sample from flask number *1* and the quenching and titration process.

Tabulate the results and draw a graph of titre of E against time from the start of the reaction. From your graph deduce the order of the reaction with respect to iodine.

Method 2

The reaction between iodine and propanone follows the overall equation

$$CH_3COCH_3(aq) + I_2(aq) \longrightarrow CH_3COCH_2I(aq) + H^+(aq) + I^-(aq)$$

The reaction is catalysed by hydrogen ions.

In this procedure an attempt is made to estimate the rate of change of concentration of iodine, r_1, at the start, the initial rate. Just after the start of the reaction, the concentration of a reactant decreases almost linearly with time, whatever the order of the reaction; it is only later that the differences begin to show significantly. In this method, therefore, the concentrations of each of the three reactants are varied systematically in the four 'runs' and a direct estimate of r_1 made from:

$$r_1 \propto \frac{\text{volume of iodine solution used}}{\text{time for iodine colour to disappear}}$$

Procedure

Make up mixtures of hydrochloric acid, propanone solution, and water according to the table, using burettes. Start the reaction by adding the appropriate volume of iodine solution, measured into test-tubes from a burette, and measure the time in seconds for the colour of the iodine to disappear.

	Run 1	Run 2	Run 3	Run 4
Volume of 2M HCl/cm^3	20	10	20	20
Volume of 2M propanone/cm^3	8	8	4	8
Volume of water/cm^3	0	10	4	2
Volume of 0.01M iodine/cm^3	4	4	4	2
Time for colour to disappear/s				
Rate (as indicated above)				

Questions

1 Why was water added to some of the reaction mixtures?

2 If you compare the mixtures for runs 1 and 2 you will see that the concentrations of propanone and of iodine are the same in both but the concentration of acid in run 2 is half of what it is in run 1.
   What is the effect on the rate of change of concentration of iodine of halving the concentration of acid?

3 What, then, is the order of the reaction with respect to hydrogen ions?

4 Using similar arguments, what are the orders of the reaction with respect to propanone and to iodine?

Kinetics and reaction mechanism

You will have seen that in experiment 14.2b the order of the reaction with respect to iodine is zero or, to put it plainly, the rate of change of concentration of iodine, r_1, does not depend on the iodine concentration at all. Putting together all the kinetic evidence we have about the reaction gives the rate equation.

$$r_1 = k \, [\text{propanone}]^1 \, [\text{hydrogen ion}]^1 \, [\text{iodine}]^0$$

Or, since any quantity raised to the power 0 is 1, we could omit the iodine from the rate equation altogether

$$r_1 = k \, [\text{propanone}]^1 \, [\text{hydrogen ion}]^1$$

At first sight it seems strange that a substance which appears in the chemical equation for the reaction does not affect its rate, whereas a substance which does not appear in the chemical equation does appear in the rate equation.

In organic chemistry, however, you have already met the idea that an organic reaction can occur in a number of successive steps. These steps are known as the mechanism of the reaction and there is no reason to suppose that all the steps take place at the same rate; in fact, it would be rather surprising if they did. When we realize this, it is clear that the whole reaction goes at the rate of the slowest of the steps in the mechanism. This slowest step is called the *rate-determining step*.

Quite a useful way of visualizing the idea of a rate-determining step is to imagine that a teacher has prepared some pages of notes and wants to collect them into sets with the help of some students. The notes are arranged in ten piles and one student collects a page from each of the piles (*step 1*). A second student takes the set of ten pages and tidies them ready for stapling (*step 2*). A third student staples the set of notes together (*step 3*).

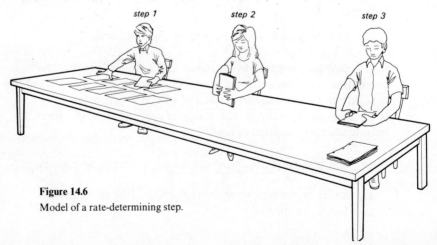

Figure 14.6
Model of a rate-determining step.

It is not hard to see that in this situation the overall rate of the process (the rate at which the final sets of notes are prepared) depends on the rate of *step 1*, the collecting of the sheets of notes, since this is by far the slowest step. It does not matter, within reason, how quickly the tidying or stapling is done; for the most part, the second and third students will be doing nothing while they wait for the first student to collect pages. The mechanism of the reaction may thus be refined to

step 1   Student 1 collects pages   (slow)
step 2   Student 2 tidies set of pages   (fast)
step 3   Student 3 staples pages   (fast)

In order to speed up the process the teacher could offer to help. It would be of no value at all if this additional help were given to either student 2 or student 3 but more people at a time collecting up the pages in *step 1* would clearly make the whole process faster. *Step 1* is the rate-determining step.

We return now to the reaction which was investigated in experiment 14.2b. Various mechanisms might be proposed for the reaction, including this one:

step 1 $H^+ + CH_3-C-CH_3 \longrightarrow CH_3-C-CH_3$ (fast)
             with O (double bond) below first, and OH^+ below second

step 2 $CH_3-C-CH_3 \rightleftharpoons CH_2=C-CH_3 + H^+$ (slow)
             with OH^+ and OH below

step 3 $CH_2=C-CH_3 + I_2 \longrightarrow CH_2I-C-CH_3$ (fast)
             with OH below left, and I above, OH below right

step 4 $CH_2I-C-CH_3 \longrightarrow CH_2I-C-CH_3 + H^+ + I^-$ (fast)
             with I above and OH below left, O (double bond) below right

As the iodine molecules are not involved in the mechanism until after the rate-determining step, the overall rate of the reaction would not depend on the concentration of iodine. This is consistent with the outcome of experiment 14.2b.

It is important to realize that the mechanistic suggestion was not deduced from the kinetic evidence; the evidence was merely shown to be consistent with the suggestion.

Another example of the use of kinetics data to decide between possible mechanisms is given by the hydrolysis of various bromoalkanes, using hydroxide ions. The general pattern of these reactions is

$$C_4H_9Br + OH^- \longrightarrow |C_4H_9OH + Br^-$$

Two mechanisms have been proposed for this type of reaction, often referred to as the S_N1 and S_N2 mechanisms.

S_N1 (substitution, nucleophilic, 1 molecule in the rate-determining step)

$$C_4H_9Br \rightleftharpoons C_4H_9^+ + Br^- \qquad \text{(slow)}$$

$$C_4H_9^+ + OH^- \longrightarrow C_4H_9OH \quad \text{(fast)}$$

S_N2 (substitution, nucleophilic, 2 molecules in the rate-determining step)

$$HO^- + \overset{\displaystyle\backslash}{\underset{\displaystyle/}{C}}{-}Br \longrightarrow \left[HO{-}{-}{-}{-}{-}\overset{\displaystyle\backslash/}{\underset{|}{C}}{-}{-}{-}{-}{-}Br \right]^- \longrightarrow HO{-}\overset{\displaystyle//}{\underset{\displaystyle\backslash}{C}} + Br^-$$

The S_N2 mechanism is effectively a one-step process in which, for an instant of time, both the incoming hydroxide ion and the outgoing bromide ion are equally associated with the hydrocarbon group. In either of the two mechanisms, water molecules can replace hydroxide ions as the nucleophiles.

Here are some kinetics data about the hydrolysis of some bromoalkanes. Determine which of the two mechanisms is the more appropriate in each case, and answer the questions, recording the answers in your notebook in such a way as to make it clear what each question was.

CASE A
The hydrolysis of 1-bromobutane

Equimolar quantities of 1-bromobutane and sodium hydroxide were mixed at 51 °C and the concentration of hydroxide ions was determined at various times, with the following results:

Time /hours	$[OH^-]$ /mol dm^{-3}	Time /hours	$[OH^-]$ /mol dm^{-3}
0.04	0.241	12.0	0.084
0.5	0.225	14.0	0.077
1.5	0.195	22.0	0.058
2.5	0.172	27.0	0.050
3.5	0.155	33.0	0.044
4.5	0.140	38.0	0.040
6.5	0.118	47.0	0.035
9.0	0.099	59.0	0.028

Questions

1 How might the results have been obtained practically?

2 From a suitable graph of the results, what is the order of the reaction?

3 Is this an overall order or an order with respect to a particular reactant?

4 Which mechanism is operating? Give reasons for your answer.

CASE B
The hydrolysis of 1-bromobutane

In this case a method was found of estimating the rate of change of concentration of 1-bromobutane, r_B, directly and the initial rate was found at various initial concentrations of hydroxide ions, and of 1-bromobutane.

Remember that the rate equation is of the general form

$$r_B = k \, [C_4H_9Br]^a \, [OH^-]^b \text{ where } a \text{ and } b \text{ are orders of reaction.}$$

Initial concentrations /mol dm^{-3}		Initial rate /mol dm^{-3} s^{-1}
[OH$^-$]	[C$_4$H$_9$Br]	
A 0.10	0.25	3.2×10^{-6}
B 0.10	0.50	6.6×10^{-6}
C 0.50	0.50	3.3×10^{-5}

Questions

1a What is the effect on r_B if the concentration of 1-bromobutane is doubled?

 b What is the order of the reaction with respect to 1-bromobutane?

2a What is the effect on r_B if the concentration of hydroxide ions is increased five times?

 b What is the order of the reaction with respect to hydroxide ions?

3 Which mechanism is operating?

CASE C and
EXPERIMENT 14.2c
The hydrolysis of 2-bromo-2-methylpropane

In this example the nucleophile is water so that the products of the reaction include a solution of hydrogen bromide. This hydrogen bromide can be titrated with standard alkali.

Full details of procedure are given but as the experiment is rather a long one it need not be done and sample results are given.

Procedure

1 Prepare a constant temperature bath at about 20 °C by running warm water into a large beaker or a pneumatic trough. The bath must be maintained at this temperature by adding small quantities of hot water as required.

2 Stand a stoppered conical flask, the reaction vessel, containing 100 cm³ of the given mixture of propanone and water in the constant temperature bath and allow it to reach the temperature of the bath.

3 While waiting for the contents of the conical flask to reach a steady temperature, fill a burette with the sodium hydroxide solution (0.05M).

4 To each of four small conical flasks add about 20 cm³ of the mixture of propanone and water from the stock solution and 5 drops of methyl red solution and cool in a mixture of ice and water.

5 Use a graduated pipette and safety pipette filler to add, as quickly as possible, about 1 cm³ of 2-bromo-2-methylpropane to the reaction vessel, mix well, and at the same time start timing.

6 After 2 minutes, withdraw 10 cm³ of the reaction mixture, using a pipette and safety pipette filler, and run it into one of the flasks containing cold solvent and indicator. Note the time at which approximately half the sample has been discharged from the pipette.

7 Titrate the sample immediately with the sodium hydroxide solution and record the result when the indicator changes colour from red to orange. *TAKE CARE:* the volume of sodium hydroxide solution required is only about 1 cm³ in the first titration.

8 Repeat the sampling and titration procedure at approximately the following times:

    5, 10, 20, 30, and 45 minutes

and record your results each time.

9 Raise the temperature of the reaction vessel by putting it in a bath of warm water (about 35 °C for about 20 minutes). Withdraw a final 10-cm³ sample and titrate as before, recording your result as a 'final' reading.

NOTE: Titrate the samples quickly; at the same time, try to avoid agitating the mixture too much. If the end point keeps fading, standardize the titration procedure so that the end point is recorded when the colour persists for about 10 seconds.

Record your results in a table of the following type. This gives sample results.

Time /minutes	Titre V_t/cm^3	$(V_{final} - V_t)$ /cm^3	
2	1.30	17.60	
5	3.10	15.80	
10	5.60	13.30	
14	7.50	11.40	
20	9.70	9.20	Temperature of water bath 20 °C
25	11.35	7.55	
30	12.45	6.45	
45	15.20	3.70	
final	18.90		

Questions

1 Explain the significance of the column headed $(V_{final} - V_t)$. The explanation is similar to one given in experiment 14.2a.

2 From a graph based on the results, what is the order of the reaction?

3 Is this an overall order of reaction or an order with respect to one reactant?

4 Do you have enough information to say which mechanism is the more appropriate? Explain how you arrived at your answer.

Here are the results from another experiment with 2-bromo-2-methylpropane which provide some further kinetics data, this time using hydroxide ions as the nucleophile.

In this experiment the concentration of 2-bromo-2-methylpropane remains the same but in successive runs of the experiment, the concentration of the hydroxide ions is increased. A few drops of an acid–alkali indicator are added to each mixture and a record is kept of the time taken for the indicator to change colour when the hydroxide ions are neutralized. The results are as follows:

Experiment number	$[C_4H_9Br] \times 10^2$ /mol dm^{-3}	$[OH^-] \times 10^3$ /mol dm^{-3}	Time /s
1a	2.5	1.25	9
1b	2.5	1.25	8
2a	2.5	2.50	17
2b	2.5	2.50	18
3a	2.5	3.75	26
3b	2.5	3.75	26
4a	2.5	5.00	37
4b	2.5	5.00	39

Questions

5 What effect does doubling the concentration of hydroxide ions have on the *time* taken for the indicator to change colour?

6 What is the effect of an increase of hydroxide ion concentration on the rate of the reaction?

7 What is the order of the reaction with respect to hydroxide ions?

8 Which mechanism is the more appropriate in this case?

There is a tendency for straight chain halogenoalkanes to hydrolyse by the S_N2 mechanism, whereas tertiary halogenoalkanes tend to hydrolyse by the S_N1 mechanism. The situation is, however, complicated in two ways:
1 some halogenoalkanes hydrolyse by a mechanism which has some characteristics of both of those described;
2 there is some competition from elimination reactions of the type

$$RCH_2CH_2Br + OH^- \longrightarrow RCH{=}CH_2 + Br^- + H_2O$$

14.3
THE EFFECT OF TEMPERATURE ON THE RATE OF REACTION

Another factor which affects the rate of reactions is temperature. We shall begin this section by obtaining some experimental data for this effect.

EXPERIMENT 14.3
The effect of temperature on the rate of the reaction between sodium thiosulphate and hydrochloric acid

The reaction between thiosulphate ions and hydrogen ions follows the equation

$$S_2O_3^{2-}(aq) + 2H^+(aq) \longrightarrow SO_2(aq) + S(s) + H_2O(l)$$

Procedure

1 Attach a piece of paper marked with a dark ink spot to the outside of a 400-cm^3 beaker. Fill the beaker with water and clamp two boiling-tubes, A and B, vertically in the beaker so that they are about half immersed. Position both the beaker and the tube A so that the ink spot is on the opposite side of the beaker from the observer and can be seen through both the beaker and the tube A.

2 Add 10 cm^3 of 0.1M sodium thiosulphate solution to tube A and 10 cm^3 of 0.5M hydrochloric acid to tube B. Use labelled measuring cylinders to measure these liquids, and make quite certain that the two are not confused. If they are, sulphur precipitates will start to form in them, making the experimental results invalid. Allow both to attain thermal equilibrium with the water and then quickly add the acid from B to the thiosulphate in A. Stir well with a thermometer and record the steady temperature. Measure the time from the mixing of the solutions to the point when the dark spot just becomes completely obscured by the formation of a precipitate of sulphur.

3 Repeat the experiment at several different temperatures, say at 20, 25, 30, 40, 50, and 60 °C, using the same concentrations of solution and the same technique of observation.

In these particular circumstances it is possible to get a satisfactory measure of the initial rate of the reaction directly from the experimental results. In each experiment the initial rate of precipitation is given by

$$\text{rate} = \frac{\text{amount of sulphur}}{\text{time to obscure spot}}$$

The quantity 'amount of sulphur' is impossible to evaluate but as it is the same in each experiment we can say that the rate of the reaction is proportional to the reciprocal of the time taken for the spot to be obscured.

You will need the following columns in your results table:

Temperature /°C	Temperature T /K	$1/T$ /K^{-1}	Time t /s	$1/t$ (*i.e.* rate) /s^{-1}	ln (rate)

Plot a graph of ln(rate) on the vertical axis against $1/T$ on the horizontal axis. The result is a straight line which indicates a relationship between rate and temperature of the form

$$\ln(\text{rate}) = \text{constant}_1 - \text{constant}_2 \, (1/T)$$

The collision theory of reaction kinetics

One model which has been used to account for the dependence of rate on temperature is the collision theory. This is most readily appreciated as applied to gas-phase reactions but its results are broadly applicable to reactions in solution too.

In the collision theory the reactant particles are pictured as moving towards each other and colliding in such a way that bonds are broken and new bonds formed. The product particles may then move away from the site of the reaction. In many gas-phase reactions it is the actual collision which controls the rate of the reaction and for an effective collision (that is, one which actually results in a reaction) the kinetic energy possessed by the colliding particles must be more than a certain minimum E.

It was seen in Topic 6 (section 6.6) that in a collection of molecules at a particular temperature T, the fraction of those molecules having an energy greater than an amount E is given by

$$\text{fraction} = e^{-E/kT}$$

where k is the Boltzmann constant.

Taking logarithms of both sides this becomes

$$\ln(\text{fraction}) = -\frac{E}{kT}$$

If E_A is an amount of energy per mole of such molecules, then

$$\ln(\text{fraction}) = -\frac{E_A}{LkT}$$

where L is the Avogadro constant.

Notice that this expression contains k, the Boltzmann constant. We shall shortly be bringing this constant into an expression involving the rate constant, which is also denoted by the symbol k. We shall therefore need to do something to avoid any confusion between these two constants.

You may recall that in Topic 3, the gas constant R, introduced in the Ideal Gas Equation

$$pV = nRT$$

was shown to be the product of k, the Boltzmann constant, and L, the Avogadro constant. In that Topic we said that we would use Lk, rather than R, in all future work. Here, however, is an exception; we have Lk in this expression, and to avoid the confusion over the two uses of k, we shall use R instead. So we have

$$\ln(\text{fraction}) = -\frac{E_A}{RT}$$

where R is the gas constant.

E_A is called the *activation energy*. In a very simple gas reaction it would have about the same value as the bond energy of the bonds that are being broken in the reaction.

The rate of the reaction would thus depend on the collision rate and on the fraction of the collisions which have an energy greater than E_A per mole. Hence

$$\text{rate} = \text{collision rate} \times e^{-E_A/RT}$$

or $\quad \ln(\text{rate}) = \ln(\text{collision rate}) - \dfrac{E_A}{R}(1/T)$

If you now compare this equation with the one that you were shown immediately after experiment 14.3

$$\ln(\text{rate}) = \text{constant}_1 - \text{constant}_2\,(1/T)$$

you will see that they are of the same form.

Comparison of both equations with the equation for a straight line

$$y = C + mx$$

where C is a constant and m is the slope of the line, indicates that the slope m is equal to $-E_A/R$. Determine the slope of your graph and, taking R as $8.4\,\text{J K}^{-1}\,\text{mol}^{-1}$, calculate a value for the activation energy E_A for the reaction in experiment 14.3. The generally accepted value is $46.58\,\text{kJ mol}^{-1}$.

The Arrhenius equation

All the quantitative parts of what we have just done depend upon the Arrhenius equation

$$\ln k = C - \frac{E_A}{R}(1/T)$$

(where k is the rate constant, and R the gas constant), or, in its exponential form,

$$k = A\,e^{-E_A/RT}$$

The quantity A (whose logarithm is the constant C) is called the *pre-exponential factor* and, in a very simple gas reaction, may be equated with the molecular collision rate, as we have seen. Generally, however, A includes other factors, the most important of which is that successful reactions can only occur between molecules which are orientated correctly at the time of collision. Sometimes this is referred to as the *steric factor*.

The Arrhenius equation shows that when the temperature rises there is a large increase in the value of k, the rate constant. This corresponds to a large increase in the number of collisions occurring with the necessary minimum energy, as can be appreciated from the distribution of energies amongst molecules at different temperatures T.

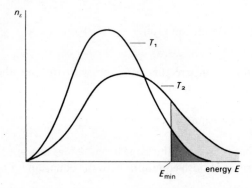

Figure 14.7

It is clear from figure 14.7 that if the necessary minimum energy is E_{min}, many more molecules have this energy at T_2 that at T_1, since T_2 is greater than T_1.

14.4
CATALYSIS

The essential feature of a catalyst is that it increases the rate of a chemical reaction without itself becoming permanently involved in the reaction. It does, however, become temporarily involved, by providing a route from reactants to products which has a lower activation energy.

From the equation

$$\ln k = C - \frac{E_A}{R}(1/T)$$

it is clear that the rate constant k, and hence the rate of the reaction, will be greater if the activation energy is lower; this assumes, of course, that the term C remains constant in both the catalysed and uncatalysed reactions.

An example of this which occurs in homogeneous catalysis is the decomposition of ethanal catalysed by iodine. Ethanal decomposes without a catalyst at about 700 K to produce methane and carbon monoxide

$$CH_3CHO(g) \longrightarrow CH_4(g) + CO(g); \qquad E_A \approx 200 \, kJ \, mol^{-1}$$

The addition of a small proportion of iodine increases the rate of this reaction by several thousand times, with the activation energy falling to $134 \, kJ \, mol^{-1}$. A series of gas reactions has occurred, involving free radicals thus

$$I_2 \rightleftharpoons 2I$$
$$CH_3CHO + I \longrightarrow CH_3CO + HI$$
$$CH_3CO \longrightarrow CH_3 + CO$$
$$CH_3 + I_2 \longrightarrow CH_3I + I$$
$$CH_3 + HI \longrightarrow CH_4 + I \qquad \text{(continuing reaction)}$$
$$CH_3I + HI \longrightarrow CH_4 + I_2 \qquad \text{(iodine regeneration)}$$

$$CH_3CHO \longrightarrow CH_4 + CO \qquad \textbf{overall reaction}$$

As a catalyst the iodine has formed some relatively stable intermediates and a reaction path or mechanism with a lower activation energy is created before the iodine is regenerated. In many homogeneous reactions, however, the intermediates are very reactive species, often present in very low concentrations, and difficult to identify.

Throughout this course there are a number of references to catalysis and when you are revising the whole course at the end you would do well to collect them all together in one place.

Apart from the short account given above, there are major references to catalysis in the following sections:

Topic 13   section 13.5 (catalysis involving enzymes)
Topic 16   section 16.5 (catalysis involving transition elements)
Topic 17   section 17.1 (catalysis in the polymerization of alkenes)

The following experiment is also about catalysis, though it treats it in an unusual way. Carry out the instructions and try to answer the questions which follow.

EXPERIMENT 14.4
A kinetic study of the reaction between manganate(VII) ions and ethanedioic acid

Potassium manganate(VII) will oxidize ethanedioic (oxalic) acid $(CO_2H)_2$ to carbon dioxide and water, in the presence of an excess of acid

$$2MnO_4^-(aq) + 6H^+(aq) + 5(CO_2H)_2(aq)$$
$$\longrightarrow 2Mn^{2+}(aq) + 10CO_2(g) + 8H_2O(l)$$

Procedure   *Method 1*

1 Prepare a reaction mixture according to the table which follows, using measuring cylinders. Some members of the class should do experiment 1 and some experiment 2. The results should then be shared. Note that ethanedioic acid (oxalic acid) is POISONOUS.

Solution	Experiment 1	Experiment 2
0.2M ethanedioic acid	$100\,cm^3$	$100\,cm^3$
0.2M manganese(II) sulphate	—	$15\,cm^3$
2M sulphuric acid	$5\,cm^3$	$5\,cm^3$
water	$95\,cm^3$	$80\,cm^3$

2 Add $50\,cm^3$ of 0.02M potassium manganate(VII) and start timing. Shake the mixture for about half a minute to mix it well.

3 After about a minute use a pipette and safety pipette filler to withdraw a $10\text{-}cm^3$ portion of the reaction mixture and run it into a conical flask.

4 Note the time and add about $10\,cm^3$ of 0.1M potassium iodide solution. This stops the reaction and releases iodine equivalent to the residual manganate(VII) ions.

5 Titrate the liberated iodine with 0.01M sodium thiosulphate, adding a little starch solution near the end point. Record the titre of sodium thiosulphate.

6 Remove further portions every 3 or 4 minutes and titrate them in the same way. Continue until the titre is less than $3 \, cm^3$.

Questions

1 What set of figures gives a measure of the reactant concentration at the various time intervals?

2 Plot an appropriate graph for each experiment. You will see that the graph for experiment 2 has a fairly conventional shape whereas the graph for experiment 1 is unusual.
What explanation can you offer for this abnormality?

Procedure *Method 2, using a colorimeter*

1 Note that ethanedioic acid (oxalic acid) is POISONOUS. Using a burette, put $10 \, cm^3$ of a solution which is 0.1M in ethanedioic ions and 1.2M in sulphuric acid into a test-tube.

2 Put $0.2 \, cm^3$ of a 0.02M solution of manganate(VII) ions into a test-tube that fits the colorimeter.

3 Adjust the meter of the colorimeter to maximum with a tube of water in place.

4 Add the ethanedioic solution to the manganate(VII) solution, shake the mixture, and start the clock.

5 Put the tube of reaction mixture into the colorimeter and take readings every 20 seconds. As the reaction proceeds you may want to take readings more frequently, but as the experiment is completed in four to five minutes, it is easy to repeat it.

You should convert the meter readings to concentration (in $mol \, dm^{-3}$) of manganate(VII) (see below) and plot a graph of concentration of manganate(VII) on the vertical axis against time on the horizontal axis.

Try to answer the following questions:

1 How does the rate of the reaction change with time?

2 Can you suggest a reason for the changes in the rate that you observe?

3 Can you suggest any experimental work that you could do to see if your suggestion is correct?

Note on the use of the colorimeter

Use of a colorimeter is a quick and easy method of determining the concentrations of solutions which are coloured. Basically the instrument consists of the components shown in figure 14.3 (see page 132).

The light-sensitive cell may either be a selenium cell which produces an e.m.f. proportional to the intensity of the light falling on it, or a cadmium sulphide cell, the electrical resistance of which is proportional to the intensity of the light falling on it. Either way the meter reading gives an indication of the intensity of light emerging from the solution.

The connection between the intensity of light emerging from the solution and the concentration of the absorbing species in the solution is

$$\lg\left(\frac{m_0}{m}\right) = \lg\left(\frac{I_0}{I}\right) \propto x$$

where I_0 is the intensity of incident, monochromatic light, I is the intensity of emergent light, m_0 and m are the meter readings, and x is the concentration of the solution.

The colorimeter is normally prepared for use by first adjusting the meter reading to maximum for I_0 by inserting a tube of pure solvent and then adjusting the intensity of light by means of a shutter placed between the bulb and the cell. The tube of solution is then put in the colorimeter and a meter reading proportional to I determined. It is *most important* to adopt this procedure for all readings if possible.

Unfortunately, the law quoted above is only obeyed accurately by certain solutions and under certain conditions. Also, the meter reading may not be accurately proportional to the intensity of light, so the instrument must be calibrated before use. This will also indicate for what range of concentration of a particular substance the colorimeter can be used.

For manganate(VII) solutions in a particular colorimeter the curve shown in figure 14.8 was obtained.

If a calibration curve is being used, it is easier to plot $\left(\frac{I}{I_0}\right) = \left(\frac{m}{m_0}\right)$ against concentration than $\lg\left(\frac{I_0}{I}\right) = \lg\left(\frac{m_0}{m}\right)$, as m_0 is normally 1 or 50 and these are useful numbers to have as denominators. Such a graph (figure 14.9) indicates that the colorimeter will not measure concentrations accurately above 0.0003M. Before these curves can be constructed, the most suitable filter must be chosen. This must select that band of wavelengths of light which are most strongly

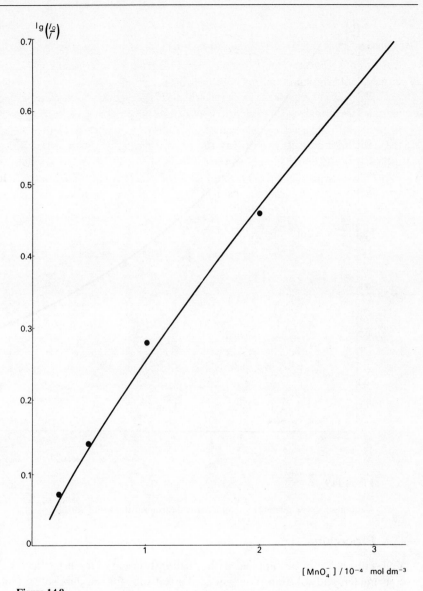

Figure 14.8

Graph of $\lg\left(\dfrac{I_0}{I}\right)$ against $[MnO_4^-]$, for a particular colorimeter.

absorbed by the solution; that is, it must be the filter which gives the lowest value of $\left(\dfrac{I}{I_0}\right)$ for a particular solution.

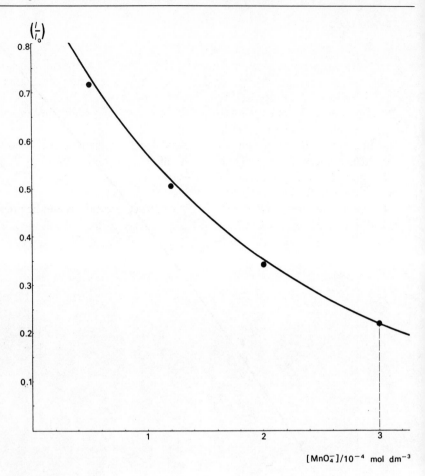

Figure 14.9

Calibration curve, relating $\left(\dfrac{I}{I_0}\right)$ for a particular colorimeter with $[MnO_4^-]$.

Procedure

You will probably be supplied with a calibration curve for a particular filter, colorimeter, and solution. You will need a test-tube for the pure solvent and an optically matched one for the solution. To take a reading, adjust the meter to maximum with pure solvent in place, then obtain the reading with solution in place. It is best then to check again with solvent in place and obtain a meter reading with solution in place once again. The value of $\left(\dfrac{I}{I_0}\right)$ so obtained can then be turned into a concentration, using the calibration chart.

SUMMARY

At the end of this Topic you should:

1 know the principal factors which affect the rate of a chemical reaction;

2 understand the quantitative significance of the terms 'rate of reaction', 'rate equation', 'rate constant', and 'order of reaction';

3 know some methods of following a reaction and be able to find the order of a reaction from experimental data;

4 understand the term 'activation energy' and how it can be evaluated from measurements of rates of reaction at different temperatures;

5 know something of the collision theory of reaction rates and its connection with the kinetic theory of matter;

6 understand more fully than before the subject of catalysis and be prepared for further exposition of this subject in later Topics.

APPENDIX I TO TOPIC 14
INTEGRATION OF RATE LAWS

For a first order reaction

$A \longrightarrow$ products

$$-\frac{d[A]}{dt} = k_1[A]$$

$$\therefore \quad -\frac{1}{[A]}d[A] = k_1 dt$$

Upon integration we have

$$-\int\frac{1}{[A]}d[A] = \int k_1 dt$$

$$\therefore \quad -\ln[A] = k_1 t + \text{constant} \qquad\qquad \textbf{1}$$

$$\left(\text{standard integral } \int\frac{1}{x}dx = \ln x\right)$$

At the start of the reaction when the concentration of A is $[A]_0$, $t = 0$, hence:

$$-\ln[A]_0 = k_1 0 + \text{constant}$$

$$\therefore \quad -\ln[A]_0 = \text{constant}$$

Substituting in **1**;

$$-\ln[A] = k_1 t - \ln[A]_0$$
$$\therefore \quad \ln[A]_0 - \ln[A] = k_1 t$$
$$\therefore \quad \ln\frac{[A]_0}{[A]} = k_1 t$$

Thus if a reaction is first order, a graph of $\ln\dfrac{[A]_0}{[A]}$ against t will be a straight line, and its gradient will be k_1. Note that absolute values of the concentrations are not required, only the ratio $\left(\dfrac{[A]_0}{[A]}\right)$, which is equal to $\left(\dfrac{V_0}{V}\right)$, where V_0 and V are titres which are proportional to $[A]_0$ and $[A]$.

The relationship between the half-life of a first order reaction and the rate constant can be obtained by inserting the condition that when $t = t_{\frac{1}{2}}$, $[A] = \dfrac{[A]_0}{2}$, into the above equation:

$$\ln\left(\frac{[A]_0}{[A]_0/2}\right) = k_1 t_{\frac{1}{2}}$$
$$\therefore \quad t_{\frac{1}{2}} = \frac{\ln 2}{k_1} = \frac{0.69}{k_1}$$

indicating that half-life is constant and independent of the concentration at the beginning of each half-life.

For a second order reaction

$A + B \longrightarrow$ products (or, if A and B are the same, $2A \longrightarrow$ products) *where the initial concentration of* A *is equal to the initial concentration of* B, *that is,* $[A]_0 = [B]_0$.

$$-\frac{d[A]}{dt} = k_2[A]^2$$
$$\therefore \quad -\frac{1}{[A]^2}d[A] = k_2 dt$$

Upon integration we have

$$-\int\frac{1}{[A]^2}d[A] = \int k_2 dt$$

$$\therefore \quad \frac{1}{[A]} = k_2 t + \text{constant} \quad \left(\text{standard integral } \int \frac{1}{x^2} dx = -\frac{1}{x} \right) \qquad \textbf{2}$$

At the start of the reaction when the concentration of A is $[A]_0 (= [B]_0)$, $t = 0$, hence:

$$\frac{1}{[A]_0} = k_2 0 + \text{constant}$$

$$\therefore \quad \frac{1}{[A]_0} = \text{constant}$$

substituting in **2**

$$\frac{1}{[A]} = k_2 t + \frac{1}{[A]_0}$$

Thus if a reaction is second order, and the two initial concentrations are the same, a graph of $\frac{1}{[A]}$ against t will be a straight line, and its gradient will be k_2. Note that absolute values of $[A]$ *are* required to obtain k_2, though titres, for instance, proportional to $[A]$ would give a straight line indicating that a reaction is second order.

The relationship between half-life of a second order reaction and the rate constant can be obtained by inserting the condition that when

$$t = t_{\frac{1}{2}}, [A] = \frac{[A]_0}{2}, \text{ into the above equation:}$$

$$\frac{1}{[A]_0/2} = k_2 t_{\frac{1}{2}} + \frac{1}{[A]_0}$$

$$\therefore \quad t_{\frac{1}{2}} = \frac{1}{[A]_0 k_2}$$

indicating that the half-life is not constant but depends on the concentration at the beginning of each half-life.

For higher order reactions

The formulae of obviously more complicated reactions can be worked out in a similar way to those above.

APPENDIX II TO TOPIC 14
ORDERS OF REACTION FROM HALF-LIFE TIMES

As a result of most methods of following chemical reactions, it is possible to arrive at some convenient measure of the concentrations C of a reactant at a series of times t from the start of the reaction. There are several ways of obtaining the order of the reaction from these results; one of them is to plot a graph of concentration C against t and read off from this the half-life time $t_{\frac{1}{2}}$ of the reactant where $t_{\frac{1}{2}}$ is the time taken for the concentration of the reactant to be reduced to half of any concentration C_0. It may readily be shown that the following simple rules apply:

Zero order – the graph is a straight line with negative gradient.
1st order – the graph is a curve for which $t_{\frac{1}{2}}$ is independent of C_0.
2nd order and above – the graph is a curve for which $t_{\frac{1}{2}}$ increases with t.

Two questions arise. 'How can we identify a reaction of the second, third, etc. order?' and 'How can we identify a fractional reaction order?' (These are sometimes mentioned in textbooks.) The following method resolves these difficulties.

$$-\frac{dC}{dt} = kC^n \qquad \text{where } n \text{ is the order of the reaction, } k \text{ is rate constant.}$$

$$\therefore \quad -\frac{dC}{C^n} = k\,dt$$

$$\therefore \quad \frac{1}{(1-n)C^{n-1}} = kt + K \qquad \text{where } K \text{ is a constant.}$$

When $t = 0$, $C = C_0$.

$$\therefore \quad K = \frac{1}{(1-n)C_0^{n-1}}$$

$$\therefore \quad kt = \frac{1}{(1-n)C^{n-1}} - \frac{1}{(1-n)C_0^{n-1}}$$

$$\therefore \quad kt = \frac{1}{1-n}\left(\frac{1}{C^{n-1}} - \frac{1}{C_0^{n-1}}\right)$$

Let $t = t_{\frac{1}{2}}$; then $C = \dfrac{C_0}{2}$.

$$\therefore \quad kt_{\frac{1}{2}} = \frac{1}{1-n}\left(\frac{2^{n-1}}{C_0^{n-1}} - \frac{1}{C_0^{n-1}}\right)$$

$$= \frac{2^{n-1}-1}{1-n} \times \frac{1}{C_0^{n-1}}$$

$$= \frac{2^{n-1}-1}{1-n} \times C_0^{1-n}$$

Taking logs of both sides,

$$\ln k + \ln t_{\frac{1}{2}} = \ln\left(\frac{2^{n-1}-1}{1-n}\right) + (1-n)\ln C_0$$

A graph of $\ln t_{\frac{1}{2}}$ against $\ln C_0$ therefore gives a straight line of gradient $(1-n)$, from which n may be identified. The graphs obtained for various orders will be as shown (see figure 14.10).

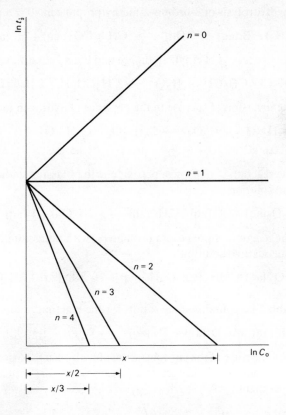

Figure 14.10

PROBLEMS

1 Devise an appropriate procedure to monitor the kinetics of each of the following reactions:

a The decomposition of 4-hydroxy-4-methylpentan-2-one (diacetone alcohol) in the presence of hydroxide ions in aqueous solution:

$$CH_3COCH_2C(CH_3)_2OH \longrightarrow 2CH_3COCH_3$$

b The effect of heat on a solution of 2,4,6-trinitrobenzoic acid –

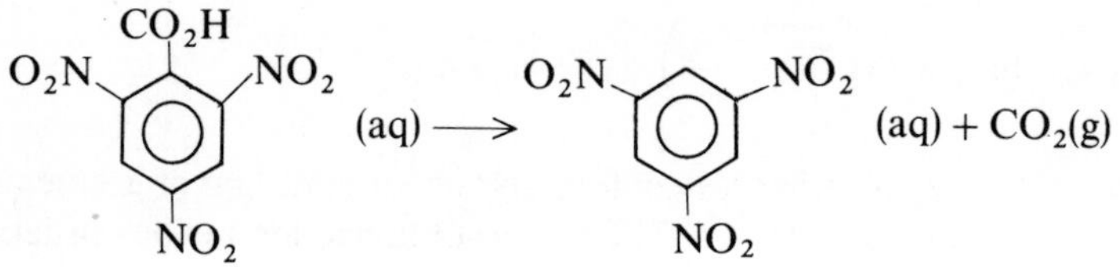

c The hydrolysis of 2-bromo-2-methylpropane in 80% aqueous ethanol –

$$(CH_3)_3CBr(aq) + H_2O(l) \longrightarrow (CH_3)_3COH(aq) + H^+(aq) + Br^-(aq)$$

d The hydrolysis of ethyl ethanoate under acid conditions –

$$CH_3CO_2CH_2CH_3 + H_2O \longrightarrow CH_3CO_2H + CH_3CH_2OH$$

e The inversion of sucrose in the presence of hydrogen ions –

$$C_{12}H_{22}O_{11} + H_2O \longrightarrow C_6H_{12}O_6 + C_6H_{12}O_6$$
$$\text{sucrose} \qquad\qquad \text{glucose} \qquad \text{fructose}$$

f The reaction of hydrogen peroxide with potassium iodide in aqueous acid solution –

$$H_2O_2(aq) + 2I^-(aq) + 2H^+(aq) \longrightarrow 2H_2O(l) + I_2(aq)$$

g The reaction of potassium bromate(v) with potassium bromide in aqueous acid solution –

$$BrO_3^-(aq) + 5Br^-(aq) + 6H^+(aq) \longrightarrow 3Br_2(aq) + 3H_2O(l)$$

2 Table 14.1 contains some results for the reaction:

$$2H^+(aq) + H_2O_2(aq) + 2I^-(aq) \longrightarrow I_2(aq) + 2H_2O(l)$$

y is a measure of the concentration of hydrogen peroxide at time t seconds. (The iodide and hydrogen ion concentrations are kept constant in this experiment.)

y	$t/10^3$ s
20.95	0
18.95	0.562
16.95	1.192
14.95	1.901
12.95	2.714
10.95	3.667
8.95	4.805
6.95	6.233
4.95	8.151
2.95	11.072
0.95	17.471

Table 14.1
(These results are taken from a famous paper published in 1867 by A. V. Harcourt and W. Esson, after whom this reaction is named.)

ai Plot a graph of y against t (y should be on the vertical axis).

 ii What is the order of the reaction?

iii Is this an order with respect to a particular reactant or is it the overall order?

bi Plot a graph of ln y against t (ln y should be on the vertical axis).

 ii Refer to Appendix I and say what can be deduced from the form of the graph.

3 Table 14.2 gives data about the decay of iodine-128.

Time /s	Counts of activity of sample/s^{-1}
1020	116.4
1740	85.2
3000	45.6
3600	35.3
4560	21.4
6300	9.7

Table 14.2

ai Plot a graph of 'counts of activity' against time.

 ii Find the half-life of iodine-128.

bi Plot a graph of ln (counts of activity) against time.

 ii Find the slope of the graph and, referring to Appendix I, find the rate constant (which, in this case, is also called the radioactive decay constant).

4 Table 14.3 contains results from an experiment concerning a reaction in which 0.0125 mole each of bromoethane and potassium hydroxide in solution are mixed, and then 20-cm^3 samples are taken and titrated against 0.05M acid.

Time/10^3 s	Titre of acid/cm^3
0	20.0
1.8	10.5
3.6	7.7
7.2	4.7
10.8	3.6
14.4	2.6
18.0	2.2
21.6	1.8

Table 14.3

a When the samples are withdrawn and are about to be titrated, is it necessary to take any action before adding indicator and proceeding with the titration? Explain how you reached your answer.

b To which reactant concentration are the titres directly proportional?

c Plot a suitable graph and identify the order of the reaction.

d What mechanism for this reaction is consistent with these results?

5 When hydrogen peroxide reacts with iodide ions in aqueous acid, iodine is liberated.

$$H_2O_2(aq) + 2H^+(aq) + 2I^-(aq) \longrightarrow 2H_2O(l) + I_2(aq)$$

The following data give some experimental results for the reaction:

Initial reactant concentrations /mol dm^{-3}			Initial rate of formation of I$_2$ /mol dm^{-3} s^{-1}
[H$_2$O$_2$]	[I$^-$]	[H$^+$]	
0.010	0.010	0.10	1.75×10^{-6}
0.030	0.010	0.10	5.25×10^{-6}
0.030	0.020	0.10	1.05×10^{-5}
0.030	0.020	0.20	1.05×10^{-5}

ai What is the rate equation for this reaction?

ii What is the value of the rate constant?

iii What are the units of the rate constant?

b A proposed mechanism for this reaction is

$$H_2O_2 + I^- \longrightarrow H_2O + IO^- \quad \text{(slow)}$$
$$H^+ + IO^- \longrightarrow HIO \quad \text{(fast)}$$
$$HIO + H^+ + I^- \longrightarrow I_2 + H_2O \quad \text{(fast)}$$

Is this mechanism consistent with the rate equation for the reaction? Give your reasons fully, discussing each of the three reactants in turn.

6 In acid solution, bromate ions slowly oxidize bromide ions to bromine.

$$BrO_3^-(aq) + 5Br^-(aq) + 6H^+(aq) \longrightarrow 3Br_2(aq) + 3H_2O(l)$$

The following experimental data have been determined.

Mixture	Volume of M bromate /cm^3	Volume of M bromide /cm^3	Volume of M H$^+$(aq) /cm^3	Volume of water /cm^3	Relative rate of formation of bromine
A	50	250	300	400	1
B	50	250	600	100	4
C	100	250	600	50	8
D	50	125	600	225	2

ai What is the rate equation for this reaction?
ii What are the units of the rate constant?
b A proposed mechanism for this reaction is

$$H^+ + Br^- \longrightarrow HBr \quad \text{(fast)}$$
$$H^+ + BrO_3^- \longrightarrow HBrO_3 \quad \text{(fast)}$$
$$HBr + HBrO_3 \longrightarrow HBrO + HBrO_2 \quad \text{(slow)}$$
$$HBrO_2 + HBr \longrightarrow 2HBrO \quad \text{(fast)}$$
$$HBrO + HBr \longrightarrow H_2O + Br_2 \quad \text{(fast)}$$

Is this mechanism consistent with the rate equation for the reaction? Give your reasons fully, discussing each of the three reactants in turn.

7 The rate constant given by

rate of change of concentration of hydrogen iodide $= k[HI]^2$

for the reaction $2HI(g) \longrightarrow H_2(g) + I_2(g)$
varies with the temperature as in the following table.

Temperature T/K	Rate constant $k/10^{-5}\,\mathrm{dm^3\,mol^{-1}\,s^{-1}}$
556	0.0352
647	8.58
700	116
781	3960

Plot a graph of $\ln k$ (on the vertical axis) against $\dfrac{1}{T}$ (on the horizontal axis) and from the gradient of this graph find the activation energy of the reaction.

8 The rate constant given by

rate of formation of nitrogen $= k[\mathrm{C_6H_5N_2Cl}]$

for the reaction

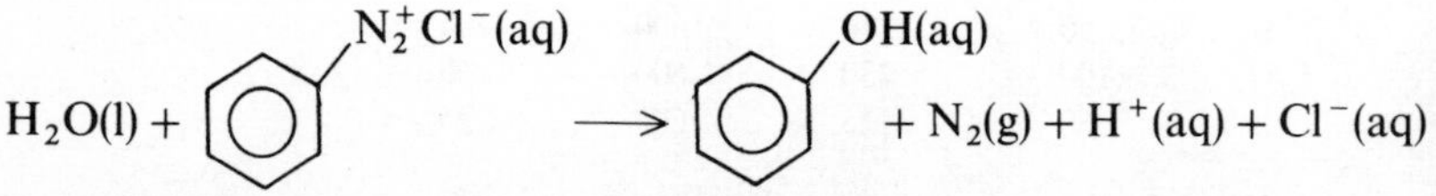

$$\mathrm{H_2O(l)} + \underset{}{\bigcirc\!\!\!\!\bigcirc}\!\!-\!\mathrm{N_2^+Cl^-(aq)} \longrightarrow \underset{}{\bigcirc\!\!\!\!\bigcirc}\!\!-\!\mathrm{OH(aq)} + \mathrm{N_2(g)} + \mathrm{H^+(aq)} + \mathrm{Cl^-(aq)}$$

varies with temperature as in the following table.

Temperature T/K	Rate constant $k/10^{-5}\,\mathrm{s^{-1}}$
278.0	0.15
298.0	4.1
308.2	20
323.0	140

Find the activation energy of the reaction.

Redox equilibria and free energy

In this Topic we shall begin with a review of what has been mentioned in previous Topics about entropy, entropy changes, and free energy. We shall then investigate some equilibria that involve the transfer of electrons (oxidation and reduction, or redox equilibria) and see how considerations of entropy changes and of free energy enable us to understand these equilibria more fully. Finally we shall investigate some examples of the use of free energy data in a number of chemical situations.

15.1
ENTROPY AND FREE ENERGY

In Topics 6, 10, and 12 we saw something of the importance of entropy changes in deciding how far a reaction will go, and how conditions such as temperature and pressure can be altered to help a reaction to take place.

The entropy of a system, S, was defined in Topic 4 as $k \ln W$, where W is the number of ways in which the molecules of the system, and their energy, can be distributed, and k is the Boltzmann constant. Because molecules don't care how they are distributed, and energy doesn't care how it is spread amongst molecules, both molecules and energy end up in the condition that is most likely by chance alone. We have seen that this is the condition that can be arrived at in the greatest number of ways, and so the value of the entropy of an isolated system left on its own becomes as large as possible.

If a change in a system can occur, there may well be an exchange of energy with its surroundings, and so the number of ways in which the energy of the surroundings can be distributed must also be considered. The total entropy change of system and surroundings that takes place during a spontaneous change (that is, during a change that takes place of its own accord), ΔS_{total}, must always be positive. This is because spontaneous changes always take place in such a way that the total value of the entropy becomes larger. This is true of the diffusion of gases, the freezing of water and the melting of ice, the manufacture of ammonia from nitrogen and hydrogen, and even of the many complex biochemical reactions occurring in your body as you read these words.

ΔS_{total} *must always be positive if a change is to occur of its own accord.*

This idea is called the Second Law of Thermodynamics.

When a change stops occurring, equilibrium has been reached. In Topic 10 we met another important statement about the total entropy change, namely:

At equilibrium, ΔS_{total} is zero

The importance of considering the *total* entropy change cannot be stressed too much. There are plenty of spontaneous changes for which the entropy of the *system* actually decreases. For example, gaseous water in the warm air of a room spontaneously changes to liquid water when it meets a cold window pane.

$$H_2O(g) \longrightarrow H_2O(l)$$

This is a change for which ΔS_{system} is negative; gaseous molecules have more ways of arranging themselves and their energy than liquid molecules, which are both more closely packed together, and possess less energy.

When we consider this change we must consider both what happens to the water molecules, and what happens to their surroundings. The entropy change of the system is negative, but the reaction is exothermic. Heat energy is given to the surroundings – the window pane and the surrounding air. This extra energy increases the number of ways the molecules of the surroundings can arrange their energy – it increases *their* entropy.

So for this change there are two entropy changes to consider,

ΔS_{system} (the system being the water) and $\Delta S_{surroundings}$; and

$$\Delta S_{total} = \Delta S_{system} + \Delta S_{surroundings}$$

When the surroundings are taken into account, the total entropy change, ΔS_{total}, is positive, because the surroundings gain more entropy than the system loses.

It turns out that many chemical reactions are like this; provided that the temperature T is comparatively low, the entropy change in the surroundings, caused by the energy the reaction gives out, is more important than the entropy change in the chemicals themselves. That is why, at ordinary temperatures, exothermic reactions are so much more common than endothermic ones; exothermic reactions increase the entropy of the surroundings, whereas endothermic reactions decrease it.

It is rather a nuisance having to bear the surroundings in mind all the time when, as a chemist, what you are really interested in is what is happening in the chemical reaction, or system. Chemists are much more interested in what is happening inside a test-tube than in the test-tube itself, let alone the surrounding air.

To make the situation easier to handle, chemists use a quantity, ΔG, first met in Topic 12. It is known as the *Gibbs free energy change*, or just the free energy change, and takes its name, and the symbol G, after the American chemist Willard Gibbs. This quantity takes account of both the system and its surroundings in a single expression.

Returning to the equation

$$\Delta S_{\text{total}} = \Delta S_{\text{system}} + \Delta S_{\text{surroundings}}$$

as $\Delta S_{\text{surroundings}} = -\Delta H/T$ for a change at constant temperature and pressure,

then $\Delta S_{\text{total}} = \Delta S_{\text{system}} - \Delta H/T$.

If the temperature is constant, then, multiplying all through by $-T$, we have

$$-T\Delta S_{\text{total}} = \Delta H - T\Delta S_{\text{system}}$$

The quantity $-T\Delta S_{\text{total}}$ is known as the free energy change, ΔG, and so we can write

$$\Delta G = \Delta H - T\Delta S \text{ (for constant temperature)}$$

For a change in a system to take place of its own accord, that is, for a spontaneous change, we have seen that, quite generally,

ΔS_{total} must be positive

Now, for a change at constant temperature and pressure,

$$\Delta G = -T\Delta S_{\text{total}}.$$

It is therefore evident that for a spontaneous change at constant temperature and pressure

ΔG must be negative.

The great value of ΔG is that it enables you to forget the surroundings. ΔH and $T\Delta S_{\text{system}}$, which together make up ΔG, can be measured or looked up in tables relating to the system (that is, the chemicals) only. ΔH and T take care of the energy which goes to (or comes from) the surroundings, and changes the entropy of the surroundings; ΔS_{system} looks after the system.

Just why free energy is called *free* will be explained at the end of this Topic. First, however, we shall investigate the equilibria that exist between metals and metal ions, and some other similar types of reactions, equilibria to which the application of free energy changes will prove to be especially useful.

15.2
REDOX EQUILIBRIA: METAL/METAL ION SYSTEMS

Earlier in the course, you have met a number of oxidation and reduction (redox) reactions and have seen, in Topic 5, that these always involve a change of oxidation number of the reacting substances. Redox reactions in which ions are involved will be studied in this and the next two sections. The principles governing these reactions provide a basis for understanding a very large number of chemical systems.

EXPERIMENT 15.2a
Some simple redox reactions

All the following reactions can be carried out in test-tubes.

1 Dip a strip of zinc foil into copper(II) sulphate solution (about 0.5M) and leave for about 30 seconds. Remove the strip and examine the zinc surface.

2 Repeat **1**, using zinc powder instead of foil, and find whether there is a temperature change during the reaction. Add zinc powder, with shaking, until there is no further change. Record any temperature change. Examine the solid and solution when reaction appears to be finished.

3 Repeat **1**, using copper foil and silver nitrate solution (about 0.1M).

4 Repeat **2**, using copper powder and silver nitrate solution.

Write ionic equations for what you have observed.

Was heat evolved or absorbed in the reactions?

Why do the changes in **2** and **4** proceed more quickly than those in **1** and **3**?

What can you say about the energy content of the products compared with the reactants for each system?

Write down the oxidation numbers of the reactants and the products in each reaction.

Which of the reactants has been oxidized and which reduced in each case?

Oxidation and reduction by electron transfer

Reactions between metals and metal ions involve the transfer of electrons from one reactant to another. This may be seen if the reactions are analysed into component reactions (or half-reactions), such as

$$Zn(s) \longrightarrow Zn^{2+}(aq) + 2e^-$$

and $2e^- + Cu^{2+}(aq) \longrightarrow Cu(s)$

In the first half-reaction, the oxidation number of zinc increases (0 to $+2$) so that an *oxidation* is involved; in the second reaction the oxidation number of copper decreases ($+2$ to 0) thus involving a *reduction*. In the complete reaction zinc is the *reductant* (or reducing agent) and copper the *oxidant* (or oxidizing agent). It will be seen that

>*loss* of electrons is an *oxidation* process
>*gain* of electrons is a *reduction* process
>species that *gain* electrons are acting as *oxidants*
>species that *lose* electrons are acting as *reductants*

Write half-reactions for the copper/silver nitrate reaction. Which species is the oxidant and which the reductant in this reaction?

From experiment 15.2a it will be seen that copper can act as either an oxidant or a reductant, depending on the conditions. This can be accounted for by treating each half-reaction as an equilibrium, so that we have

$$Cu(s) \rightleftharpoons Cu^{2+}(aq) + 2e^-$$

If electrons are added to this system, Le Châtelier's principle tells us that the equilibrium will move towards the left; removal of electrons will have the opposite effect. In the presence of a metal whose tendency to form ions in solution is greater than that of copper, the reaction moves towards the left and copper is deposited. This is the case with zinc, which dissolves to form hydrated zinc ions. With the silver system the reverse is the case and copper atoms lose electrons to become hydrated ions, while metallic silver is precipitated.

Measuring the tendency of a metal to form ions in solution

If an equilibrium is set up when a metal is placed in an aqueous solution of its ions

$$M(s) \rightleftharpoons M^{z+}(aq) + ze^-$$

we expect the metal to become negatively charged, by electrons building up on it, and the solution to become positively charged. Thus there should be an electric potential between solution and ions. If the equilibrium position differs for different metals, the potentials set up will differ also. These potentials (called absolute potentials) cannot be measured but the *difference* in potential between two metal/metal ion systems can be found by incorporating them into a voltaic cell and measuring the potential difference between the metal electrodes. Potential difference (p.d.) is measured in volts.

By using the circuit shown in figure 15.1 it is possible to explore the variation of potential difference between metal electrodes in a voltaic cell with the resistance in the circuit.

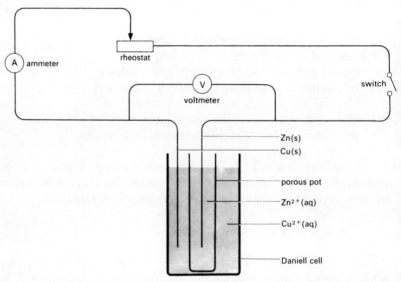

Figure 15.1

EXPERIMENT 15.2b
The variation of p.d. in a Daniell cell with change of external resistance

In this experiment a Daniell cell is used. It transfers electrical energy to the circuit as a result of the reaction

$$Zn(s) + Cu^{2+}(aq) \longrightarrow Zn^{2+}(aq) + Cu(s)$$

The Daniell cell is a combination of two electrode systems,

zinc metal immersed in zinc sulphate solution
and   copper metal immersed in copper sulphate solution.

The solutions of the two systems are prevented from mixing by a porous pot (made of unglazed porcelain); the zinc sulphate solution is usually put inside the pot and copper sulphate solution outside. Electrical contact between the solutions is established in the walls of the porous pot.

When the cell is working, electrons are transferred from the zinc electrode to the external circuit

$$Zn(s) \longrightarrow Zn^{2+}(aq) + 2e^-$$

and electrons are transferred from the circuit to the copper electrode

$$2e^- + Cu^{2+}(aq) \longrightarrow Cu(s)$$

Thus the copper is the positive terminal of the cell, and the zinc the negative terminal.

Procedure

Connect the circuit shown in figure 15.1. Use a high resistance voltmeter, and set the rheostat to the position of minimum resistance. Close the switch, gradually increase the resistance of the rheostat, and note the effect on the ammeter and voltmeter readings. Finally, open the switch and note the voltmeter reading when the only current to flow is the very small one that passes through the high resistance of the voltmeter.

The experiment shows that the potential difference between the two metal electrodes increases as the resistance increases. The p.d. is at a maximum when no current is flowing. This maximum p.d. is called the *electromotive force* of the cell (e.m.f.) which is denoted by the symbol E. When taking a measurement by a voltmeter (that is when the switch is open in figure 15.1) current must flow to operate the voltmeter. A direct voltmeter reading can therefore never give a highly accurate value for the e.m.f. of a cell. With a voltmeter of very high resistance, such as a transistorized, or solid-state voltmeter, the current taken is extremely small, however, and only a minute error is involved in measurements of e.m.f. The value of E for a cell is a measure of the relative tendencies of the electrode systems involved to liberate electrons by forming ions in solution.

Cell diagrams

It is convenient to have an agreed method of representing voltaic cells and the e.m.f. which they produce. This is called a *cell diagram*. In Britain the convention agreed by the International Union of Pure and Applied Chemistry (IUPAC) is used for such diagrams. This may be illustrated by the Daniell cell, for which the cell diagram is written

$$Zn(s)\,|\,Zn^{2+}(aq)\,\vdots\,Cu^{2+}(aq)\,|\,Cu(s); \quad E = +1.1\ V$$

The solid vertical lines represent boundaries between solids and solutions in each electrode system and the vertical broken line represents the porous partition (or other device to ensure a conducting path through the cell). The e.m.f. of the cell is represented by the symbol E, and the value is given in volts, with a sign ($+$ or $-$) preceding it which indicates the polarity of the *righthand electrode* in the diagram. In the example above the copper plate is the positive terminal of the cell. Obviously the cell diagram can be written in the reverse order but it is still the righthand electrode whose polarity is indicated. The Daniell cell can thus be written in the alternative form

$$Cu(s)\,|\,Cu^{2+}(aq)\,\vdots\,Zn^{2+}(aq)\,|\,Zn(s); \quad E = -1.1\ V$$

This is the basic pattern for all cell diagrams. Additional conventions are required for more complicated cells. These will be dealt with as they arise.

Contributions made by separate electrode systems to the e.m.f. of a cell

Measurement of the potential of a single electrode system is impossible because two such systems are needed to make a complete cell of which the e.m.f. can be measured. We can, however, assess the *relative* contributions of single electrode systems to cell e.m.f.s by choosing one system as a standard against which all other systems are measured. The standard system is then arbitrarily assigned zero potential and the potentials of all other systems referred to this value. By international agreement the hydrogen electrode is the standard reference electrode.

The hydrogen electrode

From redox reactions such as

$$Mg(s) + 2H^+(aq) \rightleftharpoons Mg^{2+}(aq) + H_2(g)$$

it is clear that an equilibrium can be set up between hydrogen gas and its ions in solution

$$H_2(g) \rightleftharpoons 2H^+(aq) + 2e^-$$

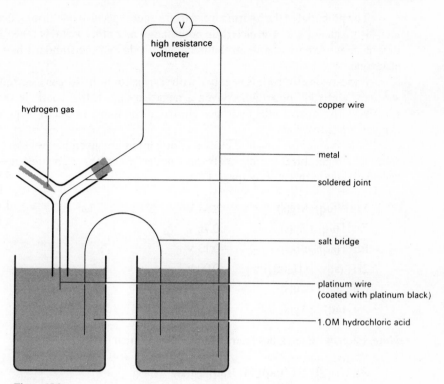

Figure 15.2

By an arrangement such as the one shown in figure 15.2, this reaction can be used in a half cell. This half-cell is called a *hydrogen electrode*. Essentially it consists of a platinum surface which is coated with finely divided platinum (usually called 'platinum black') which dips into a molar solution of hydrogen ions (usually 1M HCl(aq)). A slow stream of pure hydrogen is bubbled over the platinum black surface. Under these conditions the equilibrium

$$H_2(g) \rightleftharpoons 2H^+(aq) + 2e^-$$

is established fairly quickly. The platinum black acts as a catalyst in this process and, being porous, it retains a comparatively large quantity of hydrogen. The platinum metal also serves as a convenient route by which electrons can leave or enter the electrode system. The hydrogen electrode is represented by

$$Pt[H_2(g)] \mid 2H^+(aq) \vdots$$

The potential of the hydrogen electrode under specified conditions, detailed later, is taken as zero. The electrode potential of any other metal is taken as the difference in potential between the metal electrode and the standard hydrogen electrode.

If the metal electrode is negative with respect to the hydrogen electrode, the standard electrode potential is given a negative sign. If the metal electrode is positive with respect to the hydrogen electrode, the standard electrode potential is given a positive sign.

Some values for *standard* electrode potentials are given below. The conditions of temperature and concentration under which these are measured are specified later in this section (page 176).

$$Mg^{2+}(aq)\,|\,Mg(s) \qquad -2.37\,V$$
$$Zn^{2+}(aq)\ |\,Zn(s) \qquad -0.76\,V$$
$$Pb^{2+}(aq)\ |\,Pb(s) \qquad -0.13\,V$$
$$2H^+(aq)\ \ |\,[H_2(g)]Pt \qquad 0.00\,V$$
$$Cu^{2+}(aq)\ |\,Cu(s) \qquad +0.34\,V$$
$$Ag^+(aq)\ \ |\,Ag(s) \qquad +0.80\,V$$

Note. Each of these values refers to the e.m.f. of a real cell

$$Pt[H_2(g)]\,|\,2H^+(aq)\,\vdots\,M^{z+}(aq)\,|\,M(s)$$

From the list, values for the e.m.f. of other cells can be estimated. What would be the value of the e.m.f., and the sign of the righthand electrode for the cell

$$Pb(s)\,|\,Pb^{2+}(aq)\,\vdots\,Cu^{2+}(aq)\,|\,Cu(s)?$$

Experiment 15.2c
To measure the e.m.f. of some voltaic cells

The circuit for this experiment is shown in figure 15.3.

Several types of cell can be used. The one shown at A in figure 15.3 uses metal strips, 6 cm by 1 cm, as electrodes; they are slotted into pieces of cork held in small clamps, so that they dip into electrode solutions contained in 100 cm³ beakers. The metal strips are cleaned with emery paper before use. Each beaker contains a 'half-cell' and connection between them is made by a single strip of filter paper, about 10 cm × 1 cm, soaked in saturated potassium nitrate solution (the 'salt bridge'). Connections to the metal electrodes are made with crocodile clips. Alternative B is more portable and uses smaller volumes of solutions. The salt bridge is a plug of Polyfilla soaked in saturated potassium nitrate solution. A

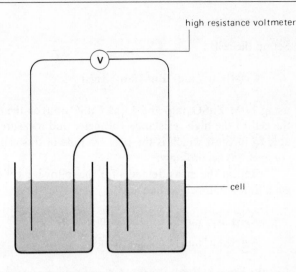

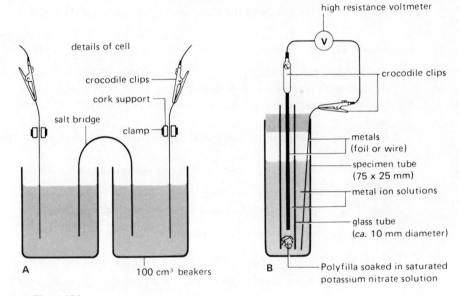

Figure 15.3

further variation can be made by using a 100×12 mm test-tube, with a hole blown in the rounded part, instead of the inner tube in version B. Small pieces of filter paper, soaked in saturated potassium nitrate solution and compressed into a plug about 1 cm thick (use a glass rod) form the salt bridge.

Procedure

Set up the cell

$$Cu(s)\,|\,Cu^{2+}(aq)\,\vdots\,Zn^{2+}(aq)\,|\,Zn(s) \qquad\qquad \textbf{a}$$

using 1.0M $ZnSO_4(aq)$ and 1.0M $CuSO_4(aq)$ as the electrode liquids. Connect the cell to the high resistance voltmeter and measure the e.m.f. You should be able to forecast which is the positive pole of this cell, and make the voltmeter connections accordingly.

Repeat the measurements for the following cells, using a fresh salt bridge each time a half-cell is changed

$$Ag(s)\,|\,Ag^{+}(aq)\,\vdots\,Cu^{2+}(aq)\,|\,Cu(s) \qquad\qquad \textbf{b}$$
$$Ag(s)\,|\,Ag^{+}(aq)\,\vdots\,Zn^{2+}(aq)\,|\,Zn(s) \qquad\qquad \textbf{c}$$

For the silver half-cell use 0.1M $AgNO_3$ with silver foil; a small piece of copper foil as backing, held in the cork but well clear of the solution, will give a more permanent contact for attaching the crocodile clip, if you are using cell A.

Questions

1 Try to relate the results of this experiment to those of the displacement reactions in experiment 15.2a. Is the system with the greatest tendency to form ions the positive or negative pole of each cell?

2 Is the e.m.f. for cell **c** what you would expect from the values obtained for cells **a** and **b**?
You could obtain the equivalent of cell **c** by connecting cells **a** and **b** together as follows

$$Ag(s)\,|\,Ag^{+}(aq)\,\vdots\,Cu^{2+}(aq)\,|\,\overbrace{Cu(s)\qquad Cu(s)}\,|\,Cu^{2+}(aq)\,\vdots\,Zn^{2+}(aq)\,|\,Zn(s)$$

The two copper electrodes would then cancel each other.

3 How do the results obtained for each cell compare with the e.m.f. values calculated from the list of standard electrode potentials given on page 172?

15.3
THE EFFECTS OF CONCENTRATION CHANGES ON ELECTRODE POTENTIALS

The value of the electrode potential of a metal depends upon the concentration of the metal ions in the solution in contact with the metal. In the next experiment we shall examine this effect, using a silver electrode.

EXPERIMENT 15.3a
To investigate the effect of changes in silver ion concentration on the potential of the $Ag^+(aq)|Ag(s)$ electrode

By applying the principle of Le Châtelier to the system

$$Ag(s) \rightleftharpoons Ag^+(aq) + e^-$$

work out the effect of altering the ionic concentration of the solution. If the solution is made more dilute, will the metal tend to form ions to a greater or lesser extent? Will this result in the electrode potential becoming more negative or more positive? Test your predictions by measuring the e.m.f. of the cell

$$Cu(s)|Cu^{2+}(aq)|Ag^+(aq)|Ag(s)$$

keeping the ion concentration in the copper electrode system constant and varying the ion concentration in the silver electrode system.

Procedure

The same circuit as that for experiment 15.2c can be used; the high-resistance voltmeter can be shared on a communal basis.

Measure the e.m.f. of the cell for various values of $[Ag^+(aq)]$, keeping the concentration of copper ion constant at 1.0M in the other electrode. Suitable silver ion concentrations are 0.01M, 0.0033M, 0.001M, 0.00033M, and 0.0001M. The concentrations other than 0.01M can be obtained by progressive dilution of 0.01M solution. Extreme care and attention to cleanliness are essential when handling very dilute solutions. A fresh salt bridge (saturated potassium nitrate solution) must be used for each new concentration of silver ions. If a cell of the type shown at B in figure 15.3 is used, each group of students could make up one cell. The cells could then be passed round so that each group measures the e.m.f. of every cell in the range quoted above.

Record the results in tabular form, using separate columns for

$$[Ag^+(aq)]/\text{mol dm}^{-3}, \ln[Ag^+(aq)], \text{ and } E/\text{volt}.$$

Plot a graph of E against $\ln[Ag^+(aq)]$, with E values on the vertical axis and $\ln[Ag^+(aq)]$ values on the horizontal axis. Choose a horizontal scale so that extrapolation to $\ln[\text{Ag}^+(\text{aq})] = -40$ is possible. This graph will be needed later in experiment 15.3b.

Questions

1 Does e.m.f. variation with ion concentration follow the predictions made by applying Le Châtelier's principle to the $\text{Ag(s)}|\text{Ag}^+(\text{aq})$ system?

2 From the shape of the graph, what can be deduced about the relation between electrode potential and ion concentration?

3 What concentration of silver ion would make the silver electrode potential exactly equal to the copper electrode potential for 1.0M $\text{Cu}^{2+}(\text{aq})$? The cell e.m.f. will then be zero. Under these conditions the two electrodes will be in equilibrium, as represented by the equation

$$\text{Cu(s)} + 2\text{Ag}^+(\text{aq}) \rightleftharpoons \text{Cu}^{2+}(\text{aq}) + 2\text{Ag(s)}$$

Calculate the equilibrium constant, K_c, for this reaction.

Standard electrode potentials

From experiment 15.3a it can be seen that the value of E for a given electrode system varies with the ionic concentration of the solution. It can also be shown that E varies with the temperature. Thus, standardized conditions of temperature and concentration must be used if electrode potential measurements are to be compared. By international agreement the conditions chosen are:

an ion concentration of one mole per cubic decimetre*
a temperature of 298 K (25 °C)

The value of an electrode potential relative to the *standard hydrogen electrode* under these conditions is called the *standard electrode potential* (or standard redox potential). By definition again, the standard hydrogen electrode consists of hydrogen gas at one atmosphere pressure bubbling over platinized platinum in a solution of hydrogen ion concentration one mole per cubic decimetre.* Standard electrode potentials are denoted by the symbol $E^\ominus$. A series of $E^\ominus$ values is given in the *Book of data*.

---

* Owing to incomplete ionization, interference between ions, and effects of 'crowding together' in fairly concentrated solution, the actual concentration of dissolved material is usually greater than this. Thus to obtain a concentration of free hydrogen ions of one mole per cubic decimetre a solution of hydrogen chloride which is 1.18M must be used. We shall ignore these effects in the simplified treatment given in this Topic.

For other ion concentrations at around room temperature, the value of E can be calculated approximately from the expression

$$E = E^{\ominus} + \frac{0.026}{z} \ln [\text{ion}]$$

where $z = $ the charge on the metal ion.

This expression applies to metal/metal ion electrodes only; other electrode systems behave differently, and will be dealt with later. The value of the constant (0.026) varies slightly with temperature. It will be seen that if [ion] is less than one, the second term on the righthand side is negative since $\ln[\text{ion}]$ will then be negative. Hence E becomes less positive, or more negative, with dilution.

The equation

$$E = E^{\ominus} + \frac{0.026}{z} \ln [\text{ion}]$$

has the general form

$$y = c + mx$$

If y is plotted against x, a straight line is obtained. The intercept on the y axis for $x = 0$ gives the value of c, and the slope of the line (calculated in terms of the units represented on the axes) gives the value of m. For our purposes, E is y, $\ln[\text{ion}]$ is x, and $\frac{0.026}{z}$ is m. The resulting graph is of the form shown in figure 15.4.

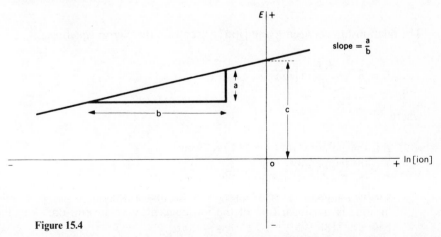

Figure 15.4

You can see whether the results obtained in experiment 15.3a fit into the above expression for calculating E. This is best done by calculating the values of E on the hydrogen scale for the electrode in which the concentration was varied. In experiment 15.3a this was the silver electrode. Thus we have

$$E_{cell} = E_{Ag} - E_{Cu}^{\ominus}$$

E_{cell} = e.m.f. of cell at various values of $[Ag^+(aq)]$
E_{Ag} = electrode potential of silver electrode
$E_{Cu}^{\ominus}$ = standard electrode potential of copper electrode (1.0M solution was used)

Rearranging the equation above, we have

$$E_{Ag} = E_{cell} + E_{Cu}^{\ominus}$$

From the *Book of data*, the value of $E_{Cu}^{\ominus}$ is 0.34 V.

So $E_{Ag} = E_{cell} + 0.34\,\text{V}.$

Calculate the values of E_{Ag} for the various concentrations of silver ion used in the experiment. Plot these values against $\ln[Ag^+(aq)]$, using the vertical axis for E values and the horizontal axis for $\ln[Ag^+(aq)]$. Draw the best straight line through the points plotted and calculate the slope of this line in terms of the units represented on the axis. The charge on the silver ion is $+1$, so the slope should be 0.026 if your results fit with the equation

$$E = E^{\ominus} + \frac{0.026}{z} \ln[\text{ion}]$$

The relationship between E and $[\text{ion}]$ arises from the *Nernst equation*.

$$E = E^{\ominus} + \frac{RT}{zF} \ln[\text{ion}]$$

where

R is the gas constant ($= 8.314\,\text{J K}^{-1}\,\text{mol}^{-1}$)
T is the temperature in K
F is the Faraday constant ($9.648 \times 10^4\,\text{C mol}^{-1}$)
z is the number of positive charges on the metal ion
$\ln[\text{ion}]$ is the logarithm of the ion concentration in mol dm^{-3} to the base e (2.718)

The Nernst equation cannot be used to calculate the value of E at different temperatures since $E^{\ominus}$ varies with temperature also. The potential of the standard hydrogen electrode, by definition, is zero at all temperatures.

By making use of known variation of ion concentration with E, the concentration of ions in very dilute solutions can be measured electrically. This method has drawbacks with concentrated solutions but can be used for solutions which would be much too dilute to be analysed by ordinary chemical methods. Some examples of the method are contained in the next section.

EXPERIMENT 15.3b
Using e.m.f. measurements to estimate small concentrations of silver ions

The cell

$$Cu(s)\,|\,Cu^{2+}(aq)\,|\,Ag^{+}(aq)\,|\,Ag(s)$$

is set up, using a standard copper electrode with 1.0M $Cu^{2+}(aq)$. The silver electrode is made up, using different mixtures of silver nitrate and other reagents. From e.m.f. measurements made on the cells which result, the silver ion concentrations in the mixtures can be found by using the graph prepared from the results of experiment 15.3a.

Procedure

This is the same as for experiment 15.3a.

Investigate as many of the following systems as you can. *For some of the measurements you may have to reverse the connections to the test cell. Use laboratory time for making measurements; the calculations can be done later.*

1 Mix one volume of 0.1M silver nitrate solution with two volumes of 0.1M potassium chloride solution. The actual volumes taken will depend on the type of cell that you are using. Insert a strip of silver foil into the mixture and measure the e.m.f. of the cell formed by this electrode and the standard copper electrode.

Calculation

Use the first graph plotted from the results of experiment 15.3a to calculate the approximate silver ion concentration in the solution of the silver electrode. This gives the value of $[Ag^{+}(aq)]_{eqm}$ in

$$AgCl(s) \rightleftharpoons Ag^{+}(aq) + Cl^{-}(aq)$$

The approximate value for the chloride ion concentration in this equilibrium can be obtained by assuming that all the silver ions used originally react with chloride ions; this is very nearly true since silver chloride is very sparingly soluble. Thus half the chloride ions added (those in one of the two volumes of potassium chloride solution used) are removed by precipitation. Since there is a total of three volumes of mixture,

$$[Cl^-(aq)]_{eqm} = \tfrac{1}{3} \times 0.1 \text{ mol dm}^{-3}$$

Calculate the approximate value of the solubility product of silver chloride

$$K_{sp} = [Ag^+(aq)]_{eqm}[Cl^-(aq)]_{eqm}$$

Compare your result with the value given in the table of solubility products in the *Book of data.*

Repeat the procedure in **1**, using the following mixtures:

2 1 volume 0.1M silver nitrate + 2 volumes 0.1M potassium bromide
3 1 volume 0.1M silver nitrate + 2 volumes 0.1M potassium iodide
4 1 volume 0.1M silver nitrate + 2 volumes 0.1M potassium iodate

BACKGROUND READING 1
The pH meter: measurement of hydrogen ion concentration by an electrical method

You will recall from Topic 12 that the hydrogen ion concentration of a solution is usually expressed as a pH number, where

$$pH = -lg[H^+(aq)]$$

Notice the use of lg, logarithms to base 10; elsewhere in this topic natural logarithms, ln, have been used. They are related in the following way:

$$\ln x = 2.3 \lg x$$

In Topic 12 you will have used pH meters to measure the pH number of solutions, and may have wondered how they work. In principle, the simplest method for measuring hydrogen ion concentration is to use a hydrogen electrode in the solution of unknown $H^+(aq)$ concentration as a half-cell. This can then be combined with a standard electrode to form a complete cell, the e.m.f. of which can be measured. The value of $[H^+(aq)]$ can then be calculated, using the Nernst equation.

For example, if the cell

$$Pt[H_2(g)] \mid 2H^+(aq) \vdots Cu^{2+}(aq, M) \mid Cu(s)$$

(Concentration
unknown)

is set up and the e.m.f. is found to be 0.43 V, then

for the copper electrode, $E = E^{\ominus} = 0.34$ V (from tables)

$\therefore$ for the hydrogen electrode $E = 0.34 - 0.43 = -0.09$ V

but, as $E^{\ominus} = 0$, $E = 0.026 \ln [H^+(aq)] = 0.06 \lg [H^+(aq)]$

$\therefore \lg [H^+(aq)] = -\dfrac{0.09}{0.06} = -1.5$

$\therefore$ pH $= -\lg [H^+(aq)] = -(-1.5) = 1.5$
also $[H^+(aq)] = $ antilg $(-1.5) = $ antilg $\overline{2}.5$
$\qquad\qquad\quad = 3 \times 10^{-2}$ mol dm^{-3}

As you will have seen, however, the hydrogen electrode is not easy to use. It is bulky when the hydrogen generator is taken into account, slow to reach equilibrium, and rather easily 'poisoned' by impurities. Alternative electrodes have therefore been sought, and, up to the present, the *glass electrode* is the most successful of these.

The glass electrode consists of a thin-walled bulb blown from special glass of low melting point. A solution of constant pH (a 'buffer' solution of which details will be given later) is placed inside the bulb with a platinum wire dipping into it. If the bulb is now immersed in a solution of unknown pH a potential is developed on the platinum wire and the whole arrangement can be used as a half-cell. When it is combined with a suitable reference electrode it is possible to make e.m.f. measurements. The resistance of the glass bulb is high (10^7–10^8 ohm) and a very sensitive voltmeter must be used to measure the e.m.f. In practice a solid-state (or transistor) voltmeter is used. The reference electrode is usually either a *calomel electrode* (a platinum wire dipping into mercury below a solution of mercury(I) chloride in saturated potassium chloride solution), or a *silver/silver chloride electrode* (a silver wire coated with silver chloride dipping into saturated potassium chloride solution).

The theory of the glass electrode is complicated but an arrangement such as

| Pt | solution A of known pH | glass bulb | solution B of unknown pH | reference electrode |

can be attached to a valve voltmeter and calibrated by using solutions of known pH in place of solution B. An instrument designed on this basis is called a *pH meter*. In commercial pH meters the glass electrode and the reference electrode may be combined in one unit which can be dipped into the solution under investigation. The voltmeter scale is calibrated to read directly in pH units.

15.4
REDOX EQUILIBRIA EXTENDED TO OTHER SYSTEMS

So far in this Topic we have been mainly concerned with redox systems involving a metal in equilibrium with its ions in solution. Earlier in the course, however, we have encountered redox systems of other kinds in which reactions between non-metals and non-metal ions, and between ions only, take place. Examples of these are

$$2Br^-(aq) \rightleftharpoons Br_2(aq) + 2e^-$$

and $$MnO_4^-(aq) + 8H^+(aq) + 5e^- \rightleftharpoons Mn^{2+}(aq) + 4H_2O(l)$$

We shall now see how reactions of this kind can be fitted into the pattern of electrode potentials which we have developed in sections 15.2 and 15.3. To begin with, we shall investigate a reaction in which there is no solid metal involved.

EXPERIMENT 15.4a
To investigate the reaction between iron(III) ions and iodide ions

To about $2\,cm^3$ of a solution which is approximately 0.1M with respect to $Fe^{3+}(aq)$ ions – iron(III) sulphate or ammonium iron(III) sulphate (iron alum) is suitable – add an equal volume of approximately 0.1M potassium iodide solution.

Test separate portions of the original solutions and the final mixture with
1 starch solution,
2 potassium hexacyanoferrate(III) solution (sometimes known as potassium ferricyanide solution).

Add potassium hexacyanoferrate(III) solution to a solution containing $Fe^{2+}(aq)$ ions – iron(II) sulphate is suitable.

The course of the reaction should be obvious from these tests, and from the colour change when the solutions are mixed. Write an equation for the reaction.

Using other redox reactions in voltaic cells

The equation for the reaction between iron(III) ions and iodide ions, studied in experiment 15.4a is

$$2Fe^{3+}(aq) + 2I^-(aq) \longrightarrow 2Fe^{2+}(aq) + I_2(aq)$$

Each element has undergone a change of oxidation number for which we can postulate two separate processes

$$Fe^{3+}(aq) + e^- \longrightarrow Fe^{2+}(aq) \quad \textit{reduction}$$

and $\quad 2I^-(aq) \longrightarrow I_2(aq) + 2e^- \quad \textit{oxidation}$

By analogy with half-reactions studied earlier we might expect two competing equilibria:

$$Fe^{3+}(aq) + e^- \rightleftharpoons Fe^{2+}(aq)$$
$$2I^-(aq) \rightleftharpoons I_2(aq) + 2e^-$$

with equilibrium positions which differ for each reaction. When the two systems are brought together the tendency for the iodide/iodine system to liberate electrons is greater than that of the iron(II)/iron(III) system. Thus the first equilibrium moves to the right and $Fe^{2+}(aq)$ ions are formed. Loss of electrons from the second equilibrium causes this to move to the right. This process will continue until the two systems are in equilibrium with each other. The equilibrium position for the overall reaction

$$2Fe^{3+}(aq) + 2I^-(aq) \rightleftharpoons 2Fe^{2+}(aq) + I_2(aq)$$

lies well over to the righthand side ($K_c \approx 10^5$ at $25\,°C$) so that we can say it is virtually complete.

It should be possible to use this reaction in a voltaic cell. The only problem is that of providing a means of allowing electrons to leave and enter the electrode systems. The solution is to use an inert electrode, as was done for the hydrogen half-cell in section 15.2, but in this case the electrode need not function as a catalyst, so smooth platinum is suitable. The following electrode systems can therefore be set up and their E values determined by using a reference electrode

$$Fe^{3+}(aq), Fe^{2+}(aq) | Pt$$

and $\quad I_2(aq), 2I^-(aq) | Pt$

We need to extend our conventions for writing cell diagrams in order to deal with systems such as these. The accepted practice is to put the *reduced* form (that in which the oxidation number is lowest) of the electrode system nearest to the electrode, and separate it from the oxidized form by a comma. (In the above examples the oxidation numbers are Fe^{3+}, $+3$; Fe^{2+}, $+2$; I_2, 0; I^-, -1.) Since iodine dissolves to a very small extent only in water and is usually obtained in solution in aqueous potassium iodide, it is sometimes represented as $I_2/KI(aq)$ in cell diagrams.

EXPERIMENT 15.4b
To measure the electrode potentials for the $Fe^{3+}(aq)/Fe^{2+}(aq)$ equilibrium and the $2I^-(aq)/I_2(aq)$ equilibrium

The circuit used is shown in figure 15.5.

Procedure

In order to allow the half-cell reactions to proceed in either direction, we must have both the oxidized and reduced forms present in the electrode systems. The lefthand electrode in figure 15.5 must therefore contain a solution in which both $Fe^{3+}(aq)$ ions and $Fe^{2+}(aq)$ ions are present for the first measurement, and $I^-(aq)$ ions and I_2 molecules for the second measurement. Keep the beakers containing these solutions for the third e.m.f. measurement. The $Cu(s)|Cu^{2+}(aq)$ system is used as a reference electrode.

Set up the following cells and measure their e.m.f.

$$Cu(s)|Cu^{2+}(aq) \vdots Fe^{3+}(aq), Fe^{2+}(aq)|Pt$$

and $$Cu(s)|Cu^{2+}(aq) \vdots I_2(aq), 2I^-(aq)|Pt$$

From the results calculate the e.m.f. of the cell

$$Pt|2I^-(aq), I_2(aq) \vdots Fe^{3+}(aq), Fe^{2+}(aq)|Pt$$

Check your calculation by measuring the e.m.f. of this cell. A second platinum electrode will be needed for this.

Concentration effects in ion/ion systems

From the results of experiment 15.4b it will be seen that ion/ion systems and non-metal/non-metal ion systems can be used in cell reactions in exactly the same way as metal/metal ion systems.

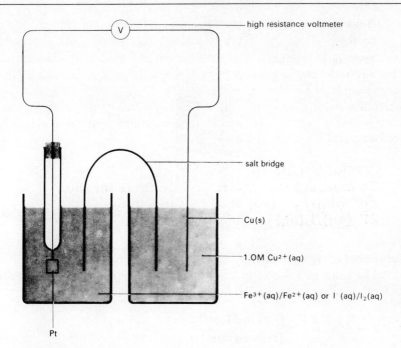

Figure 15.5

The application of either the principle of Le Châtelier or the Equilibrium Law to the iron(II)/iron(III) equilibrium indicates that the equilibrium position is affected by relative ion concentrations. The greater the relative concentration of $Fe^{3+}(aq)$ ions the more the equilibrium

$$Fe^{3+}(aq) + e^- \rightleftharpoons Fe^{2+}(aq)$$

moves to the right. This will *reduce* the absolute negative potential of the system, so that measured against the copper or hydrogen reference electrodes the difference in potential will *increase*, that is, it will become more positive.

The concentration effect is shown quantitatively by the figures given in table 15.1 on the next page.

As with metal/metal ion systems, temperature also has an effect on ion/ion equilibria. It is therefore necessary to specify both concentration and temperature conditions for standard electrode potentials involving equilibria between ions. The conditions chosen are:

equal molar concentrations of the reduced and oxidized forms of ion;
a temperature of 298 K (25 °C).

Relative concentrations /mol dm^{-3}		$\ln\dfrac{[Fe^{3+}]}{[Fe^{2+}]}$	E/volt
$[Fe^{3+}(aq)]$	$[Fe^{2+}(aq)]$		
1	9	-2.197	0.716
2	8	-1.386	0.735
3	7	-0.847	0.748
4	6	-0.405	0.760
5	5	0	0.770
6	4	$+0.405$	0.782
7	3	$+0.847$	0.792
8	2	$+1.386$	0.805
9	1	$+2.197$	0.825

Table 15.1
The variation of electrode potential with concentration for the $Fe^{3+}(aq)$, $Fe^{2+}(aq)$ electrode (E values are measured against a standard hydrogen electrode).

The symbol $E^{\ominus}$ is again used to indicate a standard redox potential

The value of E for other conditions is given by the Nernst equation in the form

$$E = E^{\ominus} + \frac{RT}{zF} \ln \frac{[\text{oxidized form}]}{[\text{reduced form}]}$$

For ion/ion systems z is the number of electrons transferred when the oxidized form changes to the reduced form. For the equilibrium

$$Fe^{3+}(aq) + e^{-} \rightleftharpoons Fe^{2+}(aq)$$

    (oxidized   (reduced
     form)   form)

$z = 1$.

If the values of $\ln\dfrac{[Fe^{3+}(aq)]}{[Fe^{2+}(aq)]}$ are plotted against E values from table 15.1, a straight line of slope 0.026 is obtained. For this system, the standard electrode potential (value of E when $[Fe^{3+}(aq)] = [Fe^{2+}(aq)]$ and hence $\dfrac{Fe^{3+}(aq)]}{[Fe^{2+}(aq)]} = 1$) is 0.770 V.

The form of the Nernst equation used earlier for metal/metal ion electrodes

$$E = E^{\ominus} + \frac{0.026}{z} \ln [\text{ion}]$$

is a special case of the general redox equation given above. For metal/metal ion systems the ion is the oxidized form and the metal the reduced form. Since the metal is a solid its concentration is constant and is thus not included in the equation. For non-metal/non-metal ion systems the non-metallic element is the oxidized form and the ion the reduced form.

Some further notes on standard potentials

You should now be in a position to appreciate all the information given in table 6.1 in the *Book of data*, the table of standard electrode potentials. The following notes may, however, be helpful in using this and other similar tables of $E^{\ominus}$ values.

1 It sometimes happens that the reduced and oxidized parts of an electrode system contain more than one chemical species (ion or molecule) taking part in the cell reaction. For example, the manganate(VII) ion generally exerts its oxidizing power in presence of hydrogen ions, and water molecules are formed amongst the products of oxidation. These ions and molecules must be included in the oxidized and reduced forms of the equilibrium mixture.

$$MnO_4^-(aq) + 8H^+(aq) + 5e^- \rightleftharpoons Mn^{2+}(aq) + 4H_2O(l)$$

The half-cell diagram for this system is written

$$[MnO_4^-(aq) + 8H^+(aq)], [Mn^{2+}(aq) + 4H_2O(l)]|Pt$$

The square brackets in this and similar diagrams do not stand for 'the concentration of' but are merely used to bracket together the oxidized and reduced forms of the equilibrium mixture.

The Nernst equation for calculating E values at other than standard concentration conditions is

$$E = E^{\ominus} + \frac{0.026}{z} \ln \frac{[MnO_4^-(aq)][H^+(aq)]^8}{[Mn^{2+}(aq)]}$$

In this equation, the square brackets have their usual significance. The concentration of water is not included since the variation of this is negligible in aqueous solutions. It will be obvious that E values for this and similar electrode systems are very sensitive to changes in hydrogen ion concentration.

2 In some tables of electrode potentials the electrode systems are given in the form

$$M^{z+}(aq) + ze^- = M(s); \qquad E^{\ominus} = \pm x \, V$$

Examples are

$$Cu^{2+}(aq) + 2e^- = Cu(s); \quad E^\ominus = +0.34\,V$$
$$Zn^{2+}(aq) + 2e^- = Zn(s); \quad E^\ominus = -0.76\,V$$
$$IO_3^-(aq) + 6H^+(aq) + 5e^- = \tfrac{1}{2}I_2(aq) + 3H_2O(l); \quad E^\ominus = +1.19\,V$$

It is easy to convert these to

$$Cu^{2+}(aq)\,|\,Cu(s)$$
$$Zn^{2+}(aq)\,|\,Zn(s)$$
$$[IO_3^-(aq) + 6H^+(aq)],\, [\tfrac{1}{2}I_2(g) + 3H_2O(l)]\,|\,Pt, \text{ if required.}$$

Note. In some older textbooks, everything is written the other way round, and all the signs are reversed, e.g. $Cu(s) = Cu^{2+}(aq) + 2e^-$; $\quad E^\ominus = -0.34\,V$.

3 Not all the $E^\ominus$ values given in tables have been obtained by direct measurements. Many of them are calculated from other experimental data.

Some uses of $E^\ominus$ values

Three important uses which can be made of tabulated $E^\ominus$ values are as follows.

1 Calculating the e.m.f. of voltaic cells

The procedure for doing this has been mentioned earlier. Write the cell diagram, reverse the sign of the $E^\ominus$ value for the lefthand electrode, and add the revised values to get the cell e.m.f. and the polarity of the righthand electrode.

As an example, we will calculate the e.m.f. of the magnesium–lead cell. The cell diagram can be written

$$Mg(s)\,|\,Mg^{2+}(aq)\,\vdots\,Pb^{2+}(aq)\,|\,Pb(s)$$

From tables

$$Mg^{2+}(aq)\,|\,Mg(s); \quad E^\ominus = -2.37\,V$$
$$Pb^{2+}(aq)\,|\,Pb(s); \quad E^\ominus = -0.13\,V$$

Reverse the sign of $E^\ominus$ value for lefthand electrode and add

$$+2.37\,V - 0.13\,V = +2.24\,V$$

The e.m.f. of the cell, under standard concentration conditions, will be 2.24 V at 25 °C and the lead electrode will be the positive pole. If other than standard ion concentrations are used, the E values for the separate half-cells must be calculated, using the Nernst equation.

2 Predicting whether a reaction is likely to take place

$E^\ominus$ values in order of increasing positive (or decreasing negative) values are also in order of decreasing tendency for the electrode system to release electrons, as shown in table 15.2.

Electrode system	$E^\ominus$/volt	
$Mg^{2+}(aq)\|Mg(s)$	-2.37	↑ increasing tendency for
$Zn^{2+}(aq)\|Zn(s)$	-0.76	electrode to release
$S(s), S^{2-}(aq)\|Pt$	-0.48	electrons
$Fe^{2+}(aq)\|Fe(s)$	-0.44	
$Sn^{2+}(aq)\|Sn(s)$	-0.14	
$2H^+(aq)[H_2(g)]\|Pt$	0.00	
$Cu^{2+}(aq)\|Cu(s)$	$+0.34$	
$I_2(aq), 2I^-(aq)\|Pt$	$+0.54$	
$Fe^{3+}(aq), Fe^{2+}(aq)\|Pt$	$+0.77$	
$Br_2(aq), 2Br^-(aq)\|Pt$	$+1.09$	decreasing tendency for
$Cl_2(aq), 2Cl^-(aq)\|Pt$	$+1.36$	electrode to release
$[MnO_4^-(aq) + 8H^+(aq)], [Mn^{2+}(aq) + 4H_2O(l)]\|Pt$	$+1.51$	↓ electrons

Table 15.2

Hence if we link two electrode systems to form a voltaic cell, the system which is higher in the series will become the negative pole (transferring electrons to external circuit) and the system which is lower will become the positive pole. This is the same as saying that, as we go down the series, the oxidizing power of the oxidized forms in the electrodes increases and the reducing power of the reduced forms decreases. This can be illustrated by the $Fe^{3+}(aq)$, $Fe^{2+}(aq)$ and $I_2(aq)$, $2I^-(aq)$ reaction studied earlier. The order of these electrode systems is

$$I_2(aq), 2I^-(aq)\|Pt; \qquad E^\ominus = +0.54\,V$$
$$Fe^{3+}(aq), Fe^{2+}(aq)\|Pt; \qquad E^\ominus = +0.77\,V$$

In a cell made from these two electrodes, the $Fe^{3+}(aq)$, $Fe^{2+}(aq)$ system will be the positive pole. $Fe^{3+}(aq)$ is a better oxidizing agent than $I_2(aq)$, and $2I^-(aq)$ is a better reducing agent than $Fe^{2+}(aq)$. When the cell is working the reaction in the upper electrode goes from *right* to *left*, $2I^-(aq) \longrightarrow I_2(aq)$, and the reaction in the lower electrode from *left* to *right*, $Fe^{3+}(aq) \longrightarrow Fe^{2+}(aq)$. This is the general situation for all reactions of this kind and can be summed up in a simple rule. For any pair of couples in the redox series, reaction will tend to go in such a way that, taking the individual species in the couples in *anti-clockwise order, starting with the bottom left in the positions that they occupy in the series, gives the reactants and products of the possible reaction.* In the redox series we have

$$\begin{array}{c} \text{--- } I_2(aq), 2I^-(aq)\,|\,Pt \leftarrow \text{----} \\ \text{above} \\ \text{--} \rightarrow Fe^{3+}(aq), Fe^{2+}(aq)\,|\,Pt \text{ --} \end{array}$$

Start from Fe^{3+}(aq) and proceed anti-clockwise

$$Fe^{3+}(aq) \xrightarrow{\ +e^-\ } Fe^{2+}(aq)$$
$$2I^-(aq) \xrightarrow{\ -2e^-\ } I_2(aq)$$

balance electron loss and gain

$$2Fe^{3+}(aq) \xrightarrow{\ +2e^-\ } 2Fe^{2+}(aq)$$
$$2I^-(aq) \xrightarrow{\ -2e^-\ } I_2(aq)$$

and add $2Fe^{3+}(aq) + 2I^-(aq) \longrightarrow 2Fe^{2+}(aq) + I_2(aq)$

A more convenient form for prediction is to write the half-equations for the electrodes, rather than the electrode systems, in a diagram (figure 15.6).

Strictly speaking, predictions made in this way refer only to standard conditions – molar solutions of reactants and products present at 298 K. In other conditions, effects of concentration or temperature may alter the equilibrium position (this is what we are really forecasting) of the opposing electrode systems and lead to a partial reversal of the change. If the difference in $E^\ominus$ values for the electrodes concerned is greater than about 0.4 V, this is unlikely to happen.

Consider another example. Will potassium manganate(VII), in acid solution, be likely to oxidize hydrogen sulphide to sulphur? From the electrode potential series we find the electrodes in the following order

$$[2H^+(aq) + S(s)], H_2S(aq)\,|\,Pt; \qquad E^\ominus = +0.14\,V$$
$$[MnO_4^-(aq) + 8H^+(aq)], [Mn^{2+}(aq) + 4H_2O(l)]\,|\,Pt; \qquad E^\ominus = +1.51\,V$$

and this can be written in a diagram as in figure 15.7.

The half-reactions should proceed as on page 192 (the equations are more complicated since more than one species is involved in the oxidized and reduced electrode systems).

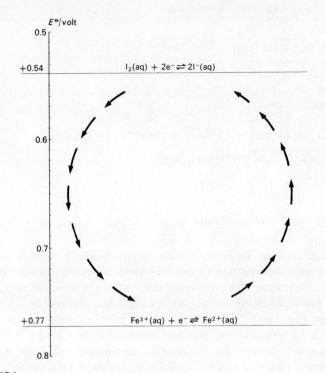

Figure 15.6

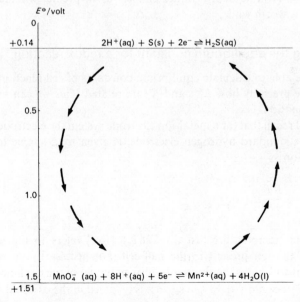

Figure 15.7

$$MnO_4^-(aq) + 8H^+(aq) \xrightarrow{+5e^-} Mn^{2+}(aq) + 4H_2O(l)$$
$$H_2S(aq) \xrightarrow{-2e^-} 2H^+(aq) + S(s)$$

Balance electron transfer

$$2MnO_4^-(aq) + 16H^+(aq) \xrightarrow{+10e^-} 2Mn^{2+}(aq) + 8H_2O(l)$$
$$5H_2S(aq) \xrightarrow{-10e^-} 10H^+(aq) + 5S(s)$$

and add

$$2MnO_4^-(aq) + 6H^+(aq) + 5H_2S(aq) \longrightarrow 2Mn^{2+}(aq) + 8H_2O(l) + 5S(s)$$

The difference between the $E^\ominus$ values is considerable ($1.51 - 0.14 = 1.37\,V$), so we should expect this reaction to proceed under all concentration conditions at ordinary temperatures. The difference in $E^\ominus$ values is, however, no guarantee that it will proceed quickly. In fact, this example is a reasonably fast reaction. Although we can use $E^\ominus$ values to tell us something about the position of equilibrium for a given change, they can never tell us how long it will take for this equilibrium to be attained. We have to experiment to find out the rate of a given reaction and, if it is slow, look for a catalyst if we want to make use of the reaction. A table of $E^\ominus$ values enables us to predict whether a search for a catalyst is worth while.

3 Finding the equilibrium constant for a redox reaction

In order to be able to calculate equilibrium constants of cell reactions, we need to know more precisely how $E^\ominus_{cell}$ and K_c are related. This we can see by using the Nernst equation.

You will recall that for a metal/ion electrode system the electrode potential, relative to the standard hydrogen electrode, is given by a simple form of the Nernst equation as

$$E = E^\ominus + \frac{RT}{zF} \ln[\text{ion}]$$

for a particular temperature (usually $298\,K$). [ion] refers to the *actual* concentration of the ion present in the half-cell (not necessarily an equilibrium concentration) and $E^\ominus$ is the standard redox potential of the electrode.

Example: Nernst equation for the Daniell cell – the cell diagram for the Daniell cell is

$$Zn(s)|Zn^{2+}(aq)|Cu^{2+}(aq)|Cu(s)$$

Figure 15.8 shows the construction of a typical Daniell cell. Inside the porous pot is the zinc half-cell which has an electrode potential, relative to hydrogen, given by the Nernst equation as

$$E_{Zn} = E^{\ominus}_{Zn} + \frac{RT}{zF} \ln [Zn^{2+}(aq)]$$

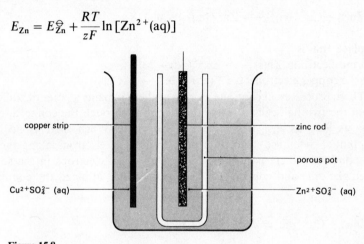

copper strip

zinc rod

porous pot

$Cu^{2+}SO_4^{2-}$ (aq)

$Zn^{2+}SO_4^{2-}$ (aq)

Figure 15.8

A diagram of a Daniell cell. The concentration of Zn^{2+}(aq) and Cu^{2+}(aq) are $[Zn^{2+}(aq)]$ and $[Cu^{2+}(aq)]$ and not necessarily 1M.

Similarly, the electrode potential of the copper electrode, relative to hydrogen, is

$$E_{Cu} = E^{\ominus}_{Cu} + \frac{RT}{zF} \ln [Cu^{2+}(aq)]$$

The e.m.f. of this Daniell cell is found by taking the difference between E_{Cu} and E_{Zn}:

$$E_{cell} = E_{Cu} - E_{Zn}$$
$$= E^{\ominus}_{Cu} - E^{\ominus}_{Zn} - \left(\frac{RT}{zF} \ln [Zn^{2+}(aq)] - \frac{RT}{zF} \ln [Cu^{2+}(aq)] \right)$$
$$\therefore \quad E_{cell} = E^{\ominus}_{cell} - \frac{RT}{zF} \ln \frac{[Zn^{2+}(aq)]}{[Cu^{2+}(aq)]}$$

This equation is a simple form of the Nernst equation for the Daniell cell. $E_{cell}^{\ominus}$ is the standard e.m.f. of the cell and has a value of 1.1 V.

Short-circuiting the cell – Suppose now that this cell is short-circuited by connecting the zinc pole to the copper pole by means of a conducting wire. What would happen?

When the cell is short-circuited, in the external circuit, electrons pass from the zinc pole along the wire to the copper pole. Inside the cell the reaction

$$Zn(s) + Cu^{2+}(aq) \longrightarrow Zn^{2+}(aq) + Cu(s)$$

takes place, that is,

at the zinc electrode: $Zn(s) \longrightarrow Zn^{2+}(aq) + 2e^-$

and at the copper electrode: $Cu^{2+}(aq) + 2e^- \longrightarrow Cu(s)$

These processes all continue as long as there remains a potential difference between the two half-cells to drive the electrons through the connecting wire, that is, as long as E_{cell} has some positive value. When there is no longer any potential difference, there is no net transfer of electrons along the wire. Correspondingly, there is no further net transfer of electrons in the solution between the zinc and copper. The cell reaction has attained the equilibrium situation:

$$Zn(s) + Cu^{2+}(aq) \rightleftharpoons Zn^{2+}(aq) + Cu(s)$$

In this situation the concentrations of $Zn^{2+}(aq)$ and $Cu^{2+}(aq)$ are the equilibrium concentrations, $[Zn^{2+}(aq)]_{eqm}$ and $[Cu^{2+}(aq)]_{eqm}$, and K_c for this equilibrium is $\dfrac{[Zn^{2+}(aq)]_{eqm}}{[Cu^{2+}(aq)]_{eqm}}$.

(In fact K_c for this particular equilibrium has a value of about 10^{37}, so that in practical terms the reaction has gone to completion, and either all the zinc or all the copper sulphate solution has been 'used up'.)

Applying the Nernst equation for the Daniell cell at equilibrium, we put in

$$E_{cell} = 0$$

and $[Zn^{2+}(aq)] = [Zn^{2+}(aq)]_{eqm}$

and $[Cu^{2+}(aq)] = [Cu^{2+}(aq)]_{eqm}$

$$\therefore \quad 0 = E_{cell}^{\ominus} - \frac{RT}{zF} \ln \frac{[Zn^{2+}(aq)]_{eqm}}{[Cu^{2+}(aq)]_{eqm}}$$

$$\therefore \quad E_{cell}^{\ominus} = \frac{RT}{zF} \ln K_c$$

In this expression the units of R are $J\,K^{-1}\,mol^{-1}$, that is $R = 8.314\,J\,K^{-1}\,mol^{-1}$. With $F = 96\,500\,C$, this gives $E^{\ominus}$ in volts.

To summarize, then, when a cell is short circuited, its e.m.f. falls to zero, the cell reaction attains equilibrium, and the Nernst equation reduces to an expression connecting $E^{\ominus}_{cell}$ and K_c for the cell equilibrium. We can therefore calculate values of K_c for cell reactions indirectly from $E^{\ominus}$ values. In fact, in using $E^{\ominus}$ values predictively we have merely been using equilibrium constants in a sort of Nernst disguise.

Table 15.3 lists some equilibrium constants for a selection of cell reactions. These K_c values have been calculated from the corresponding $E^{\ominus}_{cell}$ values, using the Nernst equation.

Reaction	$\Delta H^{\ominus}_{298}$	$\Delta G^{\ominus}_{298}$	$E^{\ominus}_{298}$	$K_{c(298)}$
$Cu^{2+}(aq) + Zn(s) \longrightarrow Cu(s) + Zn^{2+}(aq)$	-217	-212	$+1.10$	10^{37}
$Zn(s) + 2H^{+}(aq) \longrightarrow Zn^{2+}(aq) + H_2(g)$	-152	-147	$+0.76$	10^{26}
$Pb^{2+}(aq) + Zn(s) \longrightarrow Zn^{2+}(aq) + Pb(s)$	-154	-123	$+0.64$	10^{21}
$2Ag^{+}(aq) + Cu(s) \longrightarrow 2Ag(s) + Cu^{2+}(aq)$	-147	-89	$+0.46$	10^{16}
$2Tl^{+}(aq) + Zn(s) \longrightarrow 2Tl(s) + Zn^{2+}(aq)$	-164	-82	$+0.42$	10^{15}
$Cu^{2+}(aq) + Pb(s) \longrightarrow Cu(s) + Pb^{2+}(aq)$	-62.8	-89.3	$+0.47$	10^{16}
$Tl(s) + H^{+}(aq) \longrightarrow Tl^{+}(aq) + \tfrac{1}{2}H_2(g)$	$+5.9$	-31.8	$+0.34$	4×10^{5}

Table 15.3
Some energy data, $E^{\ominus}_{298}$ values, and equilibrium constants for selected reactions in solution. (Energy data are in kJ; equilibrium constants in appropriate units; standard electrode potentials in volts.)

Questions

The cell

$$Cu(s)\,|\,Cu^{2+}(aq)\,\vdots\,Br_2(aq), 2Br^{-}(aq)\,|\,Pt$$

is set up and short-circuited.

1 Write an equation for the resulting cell equilibrium reaction.

2 Calculate K_c for this equilibrium.

(Standard electrode potentials are given in the *Book of data*.)

15.5
ENTROPY CONSIDERATIONS

Entropy changes when metal ions go into solution

In the last three sections we have looked at redox reactions in cells, particularly cells involving metal/metal ion electrodes. In this section we shall consider the entropy changes that take place in these cells, and how they lead to a better understanding of cell reactions.

Consider the reaction

$$Cu(s) \longrightarrow Cu^{2+}(aq) + 2e^-$$

A simplified picture of just one Cu atom becoming a Cu^{2+} ion and entering pure water is shown in figure 15.9. Which has the greater number of possible distributions – the Cu atom in the solid lattice, or the Cu^{2+} ion in solution? There are more ways of arranging the Cu^{2+} ion, since it could go anywhere among the water molecules.

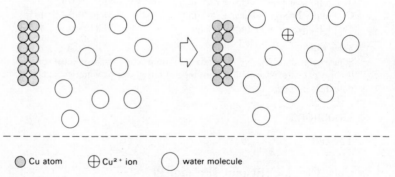

○ Cu atom   ⊕ Cu²⁺ ion   ○ water molecule

Figure 15.9

Ions in solution have a lot of freedom to move about, so the number of distributions, and therefore the entropy of ionic solutions is high – much higher than for solid lattices. In fact, ions in solution behave in many ways like molecules in gases, as is shown in figure 15.10.

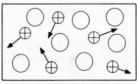

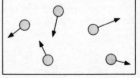

Figure 15.10   **Ionic solution**
Ions free to move anywhere
among the water molecules

Gas
Molecules free to move anywhere
in the container

Of course, this is greatly simplified. For one thing it ignores any interactions between the ions and the water molecules, whereas we know very well there are in fact strong attractions. Nevertheless, it is a useful approximate model to think of ions in solutions, especially dilute ones, as behaving like gases.

With this in mind, it is quite clear that whenever a copper atom turns into a copper ion and goes into solution, there must be an increase in the number of distributions – an entropy increase.

What if a second copper atom becomes an aqueous ion? The situation is illustrated in figure 15.11.

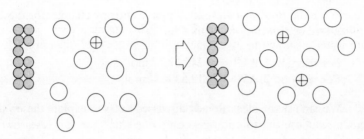

Figure 15.11

Obviously there will be a further entropy increase. But the increase will not be quite as large as last time, because this time there is an ion already present; this slightly reduces the number of arrangements available to the new ion. As more and more ions go into solution, the entropy increases each time – but each time by a little less. It is the same argument we used before with gas molecules and with energy quanta – when there are only a few molecules or quanta present, adding one more makes a big difference. When there are lots of molecules or quanta present, adding one more makes a smaller difference. It is the same with money, for that matter: when you are poor an extra pound makes a bigger difference to you than when you are rich.

So if you have a metal electrode surrounded by a solution already containing lots of ions – a concentrated solution – metal atoms turning into aqueous ions will give only a fairly small entropy increase. But if the surrounding solution is dilute, the entropy increase will be bigger.

To summarize, there are two things to bear in mind when considering metal/ion reactions.

1 **There is always an entropy increase when metal atoms go into solution as metal ions.**
2 **The more concentrated the solution of ions surrounding the metal, the smaller the entropy increase when new ions go into solution.**

Entropy changes in redox reactions

So far we have looked at the changes that occur when metal atoms turn into metal ions. This is the kind of change that occurs at a metal electrode in a cell But to make a cell you need *two* electrodes. What happens when the entropy changes at *both* electrodes are taken into account?

Consider these two simple reactions

$$2Ag^+(aq) + Cu(s) \longrightarrow 2Ag(s) + Cu^{2+}(aq) \qquad 1$$
$$Cu^{2+}(aq) + Zn(s) \longrightarrow Zn^{2+}(aq) + Cu(s) \qquad 2$$

We can use the *Book of data* to look up the entropy changes for these reactions At standard concentration ($1.0\,mol\,dm^{-3}$):

For reaction **1**, $\Delta S^{\ominus} = -245\,J\,K^{-1}\,mol^{-1}$
For reaction **2**, $\Delta S^{\ominus} = -12.5\,J\,K^{-1}\,mol^{-1}$

We can use the ideas already discussed to try to explain these values. In the Ag/Cu reaction, the solution gets only one Cu^{2+} ion but loses *two* Ag^+ ions. We can be sure that the entropy will *decrease* on this account, because aqueous ions have higher entropy than atoms in a solid. In the Cu/Zn reaction, one ion enters and one leaves. Furthermore, Cu and Zn are very alike as atoms and ions, even down to having similar mass. The entropy of the reactants and the entropy of the products are therefore very similar, so the entropy change for this reaction is small.

In this way we can rationalize the entropy change for some redox reactions In many cases though, the situation is too complex. But in any case we can *measure* the entropy change for a redox reaction quite easily, by making the reaction occur in a cell, and measuring its e.m.f.

Redox reactions in cells

The reaction

$$2Ag^+(aq) + Cu(s) \longrightarrow 2Ag(s) + Cu^{2+}(aq)$$

can easily be carried out in a test-tube. If we carry out the reaction in an insulated flask, and measure the temperature change, the reaction is found to be exothermic.

This reaction could be made the basis of a cell, which could even light a bulb (see figure 15.12).

By carrying out the reaction in this way, and not in a test-tube, nothing much has been changed. Silver ions still turn to silver atoms, and copper atoms

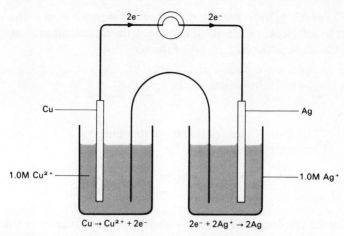

Figure 15.12

to copper ions. Heat is still evolved, but this time some of it is used to make the filament of the bulb hot, or to heat the wires a little. As before, ΔS_{system}, the entropy change for the reaction, is negative; $-245\,\text{J K}^{-1}\,\text{mol}^{-1}$.

So why does the reaction go? Because this negative entropy change is more than balanced by the *positive* entropy change that occurs when the surrounding air receives energy from the hot bulb, wires etc. Look at the figures.

$$2Ag^+(aq) + Cu(s) \longrightarrow 2Ag(s) + Cu^{2+}(aq); \quad \Delta S^{\ominus} = -245.0\,\text{J K}^{-1}\,\text{mol}^{-1}$$
$$\Delta H^{\ominus} = -146\,800\,\text{J mol}^{-1}$$

Therefore

$$\Delta S^{\ominus}_{\text{system}} \quad = \quad\quad\quad\quad\quad\quad\quad\quad -245.0\,\text{J K}^{-1}\,\text{mol}^{-1}$$

$$\Delta S^{\ominus}_{\text{surroundings}} = \frac{-\Delta H^{\ominus}}{T} = -\left(\frac{-146\,800}{298}\right) = +492.6\,\text{J K}^{-1}\,\text{mol}^{-1}$$

$$\Delta S^{\ominus}_{\text{total}} \quad = \Delta S^{\ominus}_{\text{system}} + \Delta S^{\ominus}_{\text{surroundings}} \quad = +247.6\,\text{J K}^{-1}\,\text{mol}^{-1}$$

As can be seen, $\Delta S_{\text{surroundings}}$ is more than enough to ensure that ΔS_{total} is positive. In fact, ΔS_{total} would still be positive even if much less heat were passed to the surroundings - all that is needed is to pass enough energy out to make $\Delta S_{\text{surroundings}}$ a tiny bit bigger than ΔS_{system}. That means that we could save some energy that would otherwise be passed to the surroundings, and use it to do useful mechanical work like raising a load. The maximum energy free to be used in this way is easily worked out. The method is as follows.

For ΔS_{total} to be positive the minimum energy passed to the surroundings must be sufficient to increase the entropy of the surroundings enough to compensate for the negative ΔS_{system}. So, at the very best,

$$\Delta S_{system} = \Delta S_{surroundings}$$

Now,

$$\Delta S_{surroundings} = \frac{(\text{heat passed to surroundings})}{T}$$

$$(\text{heat passed to surroundings}) = T\Delta S_{surroundings}$$
$$= T\Delta S_{system} \text{ in this case.}$$

When the reaction is carried out wastefully in a test-tube or in a short-circuited cell, all of the enthalpy change ΔH is passed to the surroundings. But all that *needs* to be passed out to ensure that ΔS_{total} is positive is $T\Delta S_{system}$. Some of ΔH is left over, free to do useful work. The amount left over after passing out $T\Delta S_{system}$ is just

$$-(\Delta H - T\Delta S_{system})$$

The minus sign arises because, while the energy has been lost by the system, it has been gained by whoever is doing the work.

As you will recall from the first section of this Topic, this is $-\Delta G$, the Gibbs free energy change.

$$\Delta G = \Delta H - T\Delta S_{system}$$

Now it should be clear why ΔG is called 'free' energy. *It is the energy available, or free, to do useful work,* once the requirement that ΔS_{total} should be positive has been met – in this case by passing enough heat to the surroundings. Of course, in practice it is always necessary to lose slightly more than the minimum theoretical amount of energy, so ΔG gives the theoretical *maximum* amount of free, useful work that is available.

$$W_{max} = -\Delta G$$

If ΔS_{system} is already positive, a positive $\Delta S_{surroundings}$ is not needed. In fact, $\Delta S_{surroundings}$ *could* be negative: ΔS_{total} would still be positive provided the magnitude of ΔS_{system} was greater than the magnitude of $\Delta S_{surroundings}$. So for reactions with a positive ΔS_{system}, heat can be *drawn in* from the surroundings and added to the energy available – the reaction can be endothermic.

Free energy and the electromotive force of a cell

In experiment 15.2b you saw how the potential difference, p.d., of a Daniell cell increased as the external resistance was increased, until it reached a maximum value, when the resistance was very high. This maximum p.d. was called the electromotive force of the cell, E_{cell}. At the same time as the p.d. increases, however, the current drops. When the p.d. reaches its maximum, E_{cell}, the current is zero.

When resistance is very high:
p.d. is a maximum, E_{cell}
but current is zero.

How much work could this cell do?
1 volt = 1 joule per coulomb.
If 1 coulomb of charge moves through a p.d. of 1 volt, it does 1 joule of work.
Thus 1 coulomb moving through v volts does v joules of work.
How many coulombs pass in this case? The electrode reactions are

$$Cu^{2+}(aq) + 2e^- \longrightarrow Cu(s)$$
$$Zn(s) \longrightarrow Zn^{2+}(aq) + 2e^-$$

For 1 mole of Cu^{2+} and Zn reacting, 2 moles of electrons pass round the circuit.
That is, $2 \times 96\,500$ coulombs of charge pass round.
So, for one mole of reactants and voltage v,

$$\text{Work done} = 2 \times 96\,500 \times v \, \text{J mol}^{-1}$$

The maximum value of v is E_{cell}.
Therefore, maximum work done $= W_{max} = 2 \times 96\,500 \times E_{cell}$.
In the general case, when z moles of electrons pass round the circuit,

$$W_{max} = zFE_{cell} \qquad (F = 96\,500 \, \text{C})$$

But we already know that $W_{max} = -\Delta G$

Thus,

$$\Delta G = -zFE_{cell}$$

So we see that there is a simple relationship between the e.m.f. of a cell, E_{cell}, and the free energy change in that cell.

For the zinc/copper cell (the Daniell cell),

$$E^{\ominus} = 1.10 \text{ V. Hence, } \Delta G^{\ominus} = -2 \times 96\,500 \times 1.10 \text{ J mol}^{-1}$$
$$= -212\,300 \text{ J mol}^{-1}$$

When a mole of zinc atoms, Zn, and a mole of copper ions, $Cu^{2+}(aq)$, react in this cell, then up to 212.3 kJ are available to do work. But remember, maximum work is only available when the p.d., $v = E_{cell}$, and for this, the resistance must be very high, compared with the resistance of the cell itself. If the cell resistance is appreciable, some of the work that could have been done outside the cell, lighting a lamp for example, must be done *inside* the cell, warming it up. The only way round this is to have a high resistance in the outside circuit – but then the current is very, very low.

Finding cell e.m.f. under non-standard conditions

So far we have only considered cells in standard conditions – concentrations of 1.0 mol dm^{-3} and temperature 298 K. What happens when the conditions are non-standard?

The table below shows E_{cell} for a Daniell cell under different conditions of concentration (all at 298 K).

Concentration of Cu^{2+}/mol dm^{-3}	Concentration of Zn^{2+}/mol dm^{-3}	E_{cell}/V
1.0	1.0	1.10
1.0	0.1	1.13
1.0	0.01	1.16
0.1	1.0	1.07
0.01	1.0	1.04

It is clear that lowering $[Cu^{2+}]$ lowers E_{cell}, while lowering $[Zn^{2+}]$ *raises* E_{cell}. Why is this?

When current is drawn from a Daniell cell, these reactions occur:

$$Zn(s) \longrightarrow Zn^{2+}(aq) + 2e^{-}$$
$$Cu^{2+}(aq) + 2e^{-} \longrightarrow Cu(s)$$

(Notice that in this case Cu^{2+} turns to copper metal. In the silver/copper cell the reverse happened. This is because copper is more reactive than silver, but less reactive than zinc.)

When zinc ions form at the zinc electrode, there is an increase in entropy, as we have already seen: ΔS is positive. We have also seen that if the zinc electrode is surrounded by a solution already containing zinc ions, the entropy change will be less than if the electrode were surrounded by water alone. *The* *more Zn^{2+} ions already there, the less the entropy will increase when further ions dissolve.* Clearly, then, increasing the concentration of zinc ions will *decrease* ΔS, and therefore decrease E_{cell}.

Exactly the opposite applies to the copper electrode. Here, copper ions are turning to copper metal, which involves a *negative* ΔS. For a copper ion being removed from a concentrated solution, ΔS will be less negative than if it were removed from a dilute solution. The overall entropy change for the cell reaction will therefore be more positive and more favourable when the concentration of copper ions is high, and this means E_{cell}, in turn, will be more positive.

The explanation just given is a *qualitative* one. A *quantitative* relation between E_{cell} and ion concentration is given by the Nernst equation.

$$E_{cell} = E_{cell}^{\ominus} - \frac{RT}{zF} \ln \frac{[Zn^{2+}(aq)]}{[Cu^{2+}(aq)]}$$

15.6
PREDICTING WHETHER REACTIONS WILL TAKE PLACE: ΔG, E_{cell}, and K_c

Chemists are often involved in predicting whether a chemical reaction will 'go' of its own accord, and if not, what can be done to make it go. We have seen that reactions are always possible if ΔS_{total} is positive – but the surroundings as well as the reacting system itself must be taken into account. A convenient way of doing this is to use ΔG instead of ΔS_{total}. A reaction can occur if it has a negative value for ΔG.

But what does it mean to say a reaction 'goes'? If you put zinc into copper sulphate solution, a reaction certainly goes:

$$Zn(s) + Cu^{2+}(aq) \longrightarrow Zn^{2+}(aq) + Cu(s)$$

The reaction appears to go to completion – equilibrium lies well over to the right. This is the Daniell cell reaction, and we have already seen that for this reaction, $\Delta G^{\ominus} = -212.3\,kJ\,mol^{-1}$, a large, negative value.

We could measure just how far to the right the equilibrium lay if we knew the equilibrium constant, K_c. This is easy enough to work out if ΔG is known.

You may remember at the end of Topic 12 we arrived at the relationship

$$\Delta G^{\ominus} = -LkT \ln K_p$$

This was derived by applying the expression

$$S = S^{\ominus} - Lk \ln p$$

to gaseous equilibria. (It will be remembered that the expression was first stated as

$$S = S^{\ominus} - Lk \ln p/p^{\ominus}$$

but it was later pointed out that if pressures are measured in atmospheres, $p^{\ominus} = 1$ and the expression reduces to

$$S = S^{\ominus} - Lk \ln p$$

accordingly.)

For ionic equilibria such as the ones we have been considering, the situation is very similar. We have already seen how ions in solution can be compared to molecules in gases – the difference is that for ionic solutions we refer to concentrations, while for gases pressures are used. In fact, pressure is proportional to concentration – after all, if a gas is compressed, a lot of it is concentrated into a small volume.

Because of this close relation between pressure and concentration, the expression

$$S = S^{\ominus} - Lk \ln p \qquad \text{for a gas}$$

can be replaced by

$$S = S^{\ominus} - Lk \ln(\text{concentration}) \qquad \text{for a solution.}$$

Furthermore,

$$\Delta G^{\ominus} = -LkT \ln K_p$$

can be replaced by

$$\Delta G^{\ominus} = -LkT \ln K_c$$

Applying the last equation to the zinc/copper reaction (for which $\Delta G^{\ominus} = -212\,300\,\text{J mol}^{-1}$) gives $K_c = 1.9 \times 10^{37}$. The reaction will certainly be very nearly complete.

A reaction is usually described as 'going to completion' if $K_c = 10^{10}$ or greater. This corresponds to a value of ΔG of about $-60\,\text{kJ}\,\text{mol}^{-1}$ or less (that is, a greater negative value). Even with $K_c = 10^{10}$, though, there will always be a very small proportion of reactants left unreacted – the reaction will never go *fully* to completion.

Similarly, if K_c has a value less than 10^{-10}, the reaction is considered not to go at all, even though there must in fact be a tiny amount of product formed (otherwise K_c would be zero). If K_c is less than 10^{-10}, ΔG must be greater than $+60\,\text{kJ}\,\text{mol}^{-1}$.

A reaction in equilibrium with equal amounts of reactants and products present has $K_c = 1$; this corresponds to $\Delta G = 0$.

Another way of predicting whether reactions can 'go' is to use $E^{\ominus}$ values. In general, if $E^{\ominus}$ for a reaction is positive, the reaction 'goes'. Now $E^{\ominus}$ is related directly to $\Delta G^{\ominus}$:

$$\Delta G^{\ominus} = -zFE^{\ominus}$$

If $\Delta G = -60\,\text{kJ}\,\text{mol}^{-1}$ $(60\,000\,\text{J}\,\text{mol}^{-1})$, E has a value of about 0.6 V.

All this can be summarized as follows:

Reaction does not go	Reaction at equilibrium with equal amounts of reactants and products	Reaction complete
$K_c < 10^{-10}$	$K_c = 1$	$K_c > 10^{10}$
$\Delta G^{\ominus} > +60$	$\Delta G^{\ominus} = 0$	$\Delta G^{\ominus} < -60$
$E^{\ominus} < -0.6$	$E^{\ominus} = 0$	$E^{\ominus} > +0.6$

A final word of warning. Many reactions with high K_c values which should apparently go to completion do not do so at room temperature. This is because the rate of reaction is very slow because of a high activation energy barrier. $\Delta G^{\ominus}$, $E^{\ominus}$, and K_c only tell you whether a reaction is feasible. They say nothing about how fast it will go.

15.7
STANDARD FREE ENERGIES OF FORMATION, AND THEIR USE

Finally in this Topic we shall look at standard free energies of formation of compounds, and see how tabulated values of these quantities can be used to calculate the standard free energy changes for reactions.

The definitions and conventions introduced here are exactly parallel to those concerning the standard enthalpy change of formation, $\Delta H_f^{\ominus}$. You may find it useful to revise section 6.1 ('Energy changes – definitions') before continuing with this.

The standard free energy change of a reaction can be regarded as an energy *difference* between free energy values which we can assign to the reactants and products of a reaction. For example in the reaction

$$C(s) + O_2(g) \longrightarrow CO_2(g)$$

we can say that

$$\Delta G^{\ominus}_{298} = G^{\ominus}_{298}[CO_2(g)] - \{G^{\ominus}_{298}[C(s)] + G^{\ominus}_{298}[O_2(g)]\}$$

where $G^{\ominus}_{298}[X]$ means 'the standard free energy of X'.

Just as in the case of enthalpies, or of potential energy, etc., absolute values of free energies are not known, so it is convenient to choose some base line, or arbitrary zero, from which to measure the standard free energies of substances. The convention chosen for free energies is the same as that for enthalpies, namely that

at   1 atm (101 kPa) pressure
     298 K
     with the elements in the physical states normal under these conditions

the standard free energies of the elements are zero.

It should be emphasized that this convention is quite arbitrary, as arbitrary as calling the potential energy of all objects at sea level zero, but it is an agreed convention. It follows necessarily from this convention that

$$\Delta G^{\ominus}_{f,298}[\text{element in physical state normal at 1 atm and 298 K}] = 0.$$

It is possible, therefore, using this convention, to tabulate standard free energies of formation of *compounds* rather than standard free energies relating to specific reactions. In fact, of course, the standard free energy of a compound does really refer to a reaction, namely the formation of one mole of the compound from its elements in physical states normal at 1 atm and 298 K. For example,

$$C(s) + O_2(g) \longrightarrow CO_2(g); \qquad \Delta G^{\ominus}_{298} = -394.4 \, kJ \, mol^{-1}$$

and $\Delta G^{\ominus}_{f,298}[CO_2(g)] = -394.4 \, kJ \, mol^{-1}$

are exactly equivalent statements.

The tabulation of standard free energy data relating to compounds rather than reactions, however, makes for very general and flexible use of the tables, as the examples later in this section will show.

How to calculate equilibrium constants from standard free energies of formation

As we have already seen (section 15.6), given the value of $\Delta G^{\ominus}$ for any chemical reaction, it is a simple matter to calculate the equilibrium constant for that reaction. The following examples show how K values can be calculated for a selection of reactions, given that values of $\Delta G_f^{\ominus}$ are available from tables.

1 The equilibrium constants of some gas phase reactions

a $N_2(g) + 3H_2(g) \rightleftharpoons 2NH_3(g)$

$\Delta G_{298}^{\ominus} = 2\Delta G_{f,298}^{\ominus}[NH_3(g)] - \Delta G_{f,298}^{\ominus}[N_2(g)] - 3\Delta G_{f,298}^{\ominus}[H_2(g)]$

Since the standard free energies of formation of elements for their standard states are zero (by convention)

$$\begin{aligned}\Delta G_{298}^{\ominus} &= 2 \times \Delta G_{f,298}^{\ominus}[NH_3(g)]\\ &= 2 \times (-16.5)\,\text{kJ mol}^{-1} \quad \text{(from the \textit{Book of data})}\\ &= -33\,\text{kJ mol}^{-1}\end{aligned}$$

By definition

$$\begin{aligned}\Delta G^{\ominus} &= -RT\ln K_p\\ \therefore \quad \Delta G_{298}^{\ominus} &= -8.3 \times 10^{-3} \times 298 \times \ln K_p\\ \therefore \quad -33 &= -2.47\ln K_p\\ \therefore \quad \ln K_p &= 13.36\\ \therefore \quad K_p &= \frac{p_{NH_2eqm}^2}{p_{N_2eqm} \times p_{H_2eqm}^3} = 6.34 \times 10^5\,\text{atm}^{-2} \text{ at } 298\,\text{K}\end{aligned}$$

thus illustrating that, if a catalyst could be found, and equilibrium could be reached, equilibrium mixtures of hydrogen and nitrogen at room temperature and pressure would be expected to contain mainly ammonia and only small amounts of unreacted hydrogen and nitrogen.

b $N_2O_4(g) \rightleftharpoons 2NO_2(g)$

The standard free energy change for this reaction is

$$\begin{aligned}\Delta G_{298}^{\ominus} &= 2\Delta G_{f,298}^{\ominus}[NO_2(g)] - \Delta G_{f,298}^{\ominus}[N_2O_4(g)]\\ &= 2 \times 51.3 - 97.8 = 4.8\,\text{kJ mol}^{-1}\end{aligned}$$

From this

$$\ln K_p = \frac{4.8}{-2.47} = -1.94 \qquad \text{and}$$

$$K_p = \frac{p^2_{NO_2eqm}}{p_{N_2O_4eqm}} = 0.14 \, \text{atm at 298 K}$$

2 Heterogeneous equilibria involving a gas

The equilibrium constant for the system

$$CaCO_3(s) \rightleftharpoons CaO(s) + CO_2(g)$$

is given by $K_p = p_{CO_2eqm}$, the partial pressures of $CaCO_3$ and CaO in the gas phase being constant and therefore incorporated in the equilibrium constant K_p. The equilibrium pressure of CO_2 over $CaCO_3$ at $25\,°C$ may be calculated as follows.

For the above reaction

$$\begin{aligned}
\Delta G^{\ominus}_{298} &= \Delta G^{\ominus}_{f,298}[CO_2(g)] + \Delta G^{\ominus}_{f,298}[CaO(s)] - \Delta G^{\ominus}_{f,298}[CaCO_3(s)] \\
&= -394.4 + (-604.0) - (-1128.8) \\
&= 130.4 \, \text{kJ mol}^{-1}
\end{aligned}$$

Now $\Delta G^{\ominus}_{298} = -2.47 \ln K_p = -2.47 \ln p_{CO_2eqm}$

so that

$$\ln p_{CO_2eqm} = -53.0 \text{ and } p_{CO_2eqm} = 1.1 \times 10^{-23} \, \text{atm}$$

As expected, calcium carbonate is highly stable at room temperature and pressure.

3 Equilibrium constants for reactions between ions in solution

Given the standard free energies of formation of the following:

	$\Delta G^{\ominus}_{f,298}$/kJ mol^{-1}
Fe^{3+}(aq)	-4.6
Fe^{2+}(aq)	-78.9
I^-(aq)	-51.6
I_2(aq)	$+16.4$

find the equilibrium constant at $25\,°C$ for the reaction

$$2Fe^{3+}(aq) + 2I^-(aq) \rightleftharpoons 2Fe^{2+}(aq) + I_2(aq)$$

(In dealing with ions, the standard free energy of formation of the hydrogen ion $\Delta G_f^{\ominus}[H^+(aq)] = 0$. This corresponds to the convention of regarding the standard potential of the hydrogen electrode as zero volt.)

The standard free energy $\Delta G_{298}^{\ominus}$ for the reaction is given by

$$\begin{aligned} \Delta G_{298}^{\ominus} &= 2\Delta G_{f,298}^{\ominus}[Fe^{2+}(aq)] + \Delta G_{f,298}^{\ominus}[I_2(aq)] \\ &\quad - 2\Delta G_{f,298}^{\ominus}[Fe^{3+}(aq)] - 2\Delta G_{f,298}^{\ominus}[I^-(aq)] \\ &= 2 \times (-78.9) + 16.4 - 2 \times (-4.6) - 2 \times (-51.6) \\ &= -29.0 \, kJ \end{aligned}$$

Using this value in the equation

$$\Delta G_{298}^{\ominus} = -2.47 \ln K_c$$

we obtain $\ln K_c = 11.74$ and

$$K_c = \frac{[Fe^{2+}(aq)]_{eqm}[I_2(aq)]_{eqm}}{[Fe^{3+}(aq)]_{eqm}[I^-(aq)]_{eqm}} = 1.26 \times 10^5$$

Now use your value for the e.m.f. of the cell based on this reaction (obtained in experiment 15.4) to calculate the equilibrium constant. How do the values compare?

BACKGROUND READING 2
Energetics in life processes

This is a field in which our knowledge of energetic relationships is rather more limited. Yet life itself depends on the continuous supply to cells of energy in the form of glucose. The human body is, among other things, a highly efficient machine driven by chemical fuels.

Within a living cell millions of chemical reactions occur each second. The reactions occur in sequences called 'metabolic pathways'. Each reaction in a metabolic pathway is characterized by a different enzyme. Enzymes are complex proteins which combine precisely with the compound they act on. They lower the activation energy required for the chemical reaction to take place, by supplying an alternative and energetically more favourable reaction mechanism. For example, the conversion of nitrogen gas, N_2, to ammonia, NH_3, in the Haber–Bosch process requires temperatures of 623–673 K, pressures of 20–35 MPa, and a finely divided iron catalyst. The enzyme complex nitrogenase, in bacteria and cyanobacteria, converts atmospheric nitrogen to ammonia at normal soil temperatures and pressures at four times the rate of the industrial process.

Let us briefly consider photosynthesis from the point of view of the energy changes involved. The overall equation can be written:

$$6CO_2(g) + 12H_2O(l) \longrightarrow C_6H_{12}O_6(s) + 6O_2(g) + 6H_2O(l)$$

The standard free energy change, $\Delta G^{\ominus}$, for this process is approximately $+2900$ kJ, so that the reaction involves a net decrease in entropy; it is apparently not feasible. Yet it is a well established process of nature, and so there must be an increase in entropy that we have ignored.

We can easily explain this difficulty if we consider more carefully what constitutes a *system* in this instance. In the absence of light, that is, in a system isolated with respect to light energy, the photosynthetic reactions in an intact cell will not occur. The energy balance adds up rather differently when the free energy associated with the incoming light is taken into account.

Experiments have shown that eight moles of photons are required to fix one mole of carbon dioxide molecules in carbon compounds in photosynthesis. These carbon compounds have a free energy content of about 480 kJ. What is the free energy content of the photons which supply the energy to make these carbon compounds?

The photosynthetic pigments absorb the photons of light. Chlorophyll a, for example, has two absorption peaks in isolation, one at 425 nm wavelength (blue light) and one at 660 nm (red light). The least energetic of the photons able to donate energy efficiently to photosynthesis has a wavelength of 680 nm. The frequency of this light $v = c/\lambda$, where c = velocity of light $(3 \times 10^8 \text{ m s}^{-1})$ and $\lambda = 680$ nm. Therefore $v = (3 \times 10^8)/(6.8 \times 10^{-7}) = 4.41 \times 10^{14} \text{ s}^{-1}$.

The energy of a single quantum of light or a photon is the product of the frequency of light and Planck's constant $(6.6 \times 10^{-34} \text{ J s}^{-1})$, *i.e.* $E = hv$. Therefore for one *mole* of photons of frequency $4.41 \times 10^{14} \text{ s}^{-1}$ (wavelength 680 nm):

$$E(\text{energy}) = (6.02 \times 10^{23}) \times (6.6 \times 10^{-34}) \times (4.41 \times 10^{14})$$
$$= 175 \text{ kJ}$$

So eight moles of quanta possess 1400 kJ of energy, 480 kJ of which would be conserved in the compounds formed by the mole of carbon dioxide molecules absorbed. Light of this wavelength seems to be 34% efficient. White light, of course, would be even less efficient. The other photons of light used in photosynthesis have more energy than photons of wavelength 680 nm. Whatever their wavelength, eight photons are still required to fix one carbon atom.

After being absorbed, the energy of the photons is spread around in packets (quanta) of much smaller energy, ultimately in infra-red radiation of longer wavelength and consequently smaller energy per quantum. The *number* of quanta increases enormously, since their energy per quantum falls. The resulting increase in entropy arises because the more quanta there are, the greater is the number of ways that they can be spread around molecules.

Much of the glucose produced is used to provide energy for cellular reactions. In most cells the glucose is oxidized in a series of reactions known as aerobic respiration:

$$C_6H_{12}O_6(s) + 6O_2(g) \longrightarrow 6H_2O(l) + 6CO_2(g); \quad \Delta G = -2900\,kJ\,mol^{-1}$$

Some of the energy released is conserved in the form of a molecule known as ATP, adenosine triphosphate, the 'energy currency of the cell'. This compound is hydrolysed in a multitude of cellular reactions. The energy released on hydrolysis is used to do mechanical or chemical work around the cell, such as the synthesis of proteins from amino acids, or the movement of muscles.

$$ATP^{4-}(s) + H_2O(l) \longrightarrow ADP^{2-}(s)(\text{adenosine diphosphate}) + HPO_4^{2-}$$

The energy released when the bonds in the products are formed is $33\,kJ\,mol^{-1}$ more than the energy required to break the bonds in the reactants.

The scale of ATP synthesis and breakdown in organisms is impressive. A human, for example, turns over 75 kg of ATP a day. Cells contain organelles known as mitochondria. They use the energy from glucose breakdown to pump some 450 moles of protons (equivalent to 45 litres of concentrated hydrochloric acid) each day, across a total area of membrane equivalent to a football field. This produces a potential of 200 mV across the extremely thin membrane, a field strength equivalent to 30 000 volts per millimetre. This potential difference is used to make ATP from its component molecules, ADP and inorganic phosphate, in a reaction which would not otherwise be energetically possible.

SUMMARY

At the end of this Topic you should:

1 be familiar with the electron-transfer aspect of redox reactions in metal/metal ion, non-metal/non-metal ion, and ion/ion systems, and be able to divide these into half-cell reactions.;

2 be able to write, and interpret, simple cell diagrams;

3 know the construction and use of the hydrogen electrode;

4 be able to determine experimentally the e.m.f. of simple voltaic cells;

5 understand what is meant by a standard electrode potential, $E^{\ominus}$ and be able to use values of $E^{\ominus}$ to calculate standard e.m.f.s of cells;

6 know that concentration changes affect electrode potentials, and be able to use the Nernst equation;

7 be able to use electrode potentials to determine the solubility of sparingly soluble salts;

8 be able to use the anti-clockwise rule to predict the likely course of redox reactions, given a table of $E^\ominus$ values;

9 be able to calculate equilibrium constants from $E^\ominus$ values;

10 have consolidated your understanding of the conclusions reached in sections 3.7, 4.5, 6.6, 10.1, 10.2, 12.6, and have extended them so that you can

a calculate $\Delta S^\ominus_{\text{system}}$ and $\Delta S^\ominus_{\text{surroundings}}$ for a proposed change and hence decide, from $\Delta S^\ominus_{\text{total}}$, whether it is likely to take place under standard conditions;

b understand that when equilibrium is reached ΔS_{total} is zero;

c appreciate the value of using ΔG, rather than ΔS_{total}, as a guide to feasibility and attainment of equilibrium, since it includes ΔS_{total}, and is often easy to measure, or to calculate from tables of data;

d appreciate that there is always an entropy increase when metal atoms go into solution as metal ions, but that this increase becomes smaller as the concentration of the solution increases;

e appreciate the difference between standard and non-standard conditions when applied to discussions of voltaic cells;

f be able to calculate the maximum work obtainable from a given voltaic cell and, hence, the value of ΔG for the cell reaction;

g be able to account qualitatively for the change in e.m.f. of a voltaic cell when the concentrations of the electrode solutions vary;

11 know what is meant by the standard free energy of formation of a compound;

12 be able to use tables of standard free energies of formation to calculate the standard free energy changes for reactions;

13 be able to comment on the possibilities of reactions occurring, given values of $\Delta G^\ominus$, $E^\ominus_{\text{cell}}$, or K_c for the changes involved.

14 be aware of some aspects of the energetics of life processes.

PROBLEMS

* Indicates that the *Book of data* is needed.

1 State which of the reactants is the oxidant and which is the reductant in each of the following reactions.

a $Fe(s) + Cu^{2+}(aq) \longrightarrow Fe^{2+}(aq) + Cu(s)$

b $Al(s) + 3H^+(aq) \longrightarrow Al^{3+}(aq) + 1\frac{1}{2}H_2(g)$

c $Zn(s) + Pb^{2+}(aq) \longrightarrow Zn^{2+}(aq) + Pb(s)$

d $2Fe^{3+}(aq) + Sn^{2+}(aq) \longrightarrow 2Fe^{2+}(aq) + Sn^{4+}(aq)$

2 Consider the following reactions and information on four elements A, B, C, D (these are not the symbols for the elements):

i $A(s) + B^{2+}(aq) \longrightarrow A^{2+}(aq) + B(s)$
ii $A(s) + 2C^{+}(aq) \longrightarrow A^{2+}(aq) + 2C(s)$
iii $B(s) + 2C^{+}(aq) \longrightarrow B^{2+}(aq) + 2C(s)$
iv $D(s) + B^{2+}(aq) \longrightarrow D^{2+}(aq) + B(s)$
v $D(s) + A^{2+}(aq)$ no action

a Are the elements likely to be all metals, all non-metals, or some of each? Briefly state the reason for your decision.
b What reaction, if any, would you expect to take place between solid D and an aqueous solution of C^{+} ions?
c Write two half-equations to represent the changes which occur in each of the reactions **i, ii, iii,** and **iv**.
d Arrange the elements in order of their relative tendency to form positive ions in aqueous solution, putting the one with the greatest tendency first. Briefly state the reasons for your decision.

*3 Calculate the $E^{\ominus}$ value and state the terminal polarity of each of the following cells (assume a temperature of 25 °C and ionic concentration 1.0M).

a $Pt[H_2(g)] \mid 2H^{+}(aq) \vdots Fe^{2+}(aq) \mid Fe(s)$
b $Ni(s) \mid Ni^{2+}(aq) \vdots 2H^{+}(aq) \mid [H_2(g)]Pt$
c $Zn(s) \mid Zn^{2+}(aq) \vdots Ni^{2+}(aq) \mid Ni(s)$
d $Al(s) \mid Al^{3+}(aq) \vdots Cr^{3+}(aq) \mid Cr(s)$

*4 $E^{\ominus}_{cell}$ is $+0.62$ volt for the cell:

$Co(s) \mid Co^{2+}(aq) \mid Cu^{2+}(aq) \mid Cu(s)$

Calculate the standard electrode potential for

$Co^{2+}(aq) \mid Co(s)$

*5 $E^{\ominus}_{cell}$ is $+1.61$ volt for the cell:

$Zn(s) \mid Zn^{2+}(aq) \vdots Hg^{2+}(aq) \mid Hg(l)$

Calculate the standard electrode potential for

$Hg^{2+}(aq) \mid Hg(l)$

*6 For each of the following cells construct the two half equations and the whole equations to represent the changes which take place when the cell terminals are connected by a conductor.

a $Al(s) \,|\, Al^{3+}(aq) \,\vdots\, Sn^{2+}(aq) \,|\, Sn(s)$
b $Ag(s) \,|\, Ag^{+}(aq) \,\vdots\, Pb^{2+}(aq) \,|\, Pb(s)$
c $Pt[H_2(g)] \,|\, 2H^{+}(aq) \,\vdots\, Mg^{2+}(aq) \,|\, Mg(s)$

*7 Arrange the following groups of ions in order of their *ability to oxidize*. Put the one with the greatest ability to oxidize first. (Assume that they are all of molar concentration.)

a $Cu^{2+}(aq)$ $Ag^{+}(aq)$ $Pb^{2+}(aq)$ $Cr^{3+}(aq)$
b $Mg^{2+}(aq)$ $Zn^{2+}(aq)$ $Fe^{3+}(aq)$ $Sn^{2+}(aq)$

8 The electrode potential of a metal (M) was measured when it was in contact with solutions of its own ions at various molarities, using a standard hydrogen electrode. The results, at 25 °C, were:

Electrode potential/volt	Molarity of solution
−0.286	0.5
−0.298	0.2
−0.307	0.1
−0.316	0.05
−0.327	0.02

a Plot a graph of electrode potential against ln [ion].
b From the graph determine the *standard* electrode potential of the metal in an aqueous solution of its own ions. Explain how you arrive at your answer.
c Determine the charge on the ions of the metal. Explain how you arrive at your answer.

9 To a solution of $10\,cm^3$ of 0.1M $AgNO_3(aq)$, $50\,cm^3$ of 0.1M potassium bromate, $K\,BrO_3(aq)$, solution were added. A piece of silver foil was put into the solution and its *electrode potential* at 25°C was found to be $+0.61$ volt. Use the graph plotted from the results of experiment 15.3a to calculate:
a The concentration of aqueous silver ions in the final solution
b The solubility product of silver bromate.

10 A certain metal (M) can exist in acidic aqueous solution with two oxidation numbers, one of which is $+2$ ($M^{2+}(aq)$). The following data give the electrode potentials at 25°C of various aqueous mixtures of the two ions of the metal, M^{2+} and M^{x+} (x is greater than 2).

Relative concentrations

$[M^{2+}(aq)]:[M^{x+}(aq)]$			E/volt
8	:	1	0.113
4	:	1	0.122
1	:	2	0.149
1	:	5	0.161
1	:	10	0.170

By a graphical method determine $E^{\ominus}$ for the electrode, $M^{x+}(aq)$, $M^{2+}(aq)\,|\,$Pt. What is the value of x? Explain how you arrive at your answers.

*11 The equilibrium constant for the reaction

$$Ag^+(aq) + Fe^+(aq) \rightleftharpoons Fe^{3+}(aq) + Ag(s)$$

as calculated from $E^{\ominus}$ values, is $3.2\,dm^3\,mol^{-1}$ at $25\,°C$.

a Use the standard enthalpy changes of formation of the aqueous ions involved to calculate the standard enthalpy change for this reaction at $25\,°C$.

b Use the equilibrium constant to calculate the standard free energy change for this reaction at $25\,°C$.

*12 When an aqueous solution of bromine is added to a solution of potassium iodide, the following reaction takes place:

$$\tfrac{1}{2}Br_2(aq) + I^-(aq) \longrightarrow \tfrac{1}{2}I_2(aq) + Br^-(aq)$$

a Write down a cell diagram for an electrochemical cell in which this reaction can also be carried out.

b Use the values of the appropriate standard electrode potentials given in the *Book of data* to calculate the standard e.m.f. of your cell.

c What is K_c for the above reaction?

*13 The oxidation of $Fe^{2+}(aq)$ ions by $MnO_4^-(aq)$ ions proceeds according to the equation

$$5Fe^{2+}(aq) + MnO_4^-(aq) + 8H^+(aq) \longrightarrow$$
$$5Fe^{3+}(aq) + Mn^{2+}(aq) + 4H_2O(l)$$

for which

$$K_c = \frac{[Fe^{3+}(aq)]_{eqm}^5[Mn^{2+}(aq)]_{eqm}}{[MnO_4^-(aq)]_{eqm}[Fe^{2+}(aq)]_{eqm}^5[H^+(aq)]_{eqm}^8}$$

a Write two half-equations, each one for a distinct half-cell reaction.
b Write a cell diagram for a suitable electrochemical cell in which this reaction could be carried out.
c Calculate the standard e.m.f. of this cell (at 298 K).
d What is the value of K_c at 298 K?

*14a Use tables to write down the standard free energy of formation of methane gas, CH_4.
b Does this value suggest that the reaction

$$C(s) + 2H_2(g) \longrightarrow CH_4(g)$$

would, from considerations of free energy, be expected to take place?
c In your experience does such a reaction take place at room temperature and atmospheric pressure?
d How do you explain any difference in your answers to **b** and **c** above?

*15a Compare the standard free energies of formation of iron(II) oxide and iron(III) oxide.
b What is the equilibrium constant for the atmospheric oxidation of iron(II) oxide to iron(III) oxide?
c Which oxide of iron is likely to be the principal naturally occurring ore of the element?

16 The relevant equation and data for the formation of the diamminosilver ion are

$$Ag^+(aq) + 2NH_3(aq) \longrightarrow [Ag(NH_3)_2]^+(aq)$$
$$\Delta G^{\ominus}_{f,298}[Ag^+(aq)] = +77.1 \text{ kJ mol}^{-1}$$
$$\Delta G^{\ominus}_{f,298}[NH_3(aq)] = +26.6 \text{ kJ mol}^{-1}$$
$$\Delta G^{\ominus}_{f,298}[[Ag(NH_3)_2]^+(aq)] = -17.2 \text{ kJ mol}^{-1}$$

a Calculate $\Delta G^{\ominus}_{298}$ for this reaction.
b Calculate K_c for the above reaction.
c Invent an experimental method for determining K_c for this reaction
i by means of an electrochemical cell
ii by some other means.

*17a Calculate both the standard free energy change, $\Delta G^{\ominus}$, and the standard enthalpy change, $\Delta H^{\ominus}$, for each of the following reactions:
i $C(s) + O_2(g) \longrightarrow CO_2(g)$
ii $N_2(g) + 3H_2(g) \longrightarrow 2NH_3(g)$
iii $CaCO_3(s) \longrightarrow CaO(s) + CO_2(g)$
iv $Zn(s) + Cu^{2+}(aq) \longrightarrow Zn^{2+}(aq) + Cu(s)$

b For which reactions do $\Delta G^{\ominus}$ and $\Delta H^{\ominus}$

i agree closely   **ii** differ significantly?

c For which *type* of reactions are ΔG and ΔH most likely to show close agreement, and for which are they most likely to differ?

d Explain why ΔH is often a good guide to whether a reaction is likely to go, but not as reliable a guide as ΔG.

*18 Consider the cell shown in figure 15.13

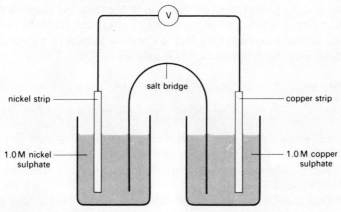

Figure 15.13

a Give the cell diagram for this cell and work out its $E^{\ominus}$ value.

b Write an equation for the reaction that occurs when current is drawn from the cell.

c Use the $E^{\ominus}$ value for the cell to calculate the standard free energy change in the cell.

d What is the maximum amount of work that could be obtained from the cell? (State your units clearly.)

e Why is this amount of work unlikely to be obtained in practice?

The Periodic Table 4: the transition elements

16.1
THE SPECIAL PROPERTIES OF THE TRANSITION ELEMENTS

In previous work you have probably used the term *transition element* to refer to those elements which come between Groups II and III in the Periodic Table. The electronic structures of the first row of these elements are given in the following table; only the outermost levels are shown:

Element	Sc	Ti	V	Cr	Mn	Fe	Co	Ni	Cu	Zn
Electronic structure	$3d^1 4s^2$	$3d^2 4s^2$	$3d^3 4s^2$	$3d^5 4s^1$	$3d^5 4s^2$	$3d^6 4s^2$	$3d^7 4s^2$	$3d^8 4s^2$	$3d^{10} 4s^1$	$3d^{10} 4s^2$

In this Topic, we shall only be concerned with these first row elements. As can be seen from the table, their electronic structures differ from each other mainly in the number of d-electrons, and for this reason these elements are often referred to as the 'd-block elements'.

True transition elements and their compounds have a number of characteristic properties, and these properties are usually a consequence of the ions of the elements having d-orbitals which are incompletely filled with electrons. This effectively removes both scandium and zinc from the list, since the only ions that these elements form are Sc^{3+} and Zn^{2+}, neither of which has incomplete d-orbitals. In future, therefore, when we refer to transition elements, we shall mean an element which contains an incomplete d-orbital in at least one compound. In the first row, this means elements Ti to Cu inclusive.

The special properties regarded as typical of transition elements are as follows.

1 Similarity of physical properties

The physical properties of the transition elements, which are all metals, show very little variation across the row. Such properties include melting point, boiling point, density, and first ionization energy.

2 Variable oxidation number

Most of the transition elements show a range of oxidation numbers in their compounds. To give an idea of how these numbers vary in the first-row transition elements, here is a chart of the oxidation numbers known to exist, with the more common ones ringed:

(Sc)	Ti	V	Cr	Mn	Fe	Co	Ni	Cu	(Zn)
				(7)					
			6	6	6				
		(5)	5	5	5	5			
	(4)	4	4	(4)	4	4	4		
(3)	3	3	(3)	3	(3)	(3)	3	3	
	2	2	2	(2)	(2)	(2)	(2)	(2)	(2)
	1	1	1	1	1	1	1	(1)	

Element   (Sc)   Ti   V   Cr   Mn   Fe   Co   Ni   Cu   (Zn)

In addition to the compounds that have transition elements with these oxidation numbers, some of the transition elements form compounds with carbon monoxide. These compounds are known as *carbonyls*, and an example is nickel carbonyl, $Ni(CO)_4$. The unusual feature of these compounds is that in them, the transition element has an oxidation number of 0.

3 Ability to form complex ions

Ions of the transition elements, and sometimes the atoms themselves, can be surrounded by, and bonded to, a number of molecules or ions called *ligands*. The result is a molecule or ion called a *complex*. For example, when copper(II) ions are in dilute solution in water, each ion is surrounded by water molecules which act as ligands but when ammonia solution is added, ammonia molecules take the place of the water molecules, giving complex ions of formula $[Cu(NH_3)_4]^{2+}$. These complexes generally have structures in which the ligand molecules or ions are arranged around the ion or atom of the transition element in one of the ways shown in figure 16.1.

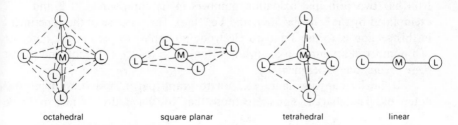

octahedral   square planar   tetrahedral   linear

Figure 16.1
Possible shapes of transition metal complexes. (M = metal; L = ligand.)

4 Colour

Many of the compounds of the transition elements are coloured, both in the solid state and in solution. This phenomenon is not, of course, confined to the compounds of the transition elements but it is relatively rare for *metal* ions outside the transition series to impart colour to their solutions.

5 Catalytic activity

Many slow reactions are accelerated by the presence of transition elements or their ions. This subject has been given a more extended treatment in section 16.5.

Personal work

Write a short introduction to the Topic in your notebook, using the account above as a basis. You should begin by explaining what a transition element is, and then list the characteristic properties. Use reference books and textbooks, together with the *Book of data*, as sources of examples.

The following sections contain practical work which is related to the characteristic properties that have just been described.

16.2
VARIABLE OXIDATION NUMBER

In this section we shall investigate some chemical reactions of compounds of the transition elements that involve changes of oxidation number. We shall begin our investigation with some reactions of compounds of iron.

The redox chemistry of iron

Iron has two principal oxidation numbers in its compounds, $+2$ and $+3$, exemplified by the ions $Fe^{2+}(aq)$ and $Fe^{3+}(aq)$. The purpose of the experiment in this section is to use standard electrode potentials to predict the outcome of a number of attempts to oxidize $Fe^{2+}(aq)$ or to reduce $Fe^{3+}(aq)$, and to test these predictions experimentally.

Begin by copying figure 16.2 on to graph paper, and stick it in your notebook. The chart is similar to those that you have already met in Topic 15.

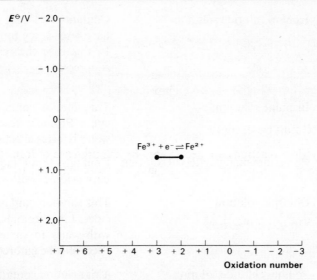

Figure 16.2

Next enter on the chart the following equilibria, in the way the Fe^{2+}/Fe^{3+} equilibrium has been entered.

a $\frac{1}{2}Br_2 + e^- \rightleftharpoons Br^-$ $\qquad\qquad\qquad\quad E^\ominus = +1.09\,V$
b $MnO_4^- + 8H^+ + 5e^- \rightleftharpoons Mn^{2+} + 4H_2O$ $\quad E^\ominus = +1.51\,V$
c $\frac{1}{2}Cl_2 + e^- \rightleftharpoons Cl^-$ $\qquad\qquad\qquad\quad E^\ominus = +1.36\,V$
d $Ag^+ + e^- \rightleftharpoons Ag$ $\qquad\qquad\qquad\quad E^\ominus = +0.80\,V$
e $SO_4^{2-} + 4H^+ + 2e^- \rightleftharpoons SO_2 + 2H_2O$ $\quad E^\ominus = +0.17\,V$
f $Zn^{2+} + 2e^- \rightleftharpoons Zn$ $\qquad\qquad\qquad\quad E^\ominus = -0.76\,V$
g $\frac{1}{2}I_2 + e^- \rightleftharpoons I^-$ $\qquad\qquad\qquad\qquad E^\ominus = +0.54\,V$

You are now ready to begin the experiment.

EXPERIMENT 16.2a
An investigation of the redox reactions of iron

You will need the following solutions, as far as possible of concentration $0.1\,mol\,dm^{-3}$.

Name	Notes
Iron(II) sulphate solution	Contains $Fe^{2+}(aq)$ ions. These ions have a tendency to react with water (hydrolysis) which eventually makes the solution go brown and become cloudy. This has been minimized by adding some sulphuric acid.

Iron(III) chloride solution	Contains Fe^{3+}(aq) ions which also have a tendency to react with water In this case the reaction has been suppressed by using hydrochloric acid.
Bromine solution	This solution contains Br_2(aq) molecules. Bromine reacts slightly with water but this effect may be ignored.
Potassium manganate(VII) solution	Contains MnO_4^-(aq) ions, and has been acidified with sulphuric acid.
Chlorine solution	This solution contains Cl_2(aq) molecules. Like bromine, chlorine reacts with water to some extent but this effect may be ignored.
Sodium chloride solution	This solution contains Cl^-(aq) ions.
Sulphur dioxide solution	This solution has been made by bubbling sulphur dioxide gas through water. The solution has a strong smell which may be harmful to those who suffer from respiratory complaints.
Silver nitrate solution	This solution contains Ag^+(aq) ions.
Potassium iodide solution	This solution contains I^-(aq) ions.

Powdered zinc is also provided.

Use the chart that you have drawn to predict the outcome of any possible reactions between the pairs of substances listed below, most of which are in solution. You should make your predictions using the 'anticlockwise rule' that was described in Topic 15.

Pairs of substances for consideration
a iron(II) sulphate and bromine water
b iron(III) chloride and zinc
c iron(II) sulphate and silver nitrate
d iron(III) chloride and sodium chloride
e iron(III) chloride and sulphur dioxide solution
f iron(II) sulphate and acidified potassium manganate(VII)
g iron(III) chloride and potassium iodide
h iron(II) sulphate and chlorine water.

For each of the pairs of substances listed, try to confirm your predictions by experiment. Do this by mixing roughly equal volumes of the two solutions in a test-tube, or by adding a spatula measure of the solid to a few cm^3 of solution.

In a number of the reactions it should be easy to tell whether a reaction has taken place or not because there are coloured reactants or coloured products. In other cases, however, it may be necessary to test the solution to find out whether the iron has changed in oxidation number. A suitable test is to add sodium hydroxide solution. Iron(II) ions give a green precipitate when sodium hydroxide solution is added, whereas iron(III) ions give a red–brown precipitate under these circumstances.

In your notebook draw up a table of results as follows:

Substances mixed	Predicted reaction, if any	Observations on mixing
(a) iron(II) sulphate and bromine water etc.		

Mention in the 'Observations' column any confirmatory tests used.

After the table of results write balanced ionic equations for the redox reactions which take place. Here is an example of how to do this. In the reaction between Fe^{2+} and Cl_2 the ions $Fe^{2+}(aq)$ are oxidized to $Fe^{3+}(aq)$ and the chlorine molecules $Cl_2(aq)$ are reduced to $Cl^-(aq)$. The equations for the half reactions are

$$Fe^{2+}(aq) \rightleftharpoons Fe^{3+}(aq) + e^-$$
$$Cl_2(aq) + 2e^- \rightleftharpoons 2Cl^-(aq)$$

The first equation involves one electron, whereas the second involves two electrons. The first equation should therefore be doubled throughout, and added to the second, so that the electrons do not appear in the final equation. This is

$$2Fe^{2+}(aq) + Cl_2(aq) \rightarrow 2Fe^{3+}(aq) + 2Cl^-(aq)$$

The oxidation numbers of vanadium

Whereas iron has only two readily accessible oxidation numbers in its compounds, vanadium has four; $+5$, $+4$, $+3$, and $+2$. In this next experiment we shall investigate this rather more complicated situation.

EXPERIMENT 16.2b
An investigation of the redox reactions of vanadium

Part 1 The reduction of vanadium(v)

The object of this part of the experiment is to start with a solution containing vanadium with oxidation number $+5$, and to select reducing agents which will reduce it to each of the other oxidation numbers.

The most convenient starting material is solid ammonium vanadate(v) which is usually given the formula NH_4VO_3. In solution the vanadate(v) ion is polymerized in a complicated way, so the formula $VO_3^-(aq)$ is a simplification. When ammonium vanadate(v) is acidified the vanadium becomes part of a positive ion, $VO_2^+(aq)$, in which the vanadium still has an oxidation number of $+5$. You are provided with such a solution. The other oxidation numbers of vanadium are included with this one in the following table:

Ion	Oxidation number of vanadium	Colour of solution
VO_2^+	$+5$	yellow
VO^{2+}	$+4$	blue
V^{3+}	$+3$	green
V^{2+}	$+2$	mauve

Copy this table into your notebook, and also the chart of electrode potentials and oxidation numbers which follows (figure 16.3).

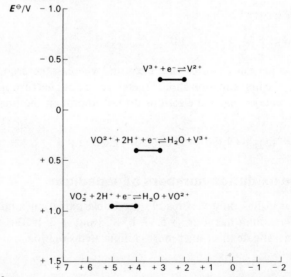

Figure 16.3

Enter on this chart the equilibria **a**, **e**, **f**, and **g** from experiment 16.2a, and also the equilibrium

$$Sn^{2+}(aq) + 2e^- \rightleftharpoons Sn(s); \qquad E^{\ominus} = -0.14\,V$$

Use the chart to select a reducing agent which should reduce vanadium from $+5$ in VO_2^+ to $+4$ in VO^{2+}, but should not reduce the vanadium any further. Then select a reducing agent which should reduce vanadium from $+5$ to $+4$ and also from $+4$ to $+3$ but no further. Finally select a reducing agent which should reduce vanadium all the way from $+5$ to $+2$. Record your selections in a table drawn up in your notebook and then try out the reactions, recording your observations in a copy of this table.

Desired final oxidation number	Selected reducing agent	Observations
$+4\,(VO^{2+})$ $+3\,(V^{3+})$ $+2\,(V^{2+})$		

If you decide to use potassium iodide solution as a reducing agent, the iodide will be oxidized to elemental iodine. This will give a colour to the solution which will mask the colour of the vanadium ion. The iodine can be reduced back to colourless iodide by reaction with sodium thiosulphate solution (see Topic 5): add only just enough to discharge the colour due to the iodine. Incidentally, it has been discovered that sodium thiosulphate will itself reduce the VO_2^+ ion. If the reaction involved is

$$2S_2O_3^{2-}(aq) \rightleftharpoons S_4O_6^{2-}(aq) + 2e^-; \; E^{\ominus} = +0.09\,V$$

what oxidation state of vanadium should result?

Keep your samples of vanadium with the four different oxidation numbers for the second part of this experiment.

Part 2 Further reactions of vanadium compounds

For the second part of this experiment, try to predict the outcome of mixing the substances listed in the table that follows. Use the standard electrode potentials given in the *Book of data* where necessary.

Using your samples of vanadium compounds obtained in part 1 of this experiment, try out the various reactions, and interpret the observations that you make. Copy the table overleaf into your notebook, and record your observations and comments.

Species to be mixed	Predicted outcome	Observations and comments
VO^{2+} and V^{2+} VO_2^+ and V^{3+} VO^{2+} and Fe^{3+} VO^{2+} and Br^- V^{2+} and Cu^{2+} V^{3+} and Fe^{3+}		

When making these predictions and interpretations you should bear in mind the appearance of the reactant solutions and that of *all* the products – that is, not only the appearance of the compounds of vanadium but that of any by-products too.

Analysis of 'iron tablets'

Potassium manganate(VII) is a well known oxidizing agent, usually used in solutions acidified with a plentiful supply of dilute sulphuric acid. Reference to the following redox potentials shows that manganate(VII) ions should oxidize iron(II) ions:

$$Fe^{3+}(aq) + e^- \rightleftharpoons Fe^{2+}(aq); \qquad\qquad\qquad E^\ominus = +0.77\,V$$

$$MnO_4^-(aq) + 8H^+(aq) + 5e^- \rightleftharpoons Mn^{2+}(aq) + 4H_2O(l); \; E^\ominus = +1.51\,V$$

Combining these two equations gives the overall equation for the reaction:

$$MnO_4^-(aq) + 8H^+(aq) + 5Fe^{2+}(aq) \longrightarrow$$
$$Mn^{2+}(aq) + 5Fe^{3+}(aq) + 4H_2O(l)$$

so that in acid solution:

1 mole of $MnO_4^-(aq)$ reacts with 5 moles of $Fe^{2+}(aq)$

Solutions containing $MnO_4^-(aq)$ ions have an intense purple colour, whereas those containing $Mn^{2+}(aq)$ ions are virtually colourless. Solutions containing $Fe^{2+}(aq)$ ions can be titrated against potassium manganate(VII) solution. The colour of the manganate(VII) is discharged, the 'end-point' of the titration being the point at which the addition of one more drop of potassium manganate(VII) gives a permanent purple colour. This titration forms the basis of an analytical technique for the estimation of iron.

EXPERIMENT 16.2c
Estimation of the percentage of iron in 'ferrous sulphate' tablets

Weigh accurately two 'ferrous sulphate' tablets. Grind up the tablets with a little M sulphuric acid, using a pestle and mortar. Through a funnel, transfer the resulting paste into a 100-cm³ volumetric flask. Use further small volumes of M sulphuric acid to rinse the ground-up tablets into the flask. During this process, you must take great care to ensure that all the particles of tablet get into the flask. When this has been done, add sufficient M sulphuric acid to make up the solution to exactly 100 cm³. Stopper the flask and shake it to make sure that all the contents are thoroughly mixed. They will not all be in solution although the Fe^{2+} ions which were present in the tablets will be dissolved.

Titrate 10-cm³ portions of the solution with 0.005M potassium manganate(VII). As explained in the introduction to this experiment, the titration is self-indicating, the 'end-point' being marked by the first permanent purple colour. Brown or red colours should not be allowed to develop; the remedy is to add more M sulphuric acid.

Write a short account of the theory of the analysis, record the practical procedure, and calculate the percentage of iron in the tablets from your results. Record these in tabular form.

Questions

1 What are the reasons for using so much M sulphuric acid during this experiment? There are two reasons, one to do with the behaviour of solutions containing Fe^{2+} ions, and the other to do with the equation for oxidations involving potassium manganate(VII).

2 Do your results agree with the analysis of the tablets given by the manufacturers on the bottle label?

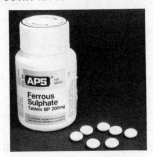

Figure 16.4
Iron is an essential element in the human diet, but our food normally contains only just enough iron in a suitable form. If necessary, the diet can be supplemented by 'iron tablets'. These contain iron(II) sulphate, generally sold under the old name of ferrous sulphate. (*Photograph, B. J. Stokes.*)

16.3
COMPLEX ION FORMATION

We shall now consider another of the characteristic properties of transition elements, namely the ability to form complex ions.

As was seen in the first section of this Topic, a complex is formed when the ion of a transition element is surrounded by ligands. These ligands are either negatively charged ions, or molecules, and in both cases contain a lone pair of electrons which is used to make the bond between the ligand and the transition element ion. As the formation of a bond using a lone pair of electrons is sometimes called *co-ordination*, complexes are sometimes referred to as *co-ordination compounds*. The number of ligand attachments around each ion of the transition element is called the *co-ordination number* of the complex.

Stability constant of copper(II) complexes

The commonest ligand is water, and aqueous solutions of simple compounds of transition elements contain complex ions with formulae such as

$$Cu(H_2O)_4^{2+} \qquad Ni(H_2O)_6^{2+} \qquad Co(H_2O)_6^{2+}$$

If a solution containing a different ligand is added to an aqueous solution containing these hydrated ions, an equilibrium is set up in which the water molecules are replaced by the new ligands. For example, in the case of the copper(II) complexes which we shall be investigating in experiment 16.3a the equilibrium in the presence of chloride ions would be:

$$Cu(H_2O)_4^{2+}(aq) + 4Cl^-(aq) \rightleftharpoons CuCl_4^{2-}(aq) + 4H_2O(l)$$

Application of the equilibrium law gives an equilibrium constant K, where

$$K = \frac{[CuCl_4^{2-}(aq)]}{[Cu(H_2O)_4^{2+}(aq)][Cl^-(aq)]^4}$$

Equilibrium constants such as this are called 'stability constants'. They enable us to compare the stabilities of complexes of a given element with different ligands; the larger the stability constant, the more stable the complex may be said to be, compared with the water complex.

For convenience, the logarithms of the values of the stability constants are given in table 16.1 for various complexes of copper(II):

Ligand		lg K
Cl^-	chloride	5.6
NH_3	ammonia	13.1

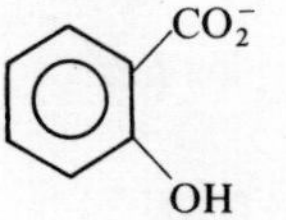

2-hydroxybenzoate   16.9

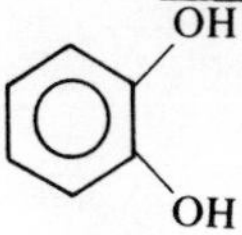

1,2-dihydroxybenzene   25.0

edta   (ethylenediamine-tetra-acetic acid)   18.8

Table 16.1
Stability constants of some copper(II) complexes. The formula of edta is given on page 231.

These stability constants will be referred to in the following experiment. They are 'overall' stability constants for the complete reaction; for example for the reaction

$$Cu(H_2O)_4^{2+}(aq) + 4Cl^-(aq) \rightleftharpoons CuCl_4^{2-}(aq) + 4H_2O(l)$$

Stability constants can also be found for each stage of the reaction, for example for the reaction

$$Cu(H_2O)_4^{2+}(aq) + Cl^-(aq) \rightleftharpoons Cu(H_2O)_3Cl^+(aq) + H_2O(l)$$

These values are given in the *Book of data*.

EXPERIMENT 16.3a
An investigation of some copper(II) complexes

As you write down the answers to the questions in the experiment, you should make it clear what each question was and to which reaction it refers.

Eye protection should be worn.

Procedure

A Put five or six drops of 0.5M copper(II) sulphate solution in a test-tube.

1 What complex ion is present?

2 What is its colour?

B   Add ten drops of concentrated hydrochloric acid drop by drop.

3   What is the colour of the solution now?

4   What ligands do you think are now present in the complex ion?

C   Keep half of the solution for D. Pour the other half into a test-tube half full of water.

5   What colour is the solution now?

6   What complex ion is now present?

7   In view of what happened in B, why do you think this reaction occurred?

D   To the solution kept from C, add concentrated ammonia solution drop by drop till there is no further colour change. Save this solution for F.

8   What colour is the solution now?

9   What ligands do you think are now present in the complex ion?

10 Qualitatively compare the relative stabilities of the various complex ions formed during these experiments: $Cu(H_2O)_4^{2+}$; $CuCl_4^{2-}$; $Cu(NH_3)_4^{2+}$.

11 Are your results consistent with the stability constants in table 16.1?

E   To 4 or 5 drops of $Cu(H_2O)_4^{2+}(aq)$ in a test-tube add a solution of the ligand edta until there is no further colour change.

12 What colour is the edta–Cu(II) complex?

13 If edta solution were added to the solution obtained in D, what do you think would happen? (Use the stability constants in table 16.1 to make your prediction.)

F   Add edta solution drop by drop to the solution obtained in D until there is no further colour change.

14 Was your prediction in question 13 correct?

G Put five or six drops of copper(II) solution in each of four test-tubes. To the first test-tube add edta solution drop by drop until there is no further colour change. To the second similarly add ammonia solution; to the third, sodium 2-hydroxybenzoate solution; and to the fourth, 1,2-dihydroxybenzene solution in 0.5M sodium hydroxide (this is an irritant: handle with care). Note that the last solution contains sodium hydroxide.

15 What are the colours of the four complexes?

16 You are going to carry out an experiment in which first a solution of ammonia, then one of sodium 2-hydroxybenzoate, then edta, then 1,2-dihydroxybenzene are added in turn to a solution of copper(II) ions. From the stability constants in table 16.1, predict what will happen.

H Check your predictions by adding 8–10 drops of concentrated ammonia solution to 4–5 drops of copper(II) solution, followed by 10–12 drops of sodium 2-hydroxybenzoate solution, 5–6 drops of edta solution, and 10–15 dops of 1,2-dihydroxybenzene solution in turn. Add the solutions drop by drop, noting the colour of the solution.

The 2-hydroxybenzoate ion, [benzene ring with CO_2^- and OH substituents]   1,2-dihydroxybenzene, [benzene ring with OH and OH substituents]

and edta, ethylenediaminetetra-acetic acid,

$$HO_2CCH_2 \diagdown$$
$$\qquad\qquad N-CH_2-CH_2-N$$
$$HO_2CCH_2 \diagup \qquad\qquad\qquad \diagup \diagdown$$

CH_2CO_2H

CH_2CO_2H

are *polydentate* ligands. That is, they can form more than one link with the metal ion. Thus in solutions in which there is an excess of ligand present, the predominating species containing the metal are:

$Cu(H_2O)_4^{2+}$ *monodentate* ligands
$Cu(NH_3)_4^{2+}$
$CuCl_4^{2-}$

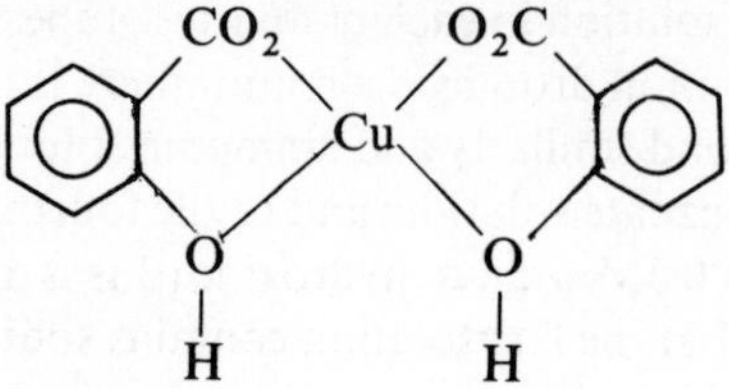

bidentate ligands

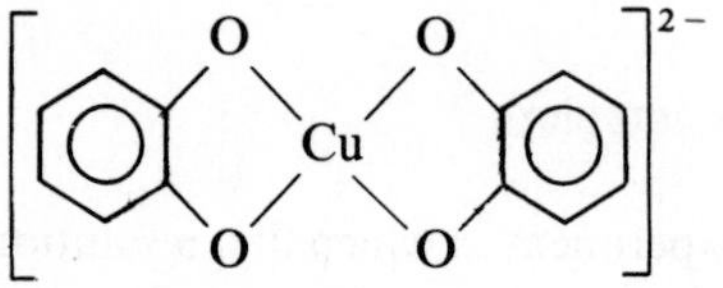

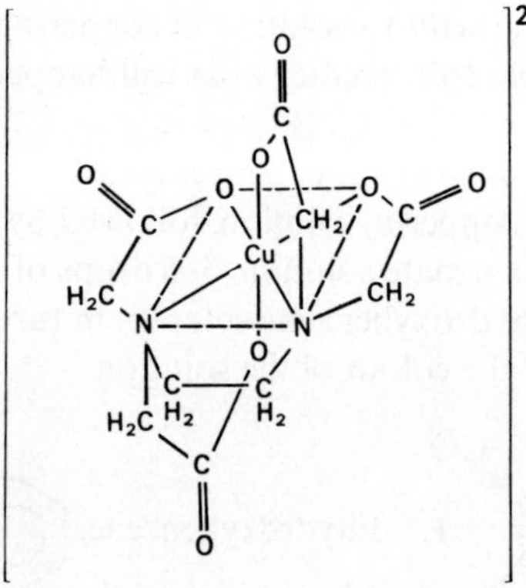

hexadentate ligand

The stoicheiometry of complexes

The word 'stoicheiometry' means 'the ratio of the constituents'. The next experiment is an example of the use of the method of continuous variation to find the stoicheiometry of some complexes, that is, the number of ligand ions or molecules associated with the ion of the transition element in each complex.

EXPERIMENT 16.3b
An investigation of the stoicheiometry of some complexes

Procedure

In the case of each complex to be investigated the method is to take eleven similar test-tubes and make mixtures in them as follows.

Test-tube number	Volume of solution A/cm³	Volume of solution B/cm³
1	0	5
2	0.5	4.5
3	1	4
4	1.5	3.5
5	2	3
6	2.5	2.5
7	3	2
8	3.5	1.5
9	4	1
10	4.5	0.5
11	5	0

Run the liquids into the test-tubes from burettes. Mix the solutions well and examine the mixtures carefully. The test-tube with the greatest amount of colour due to the complex in it has the solutions mixed most nearly to the correct ratio of ligand to transition element ion in the complex.

Try the method with some of the following examples:

	A	B
1	0.1M Ni^{2+}(aq)	0.1M edta(aq)
2	0.005M Fe^{3+}(aq)	0.005M SCN^-(aq)
3	0.1M Cu^{2+}(aq)	0.1M $\langle\!\bigcirc\!\rangle$—$NH_2$(aq)

In the case of **3** it should be borne in mind that the green complex is only sparingly soluble in water, and that a blue precipitate of 'copper hydroxide', which may also be seen, can be ignored.

If you have access to a colorimeter you could use it to make actual measurements of colour intensity.

Write an account of the method and record clearly your results and conclusions.

The preparation of some salts containing complex ions

You may be able to carry out the following preparations or some others which your teacher might suggest. If you do those described in experiment 16.3c, you should be able to do most of the practical work within one double period; but you will need to ensure that you use your time to best advantage by having both preparations going on at the same time.

EXPERIMENT 16.3c
The preparation of some compounds containing complexes

Preparation 1 Chromium(II) ethanoate, $Cr_2(CH_3CO_2)_4(H_2O)_2$

Chromium(II) ethanoate can be regarded as a neutral complex of Cr^{2+} ions with $CH_3CO_2^-$ ions and water molecules as ligands. Its structure is shown in figure 16.5. This compound is interesting because it is an example of the way in which the formation of a complex can sometimes stabilize an oxidation state which would otherwise be unstable. Chromium(II) ions are normally very readily oxidized to the chromium(III) state by the oxygen of the air.

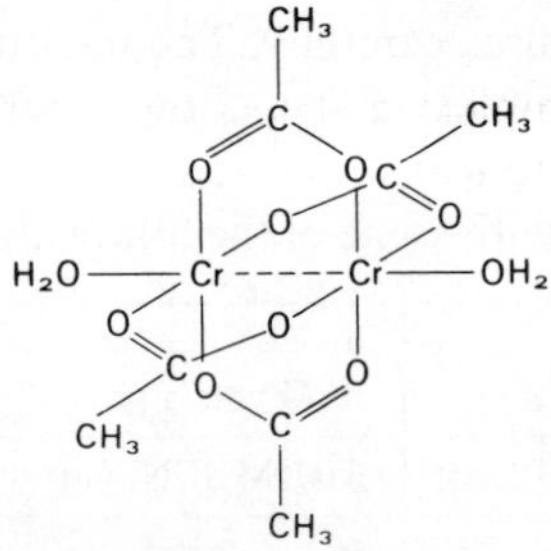

Figure 16.5
The structure of chromium(II) ethanoate.

Write an account of the following method in your notebook, and answer the questions at the end, recording your answers in such a way as to make it clear what each question was.

A set of apparatus such as that shown in figure 16.6 can be used, or one which will perform similarly.

Note that you will be using potassium dichromate(VI) which is highly irritant to skin, eyes, and respiratory system. Eye protection is essential and you must be careful not to raise any dust when handling the compound. As hydrogen is evolved, naked flames must be kept well clear.

Dissolve 1 g of potassium dichromate(VI) in $10 \, cm^3$ of water, and put it in the $50 \, cm^3$ round-bottomed flask. Add 2.5 g of granulated zinc. Put about $20 \, cm^3$ of approximately 5M hydrochloric acid (*carefully* use a mixture of $10 \, cm^3$ of concentrated hydrochloric acid with $10 \, cm^3$ of water) in the tap funnel. Put $10 \, cm^3$ of saturated sodium ethanoate solution in the boiling-tube. Add the hydrochloric acid to the mixture in the flask and leave the tap funnel open to allow the hydrogen which is vigorously generated to escape. The reduction of the chromium passes through a green (Cr^{3+}) stage and is complete when the solution is blue (Cr^{2+}). While hydrogen is still being generated close the tap

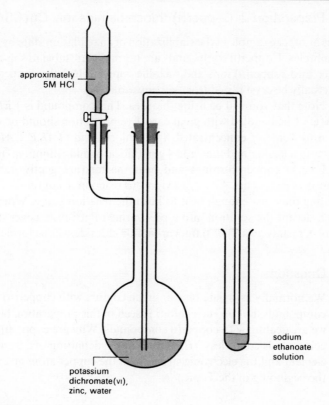

approximately
5M HCl

sodium
ethanoate
solution

potassium
dichromate(VI),
zinc, water

Figure 16.6

on the funnel so that the pressure of the hydrogen forces the blue solution over
into the saturated sodium ethanoate. A red precipitate of chromium(II) ethanoate
should be formed, and the solution will contain dissolved red chromium(II)
ethanoate. Dismantle the apparatus and pour the remaining blue solution
containing Cr^{2+}(aq) into another boiling-tube. Keep the two boiling-tubes side
by side in a rack.

Questions

1 Show, using electrode potentials, that zinc should reduce chromium
   from $+6$ in potassium dichromate to $+2$ in chromium(II) ethanoate.

2 What happens to the colour of the Cr^{2+}(aq) solution when it is
   allowed to stand open to the air?

3 Over the same period of time does the colour of the chromium(II)
   ethanoate solution also change?

Preparation 2   Copper(I) thiocarbamide ion, $Cu(CS(NH_2)_2)_3^+$

This is another example of the stabilization of an oxidation state by the formation of a complex. Copper(I) compounds are normally unstable in solution, changing at once into copper(II) ions and metallic copper. Here the copper(I) compound can actually be crystallized from aqueous solution.

Note that you will be using thiourea. This compound is *VERY* poisonous and *MUST* be handled with great care. Eye protection should be worn.

Add $4 cm^3$ of concentrated hydrochloric acid (*TAKE CARE*) to $16 cm^3$ of water in a beaker and then add 4 g of thiocarbamide (thiourea) to the mixture. Add 1.5 g of copper turnings and heat the mixture gently until a vigorous evolution of gas occurs. Allow this vigorous reaction to continue until it subsides, providing only just enough heat to keep the reaction going. When the reaction is over, decant the solution into a pre-warmed Petri dish, cover it, and allow it to cool. Crystals of copper(I) thiocarbamide chloride will be formed.

Questions

1   We normally associate blue or green colours with copper(II) compounds, but no such colour is seen in this preparation because we are dealing with copper(I) compounds. Why are copper(I) compounds colourless? (*Hint:* look at the definition of a transition element and the electronic structure of the copper atom given at the beginning of the Topic.)

2   In what way do you think the thiocarbamide molecules are arranged around the copper(I) ion?

16.4
ENTROPY CONSIDERATIONS

When bidentate ligands replace monodentate ligands in a complex, there will be an increase in the entropy of the system because one molecule of ligand is replacing two molecules, as in the following example:

$$Ni(NH_3)_6^{2+}(aq) + 3NH_2CH_2CH_2NH_2(aq)$$
1,2-diaminoethane
$$\rightleftharpoons Ni(NH_2CH_2CH_2NH_2)_3^{2+}(aq) + 6NH_3(aq)$$
ammonia

In this example there are 4 particles shown on the left of the equation but 7 particles on the right. Because the entropy of a system depends, amongst other things, on the number of particles present, the entropy of the system increases when 1,2-diaminoethane molecules replace ammonia molecules.

A similar, but larger, increase of entropy takes place when edta replaces monodentate or bidentate ligands in a complex, as, for example

$$Ni(NH_3)_6^{2+}(aq) + edta(aq) \rightleftharpoons Ni(edta)^{2+}(aq) + 6NH_3(aq)$$

Comparison of the numbers of particles involved shows a large increase, corresponding to an increase of entropy.

Because of this entropy-increasing effect, complexes with a hexadentate ligand such as edta are usually much more stable than those with a monodentate ligand. A good example of this effect is shown by the stability constants of nickel complexes:

Complex	lg K
$Ni(H_2O)_6^{2+}$	0
$Ni(NH_3)_6^{2+}$	8.01
$Ni(edta)^{2+}$	18.6

In these complexes the ligands are bonded to the nickel ion by means of lone pairs of electrons on oxygen atoms in water, nitrogen atoms in ammonia, and both oxygen and nitrogen atoms in edta. Ignoring the entropy effect, therefore, the edta complex might be expected to be comparable in stability with the ammonia complex or perhaps rather less stable. It is the increase in entropy accompanying its formation which accounts for the high stability of the edta complex compared with the other two.

16.5
TRANSITION ELEMENTS AS CATALYSTS
What a catalyst does

A catalyst increases the rate at which a chemical reaction approaches equilibrium without itself becoming permanently involved in the reaction. Nevertheless, in order to perform its task the catalyst becomes temporarily involved. It provides a new reaction path which is of lower activation energy than would be the case if it were not there.

This can be illustrated by a potential energy diagram (figure 16.7). Before the usual chemical reaction can take place the reactant molecules must be raised to a state of higher potential energy. They are then said to be *activated* or to form an *activated complex*. Reactants and products are both at stable minima of potential energy, while the activated complex is the state at the top of the potential energy barrier. The reaction follows the new path provided by the catalyst, in which the potential energy barrier (and the activation energy) is lower.

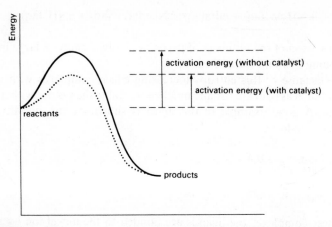

Figure 16.7

When the reaction occurs within any one phase we speak of *homogeneous* catalysis, but many catalysed reactions involve two or more phases. For example, the synthesis of ammonia involves gaseous hydrogen and nitrogen with a solid iron catalyst. This is an example of *heterogeneous* catalysis.

Heterogeneous catalysis

Important events take place at the surface of the solid catalyst. When gas molecules approach solid surfaces there is a tendency for them to interact with the surface atoms. This interaction is called *adsorption*. It is the means by which new reaction paths are formed in heterogeneous catalysis.

Adsorption To understand how heterogeneous catalysis occurs we must learn more about the process of adsorption. When new chemical bonds are definitely formed between a reactant molecule and the surface atoms of the catalyst we use the term *chemisorption*. Transition metals, for example nickel, are particularly able to chemisorb hydrogen. We may represent this process in the following simple way:

$$
\text{H}_2 + \begin{array}{c} \\ -\text{Ni}-\text{Ni}- \\ \ \end{array} \longrightarrow \begin{array}{c} \text{H----H} \\ |\quad\ | \\ -\text{Ni}-\text{Ni}- \\ |\quad\ | \end{array} \longrightarrow \begin{array}{c} \text{H}\quad\text{H} \\ |\quad\ | \\ -\text{Ni}-\text{Ni}- \\ |\quad\ | \end{array}
$$

gas   nickel   surface complex   chemisorbed
        surface atoms   hydrogen atoms

A weaker form of adsorption may also occur, called *physical adsorption*. Here the forces which attract the molecule to the surface are like those which hold molecules together in a liquid (van der Waals forces).

Ammonia synthesis catalysts When ammonia is synthesized from nitrogen and hydrogen gases, by means of an iron catalyst, the following sequence of events occurs in the reactor:

1 Nitrogen gas and hydrogen gas diffuse to the surface of the catalyst.
2 The reactant gases are then adsorbed on the catalyst surface.

$$N_2(g) \longrightarrow 2N_{adsorbed} \quad \text{and} \quad H_2(g) \longrightarrow 2H_{adsorbed}$$

3 Nitrogen and hydrogen react on the catalyst surface, forming ammonia.

$$N_{adsorbed} + H_{adsorbed} \longrightarrow NH_{adsorbed}$$

$$\xrightarrow{+H_{adsorbed}} NH_{2\,adsorbed} \xrightarrow{+H_{adsorbed}} NH_{3\,adsorbed}$$

4 Ammonia *desorbs* from the catalyst surface

$$NH_{3\,adsorbed} \longrightarrow NH_3(g)$$

5 Ammonia diffuses away from the catalyst surface.

If any one of these events is much slower than all the others, it will be the *rate-determining step*. That is to say, the rate at which this event occurs will determine how rapidly ammonia can be synthesized. Problems arising in connection with 1 and 5 are very familiar to the chemical engineer but can often be successfully overcome so that both events are made sufficiently rapid.

Iron, ruthenium, osmium, molybdenum, and tungsten can be used as catalysts for the synthesis of ammonia. Therefore, on these metals, steps 2, 3, and 4 must also take place at a reasonable rate. Nevertheless, one of these steps must be rate-determining. After many experiments and much lively discussion between American, Japanese, Dutch, and British scientists, there is a majority view that the rate-determining step in ammonia synthesis is the adsorption of nitrogen (stage 2).

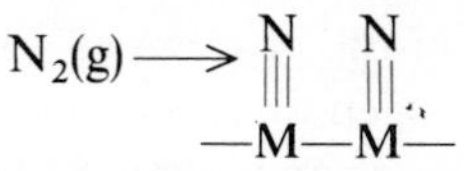

$$N_2(g) \longrightarrow \begin{array}{cc} N & N \\ ||| & ||| \\ -M & -M- \end{array}$$

(where M = a surface metal atom)

Of course, we expect catalysts for ammonia synthesis to have the capacity to chemisorb nitrogen. This is not the whole story, but it is a help to know which metals can chemisorb nitrogen and which cannot. Table 16.2 gives this information for nitrogen and for some other gases.

Metals			Gases				
			O_2	CO	C_2H_4	H_2	N_2
Ti	Zr	Hf					
V	Nb	Ta	+	+	+	+	+
Cr	Mo	W					
Fe							
Co	Ni						
Rh	Pd		+	+	+	+	−
Ir	Pt						
Al	Cu		+	+	+	−	−
Ag*							
Zn	Cd		+	−	−	−	−
In							
Sn	Pb						

* CO is known to adsorb on silver.

Table 16.2
The ability of metals to chemisorb gases.
+ means that strong chemisorption is possible.
− means that chemisorption is weak or not observed.

This information is not complete but we see that three of the metals, iron, molybdenum, and tungsten, which early workers found to be active ammonia synthesis catalysts, chemisorb nitrogen (+).

If a metal is going to be a good catalyst for the synthesis of ammonia, it must not chemisorb nitrogen too strongly, otherwise the nitrogen atoms are unreactive for the combination with hydrogen. On the other hand, if nitrogen is chemisorbed too weakly, only a few atoms will hold on to the surface at any one moment and the rate of ammonia synthesis will again be slow. The strength of chemisorption must be some suitable intermediate value if the synthesis rate is a maximum.

Now let us put these ideas together to find a good ammonia synthesis catalyst, that is, a metal which can chemisorb nitrogen but not too strongly. Figure 16.8 shows the approximate strength of nitrogen chemisorption and the rate of ammonia synthesis, plotted as a function of the position of a metal in the transition series. The ability to chemisorb nitrogen as atoms ceases to all intents and purposes to the right of iron, ruthenium, and osmium, so ruling out the metals to the right. To the left of these same metals, the strength of adsorption increases (there is a greater tendency to form bulk nitrides). There-fore we expect that iron, ruthenium, and osmium should have the greatest activity in ammonia synthesis. This is indeed the experimental result already found. It is fortunate that one of these metals, iron, is also very abundant, and therefore economically attractive as the basis of a catalyst for large scale ammonia synthesis.

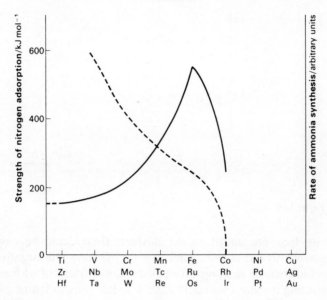

Figure 16.8
Strength of nitrogen adsorption (broken line) and rate of ammonia synthesis (continuous line) plotted against position of element in the transition series.

Homogeneous catalysis

Although heterogeneous catalysis is of more widespread significance com-mercially, homogeneous catalysis is important too and often involves transition metal ions. The ability of transition metals to change oxidation state readily is usually the key here and it is sometimes possible to identify a possible catalyst by using electrode potentials. Consider the electrode potential chart in figure 16.9.

Persulphate ions, $S_2O_8^{2-}$(aq), are capable of oxidizing iodide ions, I^-(aq), to iodine, but the reaction is a very slow one.

$$S_2O_8^{2-}(aq) + 2I^-(aq) \longrightarrow 2SO_4^{2-}(aq) + I_2(aq)$$

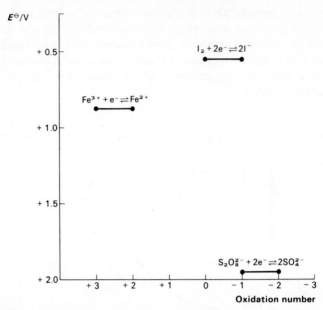

Figure 16.9

If iron(II) ions are added to the mixture, these could be oxidized by the persulphate ions rather more quickly (perhaps because the negative persulphate ions would then be reacting with positive ions instead of with negative ones). The resulting iron(III) ions could oxidize iodide ions to iodine and the iron(II) ions would be re-formed.

Clearly, in order for this catalysis to work, the electrode potential for the reaction involving the catalyst must lie between the electrode potentials involving the two reactants.

This predictive method only shows that catalysis should be possible. It does not guarantee that an improvement in rate of reaction will actually take place.

BACKGROUND READING 1
Ellingham diagrams and the iron and steel industry

Pyrometallurgy is the treatment of ores, concentrates, and metals at high temperatures. It involves many problems of chemical change and equilibrium. The reaction kinetics associated with them are complex too. Thus, most pyrometallurgical processes involve three phases, gas, slag, and liquid metal, with, consequently, three interfaces (gas-metal, slag-metal, gas-slag) at which chemical reactions take place. In such systems, factors controlling reaction rates include not only the familiar activation energies found in homogeneous kinetics but also purely physical steps involving the transport of reactants to the interfaces, for example by diffusion. Now that you have studied the Topics dealing with reaction kinetics, equilibria, and free energy you will be able to understand the following discussion.

Pyrometallurgical methods of obtaining metals from oxide ores or sulphide ores (which are first roasted to form the oxide), depend upon the action of a reducing agent. Let us begin by considering the thermodynamic stabilities of metal oxides, as measured by the free energy change $\Delta G_f^{\ominus}$, when they are formed from their elements.

Thermodynamic stability refers to the tendency for the oxide to form in a reaction such as

$$2M(s) + O_2(g) \rightleftharpoons 2MO(s)$$

where M is a metal. The equilibrium constant K_p for this process is related to the free energy change by the expression

$$\Delta G^{\ominus} = -RT \ln K_p = -RT \ln p_{O_2}$$

where p_{O_2} is the partial pressure of oxygen in the system. The greater the negative value of $\Delta G^{\ominus}$, the larger K_p becomes; that is to say, the equilibrium shifts towards the oxide.

From the relationship

$$\Delta G^{\ominus} = -RT \ln K_p$$

we obtain

$$K_p = \exp \frac{-\Delta G^{\ominus}}{R} \left(\frac{1}{T} \right)$$

A plot of K_p versus $1/T$ is of the form shown in figure 16.10a while that of K_p versus T takes the form shown in figure 16.10b. Thus we see that decomposition of the oxide into its elements is favoured at high temperatures. Figure 16.11 shows a plot of p_{O_2} versus temperature; you can see that the partial pressure of oxygen rises with increase in temperature.

Ellingham diagrams

The thermodynamic stabilities of metal oxides can be compared by determining the $\Delta G^{\ominus}$ values for reactions in which one mole of oxygen at one atmosphere

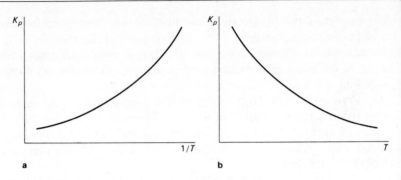

Figure 16.10
Variation of the equilibrium constant K_p with temperature.

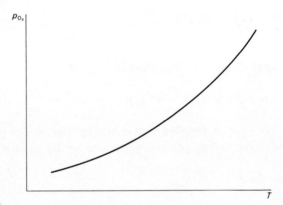

Figure 16.11
The increase in the partial pressure of oxygen with temperature, for the decomposition of a metal oxide.

pressure (101.3 kPa) is combined with the pure metal. In reactions of this kind there is a change in volume in the system since one of the reactants is a gas. The oxidation of elements such as hydrogen and carbon is also attended by volume changes. In these situations, the rate at which $\Delta G^{\ominus}$ changes with temperature is strongly affected by the volume change. Indeed at high temperatures the gradient of the graph of $\Delta G^{\ominus}$ versus T has the same sign as that of the volume change which accompanies decomposition.

An Ellingham diagram is a graphical representation of the variation of $\Delta G^{\ominus}$ with temperature. Figure 16.12 is an Ellingham diagram for the formation of water and of several metal oxides. Notice that the plots all have positive slopes, which means that as the temperature increases, $\Delta G^{\ominus}$ becomes less negative; hence K_p decreases and the equilibrium shifts to the left.

Thus, thermodynamically the oxides become less stable with respect to their elements at high temperatures; we might have here a basis for extractive metallurgy. Why not simply heat the oxide of a metal until the equilibrium moves far enough to the left to give a good yield of the element? The answer is that in most cases, uneconomically high temperatures would be needed. Iron oxides such as Fe_2O_2, would require temperatures approaching 5200 K. What do you think about the case of silver? Would simply heating the oxide be a good method of extraction for that element?

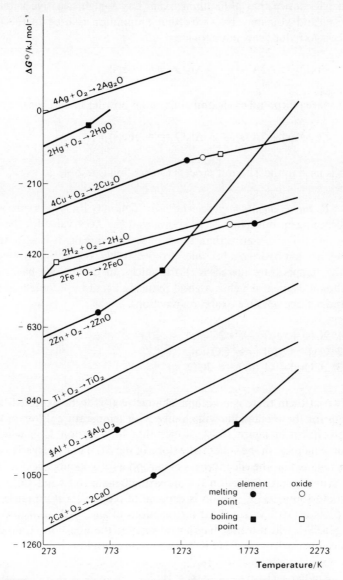

Figure 16.12
An Ellingham diagram.

If we draw a vertical line representing a fixed temperature on figure 16.1 it will intersect each $\Delta G^\ominus$ line on the diagram. The ordinates ($\Delta G^\ominus$ values) c these intersections correspond to the sequence of the thermodynamic stabilitie of the oxides. A metal which forms a more stable oxide (with $\Delta G^\ominus$ more negative is a potential reducing agent for a less stable oxide (with $\Delta G^\ominus$ less negative You will notice that aluminium oxide has a high negative value of $\Delta G^\ominus$ a 2273 K and you may be aware that aluminium is used in the Goldschmid process to reduce chromium oxide:

$$Cr_2O_3(s) + 2Al(s) \longrightarrow Al_2O_3(s) + 2Cr(s)$$

A mixture of iron(III) oxide and aluminium powder undergoes the reaction

$$Fe_2O_3(s) + 2Al(s) \longrightarrow Al_2O_3(s) + 2Fe(s)$$

This is used in the Thermit process for welding iron objects.

Reducing agents in pyrometallurgy You may think of hydrogen as a goo reducing agent but figure 16.12 shows that the $\Delta G^\ominus$ value for the formation c its oxide is less negative than those of several metals. It is also apparent fro the figure that hydrogen becomes progressively less effective as a reducing agen as the temperature increases. Fortunately, in carbon we have a cheap an abundant element which is a good reducing agent in pyrometallurgy. We mus consider three possible oxidation reactions:

1 $2C(s) + O_2(g) \longrightarrow 2CO(g)$
2 $C(s) + O_2(g) \longrightarrow CO_2(g)$
3 $2CO(g) + O_2(g) \longrightarrow 2CO_2(g)$

In two of them there is a volume change so it is to be expected that the $\Delta G^\ominus$ values for these reactions will change with temperature. Plots of $\Delta G^\ominus$ agains T are shown in figure 16.13. Notice that for reaction **1**, in which there is volume increase in the system, the slope of the $\Delta G^\ominus$ line is negative above 983 K This means that the effectiveness of carbon as a reducing agent increases abov 983 K as a result of reaction **1**. Comparing figures 16.12 and 16.13 you will als see that above 983 K carbon is capable of reducing F̶e̶O̶.̶ F̶i̶g̶u̶r̶e̶ 16.13 show the plots of $\Delta G^\ominus$ versus T for the reactions of zinc, hydrogen, and carbon wit oxygen. We will use this diagram to answer the question: 'Can zinc oxide b reduced by hydrogen at 773 K ?'. The reaction could be:

$$ZnO(s) + H_2(g) \longrightarrow Zn(s) + H_2O$$

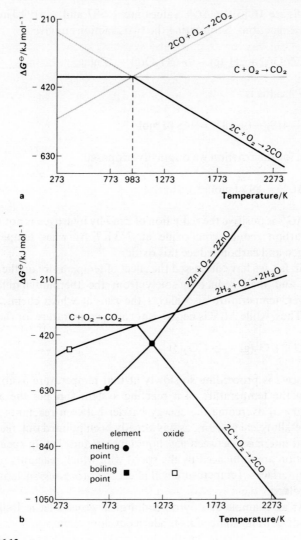

Figure 16.13
a An Ellingham diagram for the oxides of carbon.
b An Ellingham diagram for the oxides of carbon, zinc, and hydrogen.

Consider the reactions:

$$2Zn(s) + O_2(g) \longrightarrow 2ZnO(s)$$
and
$$2H_2(g) + O_2(g) \longrightarrow 2H_2O(g)$$

From figure 16.13 the $\Delta G^{\ominus}$ values are -587 and $-419\,\mathrm{kJ\,mol^{-1}}$ respective for these reactions. Subtracting the two reactions to give

$$2ZnO(s) + 2H_2(g) \longrightarrow 2H_2O(g) + 2Zn(s)$$

the $\Delta G^{\ominus}$ value is

$$(-419) - (-587) = 168\,\mathrm{kJ\,mol^{-1}}$$

so that, for the reaction we originally proposed,

$$\Delta G^{\ominus} = 84\,\mathrm{kJ\,mol^{-1}}$$

Since $\Delta G^{\ominus}$ is positive the reduction of ZnO by hydrogen is not possible at 773 Will carbon reduce zinc oxide at 773 K? At what temperature will bo hydrogen and carbon reduce this oxide?

So far, we have discussed the effect of temperature on the stability of met oxides and their reduction solely from the thermodynamic point of vie However, temperature also affects the *rate* at which chemical reactions ta place. Thus, while $\Delta G^{\ominus}$ is negative at room temperature for the reaction

$$C(s) + O_2(g) \longrightarrow CO_2(g)$$

the process is proceeding so slowly at that temperature as to be undetectab Raising the temperature of a reacting system supplies the activation ener necessary to overcome the energy barrier between reactants and products. pyrometallurgical systems, as has already been pointed out, reactions frequen occur at interfaces between gas, liquid metal, and slag. In such a situation rat of reaction are influenced by the speed with which reactants reach one anoth at the interfaces. Temperature will affect the processes, such as diffusion, whi are involved.

As an example of a pyrometallurgical process let us look at the iron a steel industry and some of its modern developments.

The iron and steel industry

Both iron and steel are required in vast quantities in industrial countries. T production of iron and that of steel are supreme examples of pyrometallurgic operations in their most awe-inspiring forms. They involve chemical process carried out on hundreds of tonnes of molten metal at temperatures of sor 1600–1900 K.

Let us begin with the production of iron itself.

Iron The metal is commonly found in the form of its oxides Fe_2O_3 and Fe_3O_4, and these oxides are reduced in a vast chemical reactor, the blast furnace. This is illustrated in figure 16.14. It may be 30.5 metres high and 14 metres across at the base; inside it iron oxides are reduced to iron, using coke as the carbon source. Limestone is added to the charge to assist in the formation of slag; this is a fusible material in which silica and other impurities are carried off from the molten iron.

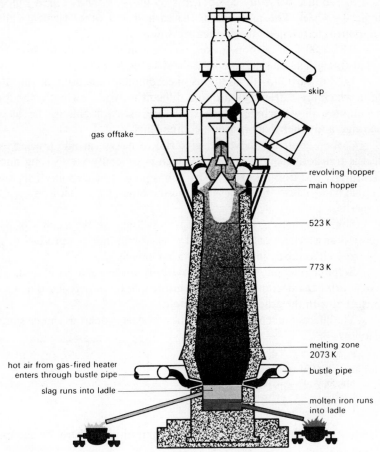

Figure 16.14
A blast furnace. The skip unloads iron ore, limestone, and coke into the top of the furnace. Molten iron and slag are tapped from different levels of the hearth.

Steel Steel is basically an iron–carbon alloy modified by the presence of other alloying elements such as manganese, silicon, phosphorus, chromium, nickel, sulphur, and oxygen.

Pure iron, as is the case with most pure metals, is both malleable and ductile. Although these are desirable properties when bending and forming railway lines, bridge girders, or car axles, all these commodities would distort and bend in use if the iron were not stiffened up in some way. The alloying elements in a steel provide the stiffeners and it is the relative proportions of these which govern the physical properties of the steel.

Steel containing 0.05% carbon provides a very ductile material. It can be used to produce car body components by pressing the required shape from a single flat sheet. The presence of the carbon and other elements allows the component to retain its shape after pressing.

0.25–1.00% of carbon makes the steel more difficult to form but gives added strength. Such steels are used for structural work.

1.0% of carbon with additions of chromium, vanadium, or tungsten gives steels which are extremely hard and difficult to work, owing to the formation of complex alloy carbides. Such steels are used for chisels, die blocks, ball bearings, and other wear-resistant components.

Carbon contents above 2% give rise to heavy carbide precipitation; the steel is transformed into cast iron, which is generally very strong and brittle. In ordinary cast iron the excess of carbon is present as graphite. This is what, on the one hand, causes brittleness but, on the other, gives cast iron many unique properties.

Since the melting point of cast iron is low, and the fluidity of its melt is high, it is an excellent casting material when attempting to reproduce an intricate pattern such as the cylinder block for a car engine.

Sulphur and phosphorus are basically undesirable in a steel. Steps are taken during the steelmaking cycle to ensure that they are reduced to a minimum consistent with the grade of steel required.

The effect of all common impurities on steel properties can be summarized as follows:

C and Mn	give strength
N, S, P, O, H	cause weakness
Mn, Si, W, V	give hardness
Cr, Ni	give corrosion resistance
Low impurities	give ductility

The control of the level of each of these impurities is therefore the essence of all steelmaking.

Steel production Considerable energy is expended in the blast furnace in producing molten iron at temperatures of 1673–1723 K, and it would be wasteful to cool it down before converting it into steel.

Heat is lost in transporting the liquid iron from the blast furnace to the steelmaking unit, and the iron enters the steelmaking unit at about 1600 K. We have already seen that the essence of steelmaking is the removal of unwanted impurities down to acceptable levels. With the exception of nickel and molybdenum, the remaining elements that are present – carbon, silicon, manganese, phosphorus, sulphur, and chromium – can be removed by oxidation. Gases may also be removed by subjecting the otherwise finished steel to a vacuum treatment either in the ladle or ingot stage. This treatment is known as *vacuum degassing.*

The rate at which steel may be produced will therefore depend upon the rate at which oxygen can be supplied to the vessel containing the liquid metal, and the kinetics of the individual oxidation reactions. These oxidation reactions may generate sufficient heat to raise the temperature of the finished steel to 1873 K, melt any cold scrap iron which may be added, flux the slag, and overcome the heat losses. External fuel may therefore not be required unless it is economical to add a high percentage of cold scrap to the molten iron.

The LD process In 1945 the Austrians nationalized the German-built open-hearth plant at Linz. This had a large tonnage liquid nitrogen plant producing liquid oxygen as a by-product. Their first experiments with oxygen consisted of injecting it into the gas space in an open hearth. The output of the open hearth increased but the roof was soon burned down and the air preheating chambers were filled with fumes. Similar experiments in an arc furnace burned the roof and electrodeholders.

At the same time the industrial chemist Durrer was working independently in Switzerland, using a Bessemer converter and an arc furnace. He injected oxygen through a water-cooled lance just above the surface of the metal. He succeeded in burning a hole in the furnace bottom and destroying the lance, but not in removing phosphorus by this technique. Joint experiments between the Linz engineers and Durrer resolved these difficulties by raising the lance and lowering the oxygen thrust, and the *LD process* was born. The phosphorus was adequately removed, the lance did not burn, and the furnace linings were not damaged. In 1952 a 35-tonne steelmaking vessel was built at Linz, and since that time LD plants have been installed all over the world. They range in size from a capacity of 35 to 400 tonnes. A typical size is 150 tonnes.

Figure 16.15 shows the layout of a modern LD system. Each unit can produce 180 tonnes of steel per hour and is equivalent to six 100-tonne open-hearth furnaces working at the rate of 30 tonnes per hour. The number of men employed and the number of cranes and wagons, etc., are all reduced. Six 100-tonne open-hearth steelmakers require 24 men per 8-hour shift; a single 150-tonne LD vessel requires 5 men per 8-hour shift. A typical charge consists of 52 tonnes of scrap steel and 120 tonnes of molten pig iron at 1623 K. The iron is charged in 3 minutes from a large lip-pouring ladle and the scrap is charged in 1 minute from a large 60-tonne capacity scrap charging machine.

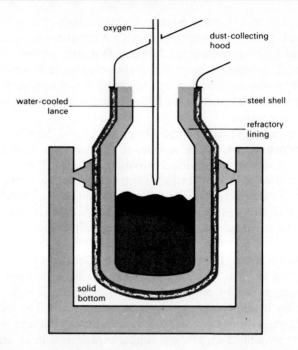

Figure 16.15
A section through an LD steelmaking vessel.

When charged, the vessel is tilted to the blowing position, and the lance is lowered to a position some 1 or 2 metres above the charge. The lance is water cooled and oxygen is injected at the rate of about 28–42×10^4 cubic decimetre per minute at $1000\,kN\,m^{-2}$; it reaches the velocity of sound.

Approximately 10 tonnes of lime and 680 kg of fluorspar are added with a little iron ore to give a fluid slag and to help to remove sulphur and phosphorus. The carbon is converted mainly to carbon monoxide which is either burnt as it leaves the vessel or collected without combustion. The total blowing time is 20 minutes and when all the required carbon has been removed the flame at the vessel mouth subsides. If proper control has been exercised the steel should then be ready for tapping. A sample is taken for analysis, and the temperature of the steel is measured. If the temperature and composition are correct the vessel is tilted further to allow the steel to flow through the tap hole into a ladle. Measured amounts of alloys are then added to the ladle to deoxidize the steel and meet the required specification for it. The finished steel may then be poured into ingots for conversion to slab, billets, reinforcing bars, and other steel shapes in the rolling mills. In most modern works a large proportion of the liquid steel is now poured through continuous casting machines which produce slabs and billets directly without the aid of a rolling mill and the yield of good steel products is 12% better than from ingots.

It is now possible to cast steel continuously in 1000–2000 tonne lots, provided the steelmaking process can be matched to the casting time and the size and quality of the product can be continuously changed to match the customer's orders.

BACKGROUND READING 2
Micronutrients

Plants take up many elements from the soil in the form of soluble compounds. The best known of these are nitrogen, phosphorus, and potassium, all usually abundant in soil and the common stuff of ordinary fertilizers. They are known as *macronutrients* because of the large quantities needed for normal growth. Other elements are required in smaller amounts, and are therefore known as *micronutrients*. Many, though not all, micronutrients are transition elements. Micronutrients are sometimes referred to as trace elements, but this term is misleading because substantially more than a trace is needed of some of them. Table 16.2 shows the average uptake per hectare of some elements needed by barley to yield 5 tonnes of grain (an average figure, about half the maximum yield) and also to produce 5 tonnes of straw, different amounts of nutrients being needed for the seed and for the leafy parts of the plant.

Uptake in:	kg ha⁻¹								g ha⁻¹				
	N	P	K	Ca	Mg	S	Na	Fe	Mn	Zn	Cu	B	Mo
To produce: 5 t ha⁻¹ grain	85	15	25	4	6	25–7	1		100	200	50		1
								2				30	
5 t ha⁻¹ straw	25	2.5	4	1.5	2.5	10–6	2.5		350	200	35		1

Table 16.2
(Separate figures were not available for iron or boron. The results for sulphur varied widely, as shown.)

The first six elements are classed as macronutrients and the last five as micronutrients, sodium and iron falling between the classes.

Other plants need different amounts and some crops need further elements such as cobalt and chlorine. In addition, plants take up elements which, although they do not seem to be needed for the growth of the plant, are vitally necessary for animals feeding on that plant. For example, mammals need all the elements so far mentioned except (as far as is known) boron, plus iodine and for some animals selenium and chromium.

Complete absence of micronutrients prevents any plant growth at all. Shortages produce stunted plants and symptoms such as chlorosis, yellowing of leaves through the absence of chlorophyll; and necrosis, decay of tissues. Animals, including humans, nourished by such plants may themselves suffer from deficiency conditions, often grave.

The position can be dramatically reversed by making good any deficiency. In Australia land which was once desert now grows crops of legumes, thanks to the application of as little as 70 grams of molybdenum per hectare (as sodium molybdate). Even more striking has been the wiping out of iodine deficiency diseases among humans living in regions lacking that element.

Soil dressing with micronutrients needs to be done only once or at most very occasionally. Excessive intake of some of these elements can cause poisoning in humans and animals.

The need for micronutrients has been shown only quite recently, since it depended on analytical techniques sensitive enough to detect, say, 1 gram of molybdenum in 5 tonnes of grain. Discoveries continue. It is hard to establish whether an element is essential to a plant in conditions resembling its natural habitat, since most of the stable elements are present, however minutely, in the rocks which weather to form soil. Nor is it easy to create an artificial growing medium completely lacking micronutrients.

Let us now consider the role of some micronutrients in more detail.

Iron

Iron was the first non-major nutrient discovered to be essential for plant growth. It is necessary for the production of chlorophyll although not present in it; deficiency results in chlorosis. In turn, uptake of iron is blocked by excessive alkalinity in the soil, a condition which affects about a third of the World's land surface. In Britain iron deficiency is a common problem in fruit trees on chalk soils.

Many animals including Man need iron for the blood protein haemoglobin: 60 to 70% of the iron in the human body is found here, most of the remainder in other proteins. The structural unit of haemoglobin which contains iron(II) is shown in figure 16.16. Haemoglobin contains only about 0.34% iron by mass, but this iron is the key to the blood's transport of oxygen. When red blood corpuscles come to the end of their usefulness and are broken up, their haemoglobin is destroyed but about 90 per cent of the iron is liberated and re-used. Therefore the daily iron requirement of a healthy animal is small. Deficiency causes anaemia, which is most often seen in rapidly growing sucklings since milk contains little iron. Piglets raised in pens without access to soil or pasture often suffer from anaemia. They need about 7 milligrams of iron a day of which only 1 mg comes from sow's milk. The rest must be provided by dosing or injection.

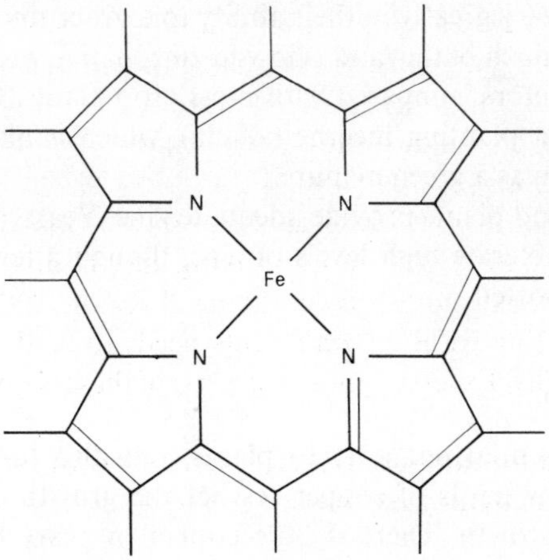

Figure 16.16
The basic porphyrin unit of haemoglobin containing four-coordinated iron(II) ions.

Manganese

Manganese is needed in several enzyme systems, for example by plants in the reduction of nitrates to amino acids. Uptake of manganese from the soil is linked to that of iron, and is affected by soil pH. In alkaline soils little manganese enters the plant. In highly acid soils too much is taken up, blocking the absorption of iron. Manganese deficiency in plants tends to occur on naturally acid soils which have been limed to over pH 7. The usual symptom is chlorosis. In *Brassica* (the cabbage genus) severe deficiency causes the leaves to become almost bleached except for the veins which remain green.

Manganese deficiency in animals causes sterility and crippling deformities of the skeleton. However, the element is widely found in food plants. Most pastures contain 40 to 200 milligrams of manganese per kilogram of dry matter. There is a wide margin of safety. Hens have been given feed with levels as high as $1 \, g \, kg^{-1}$ dry matter without apparent ill effects.

Zinc

Zinc is used by plants in the production of both carbohydrates and fats. Plants lacking zinc accumulate high levels of phenolic compounds such as 1,2-dihydroxybenzene (catechol) in their leaf cells at the expense of substances normally present. Again, the most usual visible symptom is intervenal chlorosis (between the leaf veins).

In animals zinc is found in every tissue. Like many micronutrients it is stored mainly in the liver.

Plants vary greatly in their ability to extract zinc from the soil. In Florida, USA, where much of the land is low in zinc, native weed species are particularly efficient extractors compared with most crop plants. Crops can be successfully grown by first planting lucerne (alfalfa), which is also a good extractor, and ploughing it in as a green manure.

Most food plants provide adequate zinc. Yeast, too, is a rich source. Most animals can tolerate high levels of zinc, though a few cases of zinc poisoning have been reported.

Copper

Copper is essential not only to plants, but also to animals in many bodily processes. Low levels of copper restrict the growth of plants; very low levels prevent any growth. There is little copper in peaty heathlands such as those of East Anglia, or in recently ploughed old grassland on chalk soils as in Wiltshire. Deficiency symptoms are different in each place, but in both small dressings of copper compounds give striking improvements in crop yields.

It has been known that copper is needed in the diet of animals since 1924 when experiments with rats showed that it has a role in haemoglobin formation. In some molluscs such as snails oxygen is carried by a blue substance, haemocyanin, in which copper plays a role like that of iron in haemoglobin.

Copper is a constituent of several enzymes including ascorbic acid oxidase, lactase, and tyrosinase. Another copper-containing enzyme, amine oxidase, catalyses the change of the amino acid lysine to desmosine. Desmosine provides cross-links in elastin, the main protein in the wall of the aorta and other blood vessels, to which they owe their elasticity and toughness. Copper has been found to influence the rate of growth of pigs and other animals.

Deficiency of copper leads to anaemia and bone disorders. Low copper levels of $2-4\,mg\,kg^{-1}$ dry matter in some Australian pastures have caused lambs to suffer from enzootic ataxia, a nervous disorder involving lack of co-ordination or even complete inability to stand up. A British term for a similar disorder of lambs is 'swayback'. In Somerset, soils locally called 'teart', found on the alkaline Lias Clay, have unusually high levels of molybdenum which give pasture a level of $20-100\,mg\,kg^{-1}$ dry matter compared with $3-5\,mg$ on normal soils. The excess of this element reduces the availability of copper, so that cattle suffer from 'scouring', a severe diarrhoea. Horses and pigs are not usually affected.

Copper accumulates in the liver. Too much can be poisonous. Sheep are particularly susceptible, so much so that dosing with copper compounds to balance a copper deficiency in pasture can easily kill them. Chronic copper poisoning has occurred naturally among sheep in parts of Australia where the copper content of pasture is high.

Boron

Boron is the only element thought to be necessary for plants but not for animals. The range between deficiency and excess is uniquely narrow and unfortunately varies from one plant to another. Boron is essential to cell division in the meristem tissues – the budlike growth point of a plant – and plays a role in the synthesis of proteins. Deficiency first halts growth, then kills the meristem tissues so that no further growth is possible.

Sugar beet grown on boron-deficient soil is routinely treated with boronated NPK fertilizer. Sometimes the same fertilizer has accidentally been applied to barley grown on the same farm. Barley has a much lower tolerance of boron, and the result has been a lowering of yield.

Molybdenum

Molybdenum's function in plants appears to be in the reduction of nitrates to amino acids, for lack of it causes nitrate ions to accumulate in the plants. Where nitrogen is supplied as ammonium rather than nitrate ions, much less molybdenum is needed. Legumes require molybdenum to allow their root nodules to fix atmospheric nitrogen. General signs of deficiency in plants are pale chlorotic or decayed necrotic areas in the leaves.

Molybdenum is also needed by animals. It is a constituent of the enzyme xanthine oxidase which plays a part in the metabolism of purines; these make up two of the four bases of the DNA code. In practice the live mass of young lambs has been increased by adding molybdenum (as molybdate) to a diet known to be low in the element. It is thought to have worked by stimulating the breakdown of cellulose by micro-organisms in the rumen, part of the animal's complex series of 'stomachs'.

Cobalt

Cobalt is not essential to the higher plants, but is certainly needed by at least some animals including sheep. A wasting disease known as 'pining' which affects sheep grazing on pasture on the granitic soils of the West of England has been traced to lack of cobalt in the soil and plants. Pining has also been observed in New Zealand, Australia, and the USA. The condition is due to lack of cobalt preventing micro-organisms in the rumen from synthesizing vitamin B_{12} (a compound containing cobalt). It is corrected by implanting a slow-release cobalt 'bullet' (90% cobalt(III) oxide) in the reticulum (another 'stomach'). In Man, a deficiency of vitamin B_{12} in the diet is a cause of pernicious anaemia.

Cobalt can be toxic but safety margins are high, for an excess is soon excreted. In cattle no harm is done unless intake exceeds 9–11 mg cobalt per 10 kg body mass daily.

Iodine

Apparently no plants need iodine except perhaps seaweeds. However, for animals including humans, small amounts of iodine are essential for the production of the hormone thyroxine by the thyroid gland in the neck. Adult animals contain less than 0.60 mg iodine per kg body mass. Thyroxine is a vital regulator of the metabolic rate.

The structure of thyroxine

Deficiency of iodine causes several conditions. Breeding animals give birth to hairless, weak, or stillborn young. Human diseases are goitre, a morbid swelling of the thyroid gland often to enormous size; and cretinism, a severe form of mental and physical retardation in children. Both were once common in mountainous districts of Europe and America, and also in Derbyshire, all places where the soil had a very low iodine content. They have now been completely overcome by adding iodine to the diet, for example by the sale of iodized salt. In fact, most foods contain traces of iodine, so there is danger only when diet is very restricted.

SUMMARY

At the end of this Topic you should:

   1 know some of the characteristic properties of the transition elements;

   2 be able to predict from data on electrode potentials which redox reactions are likely between transition elements and common reagents;

   3 know some of the reactions of iron ions and vanadium ions;

   4 be able to estimate concentrations of iron(II) ions, using potassium manganate(VII);

   5 be able to apply stability constants to predict the outcome of attempting to change the ligand in a complex, and understand the role of entropy change in complex formation;

   6 know a method of finding the stoicheiometry of a complex experimentally;

   7 know something of the role of transition elements as catalysts;

   8 be aware of the processes used in the iron and steel industry and of the application of Ellingham diagrams in extraction metallurgy;

   9 be aware of the role of some transition elements in micronutrients.

PROBLEMS

* Indicates that the *Book of data* is needed.

***1** This question concerns the following electrode potentials for manganese:

A $MnO_4^- + e^- \rightleftharpoons MnO_4^{2-}$ $\quad\quad\quad\quad E^\ominus = +0.56\,V$

B $MnO_4^{2-} + 2H_2O + 2e^- \rightleftharpoons MnO_2 + 4OH^-$ $\quad E^\ominus = +0.60\,V$

C $MnO_2 + 4H^+ + e^- \rightleftharpoons Mn^{3+} + 2H_2O$ $\quad\quad E^\ominus = +0.95\,V$

D $Mn^{3+} + e^- \rightleftharpoons Mn^{2+}$ $\quad\quad\quad\quad\quad E^\ominus = +1.15\,V$

E $MnO_4^- + 8H^+ + 5e^- \rightleftharpoons Mn^{2+} + 4H_2O$ $\quad E^\ominus = +1.51\,V$

a Referring to these equilibria by their letters (A to E), which of these electrode potentials are independent of pH?

b i Explain, using the electrode potentials, why a neutral solution of the green manganate(VI) ion changes, on standing, to a purple solution and a black precipitate.

ii Using the Nernst equation $E = E^\ominus + \dfrac{0.026}{z}\ln\dfrac{[MnO_4^{2-}]}{[OH^-]^4}$ explain why this reaction does not occur in very alkaline solutions.

iii Hence, suggest a method of obtaining a solution containing manganese in oxidation number +6.

iv What name is given to a reaction like this in which an element changes both upwards and downwards in oxidation number?

c What would you expect to happen to the red ion Mn^{3+} in neutral solution?

d Solutions of potassium manganate(VII) can be used to estimate the concentrations of other ions by titration.

i Which two of the following ions cannot, for reasons related to electrode potentials, be estimated using potassium manganate(VII)?

$Sn^{2+}, Zn^{2+}, Co^{2+}, Tl^+$

ii Why might it be very difficult to estimate the concentration of $Cr^{2+}(aq)$ accurately by titration with potassium manganate(VII)?

iii What colour change would mark the end-point if potassium manganate(VII) were used to titrate a solution containing the ion $V^{3+}(aq)$?

***2** 2.41 g of a certain salt containing iron(III) ions were dissolved in dilute sulphuric acid and zinc was added to reduce the $Fe^{3+}(aq)$ to $Fe^{2+}(aq)$. The resulting solution was made up to $100\,cm^3$ with dilute sulphuric acid and $10\,cm^3$ portions were titrated with 0.02M potassium manganate(VII). Exactly $10.0\,cm^3$ of 0.02M $KMnO_4$ were required.

a Show by using electrode potentials that zinc should reduce $Fe^{3+}(aq)$ to $Fe^{2+}(aq)$.

b Why was the solution made up to $100\,cm^3$ by using dilute sulphuric acid rather than water?

c What is the percentage of iron by mass in the original salt?

*3 When iron(III) ions, $Fe^{3+}(aq)$, are added to iodide ions, $I^-(aq)$. iodine is produced, but if sodium fluoride is first added, the reaction does not take place. The complexes FeF_6^{3-} and FeF_6^{4-} both exist.

a The atomic number of iron is 26. Give the ground-state electronic structure of the ion Fe^{3+}.

b Sketch the shape you would expect the complex FeF_6^{3-} to have.

c What is the oxidation number of iron in FeF_6^{4-}?

d Show by means of electrode potentials that iron(III) ions should oxidize iodide ions.

4 What factors would you consider affect the price of steel? Does a consideration of these factors help in any way to explain the decline of the British steel industry?

5 In the Ellingham diagram in figure 16.10 the slopes of most of the lines are positive (that is the lines go up to the right). The slope of the Ellingham diagram for carbon burning to carbon dioxide (figure 16.13a) is virtually zero (horizontal) and the slope of the diagram for carbon burning to carbon monoxide (figure 16.13b) is negative (down to the right).

Explain in terms of entropy changes why these differences of slope exist. Why is it that carbon is so widely used as a reducing agent rather than, say, sulphur or hydrogen?

6 Consider the following stability constants for cobalt(II) complexes which are octahedral:

Ligand	lg K (overall)	Colour
Cl⁻	No complex	–
H_2O	–	pink
NH_3	4.39	green
edta	16.3	pink

a What would you expect to see happening if ammonia solution were added to a solution of cobalt(II) chloride?

b What would you expect to see happening if the solution resulting from
   a were added to hydrochloric acid? Give a reason for your answer.
c What would you expect to see happening if edta solution were added
   to a solution of cobalt(II) chloride? Give a reason for your answer.
d Complexing by edta involves nitrogen atoms and oxygen atoms such as
   those in ammonia and water, so why is the stability constant of
   Co(edta) so much larger than that of $Co(NH_3)_6^{2+}$?

TOPIC 17
Synthesis: drugs, dyes, and polymers

This last Topic on organic chemistry is mainly concerned with the application of organic reactions to the industrial preparation of useful materials. Without the successful research of organic chemists working in industry, the Industrial Revolution would not have progressed much beyond what is possible with steel and cement. Although the successes are impressive, progress has not been uniform or easy. Previous ages were very casual about where and how they disposed of their waste materials and legal provisions over the control of waste disposal still vary greatly from one country to another. Furthermore, some products which once seemed safe have subsequently proved to be very serious hazards; examples include benzidine (biphenyl-4,4′-diamine), NH_2—⟨O⟩—⟨O⟩—NH_2, once used in the production of dyes but now known to be a cause of cancer, and the drug thalidomide which was found to have disastrous side effects.

The manufacturing process itself can be hazardous, as was seen when cyclohexane, being processed for the manufacture of nylon, escaped from piping at Flixborough in 1974 (figure 17.1) and the resulting explosion devastated the area and killed 28 workers at the plant; or again at Seveso in 1976 in Italy, when a reactor went out of control and poured its toxic contents into the atmosphere.

Society has to decide for each chemical application whether the benefits to be gained outweigh the risks; but chemists will always have a special responsibility to consider the risks as well as the benefits and to see that decisions are based on the best information that is available.

17.1
HIGH POLYMERS

There are many examples of polymers in living tissue, proteins and cellulose, for example. Some of these, such as wool, cotton, tortoiseshell, and horn, have been used as fibres and plastics since the dawn of history. The development of synthetic polymers, however, is more recent. Even so the first public display of articles made of plastics was seen in London over 100 years ago, at the International Exhibition in 1862. The material on show then was Parkesine, which was a modified natural polymer – nitrated cellulose, in fact. It was some years later in 1907 that Leo Baekeland was granted a patent for a completely synthetic polymer, Bakelite, made by condensing phenol with methanal.

Figure 17.1
Cyclohexane explosion at Flixborough in 1974.
Photograph, Keystone Press Agency Ltd.

Nowadays, polymers are found in an enormous diversity of applications. This is partly because they can be produced in a variety of forms – as thin films, as solid mouldings, as lightweight foams (figure 17.2), or as fibres – and partly because they can have such a wide range of properties – rigid, flexible, or elastic. The diversity of applications for polymers has led to their manufacture on a very large scale. From 3 million tonnes per year in 1950, worldwide production has grown almost continuously, reaching 8 million tonnes in 1960, 32 million tonnes in 1970, and 86 million tonnes in 1980. Manufacture on this scale means that production of polymers significantly exceeds the production of many metals.

The conditions that are required for the chemical synthesis of polymers were first described by the American chemist, W. H. Carothers, who worked for the Du Pont company from 1928 until his premature death in 1937. Carothers's success was based on an accurate understanding of precisely defined

Figure 17.2
Polystyrene foam blocks being used in Norway as the foundation for a road.
Photograph, Svein Alfheim/Norwegian Road Research Laboratory.

objectives – that the '... primary object was to synthesize giant molecules of
known structure by strictly rational methods' or, more informally, that 'It would
seem quite possible to beat Fischer's [molecular mass] record of 4200'. His
early papers clearly state the reasoning behind his work. Thus, polymerization
is described as '*any chemical combination of a number of similar molecules in
which they form a single molecule.* A polymer, then, is any compound that can
be formed by this process or degraded by the reverse process: rubber can be
formed by the reaction of isoprene with itself, and cellulose can be hydrolysed
to glucose.'

Furthermore he distinguished two types of polymerization reaction.

1 *Addition polymerization* produces a polymer with the same empirical
formula as the monomer from which it was made, for example, poly(chloroethene), first manufactured in the 1930s from chloroethene. The polymer is
generally known as PVC (from its non-systematic name, polyvinyl chloride).

$$nCH_2{=}CH \longrightarrow \left[CH_2{-}CH \right]_n$$
$$\quad\quad\; | \quad\quad\quad\quad\quad\; |$$
$$\quad\quad Cl \quad\quad\quad\quad\quad Cl$$

monomer   polymer

2 *Condensation polymerization* produces a polymer from monomer molecules by elimination of simple substances such as water or hydrogen chloride, for example, nylon 66, first described by Carothers in a 1935 patent.

$$nHO_2C{-}(CH_2)_4{-}CO_2H + nH_2N{-}(CH_2)_6{-}NH_2$$
$$\longrightarrow [OC{-}(CH_2)_4{-}CO{-}NH(CH_2)_6{-}NH]_n + (n-1)H_2O$$

EXPERIMENT 17.1a
The preparation of some synthetic polymers

A selection of experiments is included; you should attempt only one or two that are of interest, depending on the availability of time and materials.

C A U T I O N: These experiments should be carried out in a fume cupboard, as unpleasant fumes are evolved. Eye protection must be worn.

1 *Poly(methyl 2-methylpropenoate), Perspex*
The behaviour of catalysts used to bring about polymerization reactions is such that they are better described as 'initiators' for they participate in the reaction and are used up in the course of the reaction.

In a test-tube, mix $5\,cm^3$ of methyl 2-methylpropenoate (methyl methacrylate: T_b 100 °C) and 0.1 g di(dodecanoyl) peroxide. *TAKE CARE:* the liquid is highly flammable and has a harmful vapour. Stand a wooden splint in the reaction mixture as a means of testing the viscosity of the liquid.

$$\quad\quad\quad CH_3$$
$$\quad\quad\quad\; |$$
$$CH_2{=}CH$$
$$\quad\quad\; |$$
$$\quad\quad CO_2CH_3$$

methyl 2-methylpropenoate

Heat a water bath to boiling, remove the Bunsen burner, and place the test-tube in the water bath. Allow the test-tube to stand in the slowly cooling water bath and try stirring the mixture with the wooden splint every 5 minutes.

Write an equation for the polymerization of methyl 2-methylpropenoate. Was the reaction an addition or a condensation polymerization?

2 *Poly(propenamide)*

CAUTION: Propenamide is a skin irritant. You should wear eye protection and protective gloves for this experiment. Caution is also necessary because the reaction is exothermic.

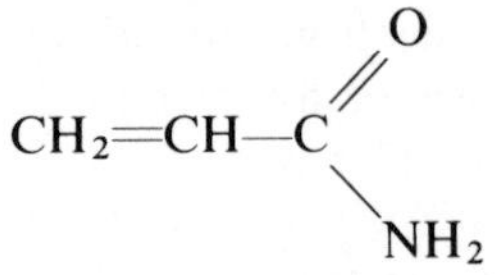

$$CH_2{=}CH{-}C{\overset{\displaystyle O}{\underset{\textstyle NH_2}{}}}$$

propenamide

Make a solution of 10 g of propenamide (acrylamide) in 50 cm^3 of water and warm in a 250-cm^3 beaker to *not more than* 85 °C. Pour the solution into a throw-away container (such as a tin can) on a heat-resistant mat and add about 0.1 g of potassium persulphate to initiate the polymerization.

Was the reaction an addition or a condensation polymerization?

Write an equation for the reaction. Use the table of bond energies in the *Book of data* to estimate the enthalpy change of polymerization.

3 *Polyester resin*

Wearing safety glasses and protective gloves, mix 3 g of benzene-1,2-dicarboxylic anhydride (phthalic anhydride) with 2 cm^3 of propane-1,2,3-triol (glycerol) in a test-tube. Measure out the propane-1,2,3-triol with a dropping pipette and allow the pipette plenty of time to drain.

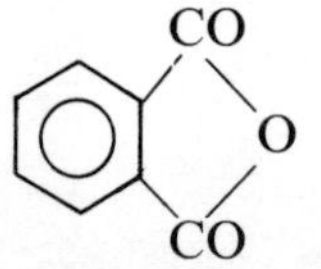

benzene-1,2-dicarboxylic anhydride

Heat to 160 °C and then more slowly to 250 °C in a fume cupboard. When the mixture ceases to bubble allow it to cool. Test the viscosity of your product.

Write an equation for a possible polymerization reaction. Was the reaction an addition or a condensation polymerization?

4 *Phenolic resin (a form of Bakelite)*

Dissolve 0.2 g of sodium hydroxide in 1 cm³ of 40% aqueous methanal solution and then add 3 g of phenol (*CAUTION:* it is corrosive). Heat the mixture in a fume cupboard until it becomes viscous. Allow the mixture to cool, when it should set to a brittle solid.

Was the reaction an addition or a condensation polymerization?

Suggest a structure for Bakelite based on methanal reacting with the 2, 4, and 6 positions of the phenol molecules in the polymer.

EXPERIMENT 17.1b
An examination of the physical properties of some polymers

CAUTION: Do not use polymer samples of unknown composition. Use only poly(ethene), nylon, Perspex, PVC, Bakelite, and rubber in these tests. Wear safety glasses.

1 *Density* Drop polymer granules into water and note which float and which sink. A drop of detergent will help to break up air bubbles.

Why does poly(ethene) vary in density?

2 *Melting point* In a fume cupboard, arrange single granules of different polymers on a heavy metal plate (or tin lid) and heat gently. Try to prevent the samples from catching fire. Note whether they melt, or soften and slowly become liquid, or char without becoming liquid.

What is the difference in structure between polymers that have a definite melting point and those that soften or char and therefore have no definite melting point?

3 *Elongation* Cut several strips, 2 cm × 10 cm, of poly(ethene) film. Strips of film should be cut with a sharp knife; otherwise, the film may tear rather than stretch when under stress. It is worth while cutting two strips, one parallel to the edge of the film and the other at right angles, as the film will have been partially stretched when being processed.

Pull the strips gently so as to stretch the film. The experiment can be repeated with nylon film if available.

Does poly(ethene) show the same behaviour as rubber when stretched?

Is the behaviour of nylon similar?
Is a greater stress needed to stretch nylon?
What do you think happens to the molecular chains in
poly(ethene) when it is stretched?

4 *Forming fibres* In a fume cupboard, melt some granules of poly(ethene)
on a heavy metal plate, removing the Bunsen burner when the polymer is
molten. Dip a metal spatula into the molten polymer and pull out a long strand.
Attempt to stretch the strand of polymer. The experiment can be repeated with
Perspex and nylon.

Which strands are brittle and which are elastic?
How much stress is needed to break the strands?
Which polymers are usually processed to form fibres?

Choosing the right fibre for a particular application

Synthetic polymers in one important group find application as fibres because
they can be drawn out into threads which have good strength and resistance
to wear and tear. The production of natural fibres needs land that could otherwise
be used for growing food. To produce 1 tonne of wool needs 160 000 m^2 of
grazing for sheep and 1 tonne of cotton needs 45 000 m^2 for cotton plants; but
50 000 tonnes of synthetic fibre nylon can be produced each year in a factory
occupying 45 000 m^2. Each batch of natural fibres will vary in properties, whereas
synthetic fibres can be made to a specified length and diameter. Nevertheless,
every fibre, natural or synthetic, has properties which make it particularly
suitable for particular applications.

The manufacturer of a fabric must evaluate the characteristics of both
natural and synthetic fibres before a proper choice can be made. If the wrong
choice is made, the fabric may shrink when washed or fail in some other way.
To meet as many applications as possible, many grades of each basic polymer
are available commercially and all have been rigorously tested for chemical and
mechanical stability. Thus, 'accelerated wear' tests are used to gauge the effect
of longterm use; for example, exposure to ozone is used to estimate more quickly
the effect of atmospheric oxygen, and ultra-violet irradiation is used to estimate
the effect of sunlight.

Consider four very different but important applications for fibres:

1 boat sails
2 underwear
3 rope
4 tights

Properties of fibres

	Density /g cm^{-3}	Dry strength /g dtex^{-1}*	Wet strength /g dtex^{-1}*	Stretch before breaking/%	Force needed to stretch /'g dtex^{-1}*	Elastic recovery /%**	Moisture absorbance /%	Handling and wear
A	1.31	1.1	0.9	30–45	29	100	16	Warm 'feel', can shrink a lot, moderate wear, can irritate skin.
B	1.55	3.2	3.4	5–10	53	70	9	Firm, comfortable, tough, can shrink a little, may crease.
C	1.34	3.6	2.6	20–25	63	90	11	Crisp attractive 'feel', very poor resistance to wear.
D	1.52	2.4	1.3	17–25	64	60	13	Soft, very good resistance to wear, crease-resistant.
E	1.30	1.2	0.8	25–28	36	85	3	Crisp sharp 'feel', gives permanent creases, crease-resistant.
F	1.14	4.5	3.8	23–42	20	100	5	Soft, hard-wearing, crease-resistant, very tough.
G	1.38	4.1	4.1	25–35	60	80	0.5	Hard-wearing, gives permanent creases, very tough.
H	1.16	2.4	2.0	25–45	40	90	2	Soft, warm 'feel', non-shrink, tough, crease-resistant.

* Grams per decitex: a measure of force per unit cross section; a decitex is the mass in grams of 10 km of yarn.

** the extent to which the fibre returns to its original shape after stretching.

Table 17.1
Selected properties of some fibres.

Discuss in a small group which of the properties of fibres listed in table 17.1 are the important properties for each of the four applications. Make a note of your conclusions and then choose which fibre, **A** to **H**, seems most suitable for each application. The list **A** to **H** includes both natural and synthetic fibres. When you have finished your discussion, compare your conclusions with those of other groups. Your teacher will identify the fibres **A** to **H**. Have you chosen the fibres commonly used for the four applications?

The nature of polymers

Some polymers, when they are put under stress, will break in a brittle fashion. In contrast, other polymers can be stretched to several times their original length and will recover completely and rapidly when the stress is removed – they are elastic. A third group of polymers will recover when the stress is removed if they are only stretched a little, but they will not recover their original size if stretched to even a moderate extent.

In the first group, covalent chemical bonds link together the molecular chains of the polymer. A large number of such 'cross links' will effectively prohibit chain movement so the polymer is brittle; nor will it soften and melt, even at high temperature. Polymers which behave in this way, with the cross links formed by the action of heat, are known as *thermosetting* (figure 17.3).

Figure 17.3
Possible shapes of the molecules of thermosetting polymers.

Polymers in the second group are referred to as *elastomers*. When a sample is stretched, there is some straightening of the long chain molecules of the polymer. The favoured state for a flexible long chain molecule is for it to be folded and twisted so that it is only a fraction of its fully extended length. The applied stress provides energy for partial chain straightening to occur; the flexibility of the chain arises from a rotation about carbon–carbon bonds. If the stress is removed, the molecular chains return to their original folded and twisted pattern and the sample recovers its original size. Complete recovery from considerable stretching will occur because a small number of cross links will restrain the molecular chains from slipping past each other (figure 17.4).

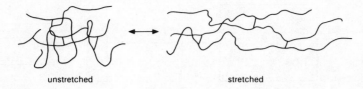

unstretched   stretched

Figure 17.4
Arrangements of the long-chain molecules of an elastomer.

In the third group of polymers the neighbouring chains are not held together by cross links, but in other ways such as hydrogen bonding, as between the —CONH— groups in nylons, or the forces of attraction associated with crystalline regions. Many polymers are able to form crystalline regions (figure 17.5), in particular those polymers of regular structure without bulky side groups or side chains.

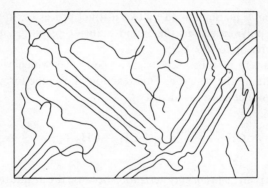

Figure 17.5
Crystalline packing in a plastic polymer.

The crystalline regions are separated by amorphous regions and any one polymer chain may be locked into several regions of crystallinity, thus tending to keep these regions together and keeping the polymer in the solid state. When the material is heated the molecular chains become increasingly flexible and the material becomes progressively softer as the amorphous regions are disrupted and the crystalline regions 'melt'. Materials which soften on heating are known as *thermoplastic* materials. When a sample of a thermoplastic such as poly(ethene) is stretched, the material thins at one place and a 'neck' travels along the sample. You should have been able to observe this phenomenon of 'cold drawing' when you tried experiment 17.1b, part **3**. What happens is that the different crystalline regions which are normally aligned in a random fashion become aligned along the direction of stretch (figure 17.6).

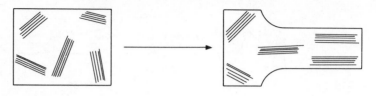

Figure 17.6
The orientation of crystalline regions in a polymer by 'cold drawing'.

Such drawn materials exhibit a considerable increase in tensile strength in the direction of draw, at the expense of strength at right angles to this direction. For fibre applications, such strength is an essential property and strength across the fibre is of far less importance.

Commercial manufacture of some synthetic polymers

Poly(ethene) is manufactured by two distinct processes. In the original process, ethene is polymerized in the liquid state at pressures up to 3000 atm and temperatures up to 300 °C.

A variety of substances such as peroxides which yield free radicals are used to initiate the reaction. This reaction allows the formation of side chains and the product has a low density.

$$n CH_2{=}CH_2 \xrightarrow{\text{free radicals}} \left[CH_2{-}CH_2 \right]_n$$

This is an *addition polymerization* because the unsaturated monomer reacts to give the polymer as the only product. In the 1950s Karl Ziegler developed a process in which ethene could be polymerized at a relatively low pressure and temperature: less than 50 atm and below 100 °C. In the Ziegler process the polymerization takes place on the surface of catalyst particles to give straight unbranched chains and the product has a higher density. The catalysts used are organo-metallic complexes, often based on chromium or vanadium.

The variation in density of poly(ethene) is related to a variation in the degree of crystallinity: high density poly(ethene) has a density up to $0.96 \, \text{g cm}^{-3}$ and is about 95% crystalline, while low density poly(ethene) has a density in the range 0.91 to $0.93 \, \text{g cm}^{-3}$ and is about 50% crystalline. The ability to crystallize is related to the regular shape of the molecular chain in poly(ethene) (figure 17.7).

Poly(chloroethene), PVC, is usually prepared as a suspension in water. Typically, equal quantities of water and chloroethene and a small quantity of a surface active agent are stirred together in an autoclave to produce a suspension of chloroethene in water. An initiator, such as potassium peroxodisulphate(VI),

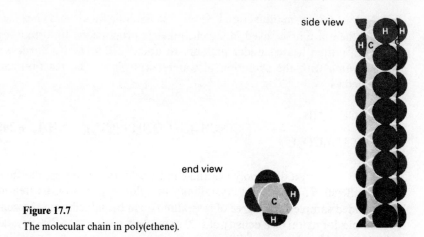

side view

end view

Figure 17.7
The molecular chain in poly(ethene).

is added and the temperature raised to about 60 °C. The droplets polymerize to form solid particles of poly(chloroethene) which can be recovered by filtration. The water acts as the diluent and makes it possible to remove reaction heat.

$$n\text{CH}_2\!\!=\!\!\text{CHCl} \xrightarrow{\text{free radicals}} \text{+CH}_2\!\!-\!\!\text{CHCl+}_n$$

The pure polymer is a rigid material, rather unstable to heat, and cannot be readily processed. Because of this, a complex variety of additives are usually present in commercial PVC. By the addition of suitable compounds, known as plasticizers, interaction between the polymer chains is reduced and a flexible form of PVC can be produced. The development of a form of PVC suitable for the manufacture of gramophone records is described in the Background reading on page 276.

Nylon: there are a number of different *nylon*-type polymers but they all have an amide linkage and they differ in having different numbers of carbon atoms between each (—NH—) group. Only nylon 6 and nylon 66 are manufactured in quantity.

Nylon 6 is manufactured by the polymerization of 6-hexanolactam (caprolactam) at about 250 °C with a little water

$$n\begin{array}{c} \text{CH}_2 \\ \text{CH}_2 \quad \text{CH}_2 \\ | \qquad | \\ \text{CH}_2 \quad \text{CH}_2 \\ \text{CONH} \end{array} \longrightarrow \text{+CO—(CH}_2)_5\text{—NH+}_n$$

Nylon 66 is manufactured from hexanedioic acid and hexane-1,6-diamine; these are first mixed in stoicheiometric proportions to form a 'nylon salt'. This is then heated under pressure to about 270 °C and a *condensation polymerization* with the evolution of water takes place. The reaction can be represented as

$$nNH_2(CH_2)_6NH_2$$
$$+ \longrightarrow \text{\textlbrackdbl}CO(CH_2)_4\text{---}CONH\text{---}(CH_2)_6\text{---}NH\text{\textrbrackdbl}_n + 2nH_2O$$
$$nHO_2C(CH_2)_4CO_2H$$

Nylon is a crystalline polymer and can be cold drawn in a similar manner to poly(ethene). The degree of crystallinity in nylon depends on its treatment, For moulded samples the degree of crystallinity can be judged from the density: for example, for nylon 66 a density of 1.20 g cm^{-3} corresponds to 75% crystalline and 1.12 g cm^{-3} to 20% crystalline.

The names of polymers

The names of polymers are more confusing than those of most other compounds. Each polymer has a systematic chemical name, often a trivial (non-systematic) chemical name, and usually a variety of trade names. Some of the more common polymers are listed in table 17.2, together with some of their alternative names.

Monomer	Polymer	Systematic chemical Name	Other names (trade names begin with a capital letter)
$CH_2{=}CH_2$	$\text{\textlbrackdbl}CH_2\text{---}CH_2\text{\textrbrackdbl}_n$	poly(ethene)	polyethylene polythene Alkathene
$\underset{\overset{\|}{CH{=}CH_2}}{CH_3}$	$\left[\underset{\overset{\|}{CH\text{---}CH_2}}{CH_3}\right]_n$	poly(propene)	polypropylene Propathene
$\underset{\overset{\|}{CH{=}CH_2}}{Cl}$	$\left[\underset{\overset{\|}{CH\text{---}CH_2}}{Cl}\right]_n$	poly(chloroethene)	polyvinyl chloride PVC, Corvic
$\underset{\overset{\|}{CH{=}CH_2}}{CN}$	$\left[\underset{\overset{\|}{CH\text{---}CH_2}}{CN}\right]_n$	poly(propenenitrile)	polyacrylonitrile acrylic fibre Courtelle
CH=CH₂ (phenyl)	[CH—CH₂]ₙ (phenyl)	poly(phenylethene)	polystyrene

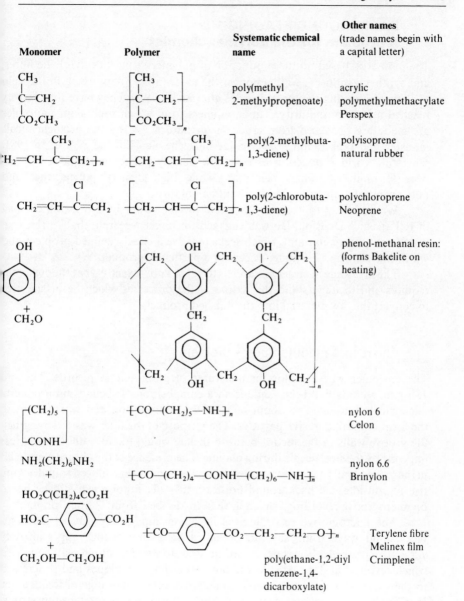

Table 17.2

BACKGROUND READING 1
A problem for the polymer chemist

The industries based on the production of polymers – plastics, synthetic fibres, and synthetic rubber – were amongst the most rapidly developing in the field of chemistry during the three decades after about 1950. They have now largely reached a state of maturity. The monomers from which such giant molecules are made are produced from crude oil or natural gas by the petrochemicals industry and rank amongst the major bulk chemicals of the World. In 1981, the total World production of both low density poly(ethene) and PVC was over ten million tonnes per year, while high density poly(ethene) and poly(propene) each amounted to about half of this.

We are going to examine an actual piece of work by a research chemist of ICI Plastics Division. He was engaged on target research, that is, research which has a particular aim, which seeks to solve a clearly defined problem and which must satisfy a customer's need for a particular product.

The first stage in target research is to get to know exactly what the customer requires and to understand the nature of the processes which he or she uses to convert the raw material into the finished product.

The record manufacturer's problem

The customer was a major manufacturer of gramophone records. The old 78 r.p.m. records had been made from a composition of shellac and a mineral filler which amounted to about 60% of the composition and was present in the form of hard, abrasive particles. The purpose of the filler was to strengthen the groove walls in the record because shellac would fracture under the stress imposed by the steel needle during playing. The presence of the abrasive particles in the groove walls had a most undesirable effect on the reproduced sound, giving rise to considerable background noise or 'hiss'. This noise can still be heard on microgroove recordings which have been 'dubbed' from 78 r.p.m. originals.

The microgroove recording, as its name implies, reproduces the complex sound wave from, say, an orchestra, in a tiny groove approximately 7.5×10^{-3} cm wide. It was developed in order to provide a long playing record which revolved at $33\frac{1}{3}$ r.p.m. At the time, recording techniques had progressed to the stage where it was possible to record a very wide range of frequencies, from the bass notes of the organ to the high-pitched sounds produced by strings and percussion instruments such as the cymbal and triangle (from about 30 to 2×10^4 Hz). In order to get the maximum benefit from this, as a realistic sound from the gramophone, 'hiss' must be eliminated from the record. Consequently, filled shellac is ruled out as a possible material. It must be replaced by one of smooth texture which is strong enough to withstand the forces exerted by the

Figure 17.8
The automatic pressing of a record.
Photograph, EMI Records UK.

needle on the groove walls. This is one of the fundamental characteristics of the polymer required by the customer.

This situation incidentally provides a good example of the manner in which the chemist must respond to developments in other sciences. Society cannot benefit from the skill of the recording engineer until the chemist has produced a material on which an extended frequency range can be recorded.

When a record is made, the recording done in the studio is converted, by processes which need not concern us here, into a metal 'master'. Masters corresponding to the two sides of the record are placed in a press, one forming the upper and the second forming the lower plate. Figure 17.8 illustrates such a press. The operator places a disc of material, known as the 'biscuit' or 'patty' in the centre of the press, which is heated to 150 °C. When the press closes, the biscuit material flows across the master surfaces and fills the tiny grooves of the recording. This sounds very simple, but consider for a moment what is implied. The biscuit material must flow very freely in the press and must fill completely and accurately the complex twists and turns of the modulations in the grooves of the master. Figure 17.9 is a greatly enlarged photograph of grooves

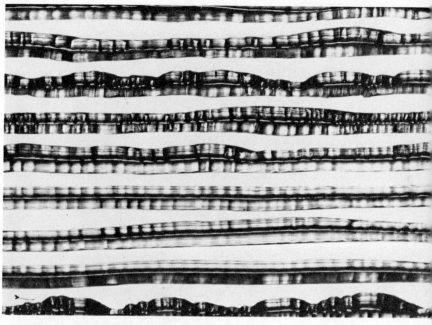

Figure 17.9
Stereophonic record grooves, magnified × 140.
Photograph, EMI Records UK.

in a stereophonic record; you can see for yourself the modulations in the walls
which correspond to the original sounds produced in the studio. When the record
is removed from the press, the free flowing melt must have hardened to produce
a record with strong groove walls which will stand up to repeated playings
without showing signs of wear. You, as the ultimate customer, will certainly
appreciate the latter point.

Even now, however, the record manufacturer was not satisfied – he wanted
a virtually unbreakable record.

The requirements for the record material which we have just outlined add
up to a formidable specification. Let us see how this specification was successfully
met and how a polymer was designed which is still in use today.

The first course of action was to see if any polymer which ICI was
producing at the time would be suitable. Cost is always an important aspect
of most pieces of target research, especially where several firms may be competing
for the same contract. The stimulus provided by the search for low cost materials
can often lead to the development of new synthetic routes.

Table 17.3 lists a number of polymers which were considered as possible
candidates for the record material, together with comments on their suitability.

Polymer	Suitability for record material
poly(ethene)	too soft and flexible
poly(methyl 2-methylpropenoate)	rigid but expensive
poly(phenylethene)	too brittle and not manufactured by ICI
poly(chloroethene), PVC	physical properties in the right range and cheap to produce

Table 17.3

The obvious choice was PVC which was already in production and could be manufactured at a competitive price. The company had to consider whether any existing grades of PVC which it was producing would be suitable, but none had sufficiently good flow properties to meet the stringent requirements of the press. The problem went to the research chemist and the piece of target research was under way.

The problem was defined like this: 'We need a polymer which, in the melted form, will flow easily so as to fill accurately the tiny grooves in the master yet which, on cooling, will be neither too brittle nor too soft.'

Solving the problem: stage 1 – control of molecular mass

What do we know about the factors which influence flow in molten polymers, and how can we satisfactorily express 'flow' in the laboratory?

We have a starting point for an attempted solution to the problem: control of viscosity by control of molecular mass. The viscosity of a liquid is a measure of its resistance to flow and it is known that the viscosity of a polymer is related to its molecular mass. The lower the relative molecular mass, the lower the viscosity will be. To understand how the chemist can control the molecular mass of the polymer he is synthesizing we shall need to look at the chemistry of the polymerization reaction itself. PVC is made by addition polymerization, that is to say, the polymer molecule grows by addition of monomer units to the chain. The mechanism of the reaction can be divided into three stages; initiation, propagation, and termination.

Initiation   Here, free radicals are formed from stable monomers by heat, by electromagnetic radiation, or by the employment of an initiator. The first two may be represented by

$$CH_2{=}CHCl \xrightarrow{\text{heat}} \cdot CH_2{-}CHCl\cdot$$

$$CH_2{=}CHCl \xrightarrow{\text{radiation}} \cdot CH_2{-}CHCl\cdot$$

The action of an initiator depends on the low temperature formation
free radicals from the initiator itself. Let us have a look at a typical or
di(benzoyl) peroxide $(C_6H_5CO_2)_2$. This molecule decomposes at relatively lc
temperatures to give $C_6H_5\cdot$ free radicals, thus

$$(C_6H_5CO_2)_2 \xrightarrow{\text{heat}} 2C_6H_5\cdot + 2CO_2$$

In the presence of a monomer, this is followed by

$$C_6H_5\cdot + CH_2{=}CHCl \longrightarrow C_6H_5CH_2{-}CHCl\cdot$$

so that a new free radical is formed. It is this radical which then enters in
the second stage of the reaction mechanism.

The initiator used in the manufacture of PVC is a closely guarded secr
and we cannot reveal it here.

To simplify the writing of equations in our discussion of the polymerizatic
mechanism, let us represent the free radical formed from the initiator by F
The reaction with the monomer is then

$$R\cdot + CH_2{-}CHCl \longrightarrow RCH_2{-}CHCl\cdot$$

Propagation   The free radicals formed from the monomer during initiatic
proceed to react with more monomers to build up the chain:

$$RCH_2{-}CHCl\cdot + CH_2{=}CHCl \longrightarrow RCH_2CHClCH_2{-}CHCl\cdot$$

or, in general

$$R(CH_2CHCl)_nCH_2{-}CHCl\cdot + CH_2{=}CHCl$$
$$\longrightarrow R(CH_2CHCl)_{n+1}CH_2{-}CHCl\cdot$$

This process is a typical chain reaction in which each step produces a fr
radical capable of further reaction with the monomer. You will notice that tl
sequence of addition reactions in the propagation stage is building up a linea
or 'head to tail' polymer.

Termination   The polymerization stops when the free radicals or growir
chains, as we might call them now, undergo reactions which produce stab
polymer molecules. There are a number of ways in which this can happen. I
the polymerization of chloroethene the most important termination process
one in which the odd electron is transferred from a growing chain to a monome
that is to say, the polymer reacts with the monomer in a different way fro

the normal propagation reaction. This may be represented as follows:

$$P\cdot + M \longrightarrow P + M\cdot$$

You will notice that this reaction, which is called 'chain transfer', gives a stable polymer molecule but at the same time another active monomer molecule is formed which can start a new chain. In this way we can produce more polymer molecules than would arise from the number of initiator molecules put into the system and polymer molecular mass is therefore not dependent on initiator concentration. Now we are in a position to consider what determines the chain length and relative molecular mass of the polymer formed in the reaction. A little thought will show that it must be the relative efficiencies of the propagation and termination steps in the mechanism. We can express these efficiencies in terms of the rates of the propagation and termination reactions and arrive at the conclusion that, for long chains,

$$\frac{\text{rate of propagation}}{\text{rate of termination}} \gg 1$$

A detailed treatment of the kinetics of a polymerization reaction is too advanced to give here, but it can be shown that the rate of termination is increased, relative to that of propagation, by raising the temperature at which the reaction proceeds. Thus, we control the relative molecular mass of the polymer by controlling accurately the temperature at which polymerization occurs. The effect of temperature on the relative molecular mass of PVC is shown in figure 17.10a and in figure 17.10b we see the effect of temperature of polymerization on the viscosity of PVC.

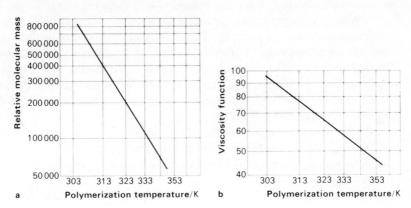

Figure 17.10

a The molecular mass of poly(chloroethene) plotted against polymerization temperature.

b The viscosity function of poly(chloroethene) plotted against polymerization temperature.

We cannot solve the problem of producing PVC with low viscosity simpl
by reducing the relative molecular mass of the polymer. A polymer of lov
relative molecular mass and the correct flow properties was found to be too
brittle to meet the requirements of the customer. The record manufacture
would have been quite satisfied if his records could be folded in two, like sheet
of paper, without shattering, which, in fact, was the test which was applied t
the PVC samples.

What else can be done to increase the flow properties without incurrin
brittleness?

Solving the problem: stage 2 – plasticizers

There are two lines of attack: firstly we can attempt to use a plasticizer an
then, if this proves unsatisfactory, we can modify the structure of the polyme
molecule itself.

The properties of polymers can be explained by a knowledge of molecula
structure. To understand how plasticizers work, we shall have to consider th
relationship between the properties of a polymer and the structure of the polyme
molecules themselves.

We have pictured polymer molecules as being made up of very long chain
of simple chemical units, usually 1000 or more of them in an actual sample. I
we could take a photograph of the polymer, the picture might look like a tangle
ball of many long separate pieces of string. One of the important factors tha
influences the properties of polymer materials is the existence of intermolecula
forces of attraction between long-chain molecules. These forces are much weake
than the chemical ones which bind atoms together to form molecules, and the
are really effective only when the chains are very close together.

When a polymer such as poly(ethene) is heated, the molecules gain energy
and move about more vigorously. The chains therefore become further separate
and the intermolecular attractive forces are weakened. As a result, th
poly(ethene) becomes softer and more flexible, and on further heating turns int
a viscous liquid. When the poly(ethene) is cooled, the molecular chains com
closer together again and attract one another again. The material solidifies an
becomes stronger and stiffer.

Polymers such as poly(ethene), PVC, and nylon can be repeatedly softene
on heating and hardened on cooling, provided the polymer is not heated s
strongly that it decomposes. Plastics which have this property are known a
thermoplastics.

The arrangement of the molecules in poly(ethene) is not always completely
disordered and random. In some samples, several chains or parts of chains ten
to run parallel to one another for all or part of their length, giving a mor
orderly or closely packed assembly. These regions of orderly alignment of th

chains are described as being *crystalline*, in contrast to the regions of disorder which are called *amorphous*. In the crystalline regions the attractive forces between the molecules are greater because the chains lie closer together over a longer distance. Therefore the molecules cannot move as freely over one another and the mechanical strength of the material is greater.

Let us consider the molecular structure of PVC, and the reason for the toughness of this polymer. In order to make PVC we polymerize a monomer, chloroethene, which, as it polymerizes, gives rise to one chiral centre in the molecule of the polymer for every molecule of monomer which joins. During the propagation steps, these chiral centres form on the growing chain to give three possible arrangements in space which differ in the manner in which the chlorine atoms are orientated with respect to other atoms in the chain.

Atactic PVC   The chlorine atoms are orientated in a random manner on both sides of the chain.

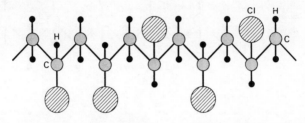

Syndiotactic PVC The chlorine atoms are arranged alternately on opposite sides of the chain.

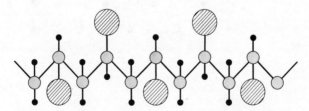

Isotactic PVC   The chlorine atoms are all on the same side of the chain.

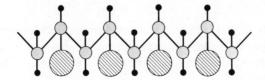

If models of these chains are made, in which the tetrahedral arrangement of the carbon atom bonds is adhered to, the difference between the three forms of PVC will become apparent.

When PVC is manufactured, it is mainly the atactic form with some syndio tactic form present, as well as some chain branching. The molecules in such specimen do not readily lie together to form crystalline regions and the polyme is said to be amorphous. Nevertheless, it is tough and rigid like high densi poly(ethene). Although there is no crystallinity in PVC, the intermolecular force between the chains are quite strong. Figure 17.11 illustrates the hydroge bonding in PVC. The C—Cl bond in PVC is polarized and the slightly negativ chlorine atoms attract hydrogen atoms on an adjacent chain to form hydroge bonds. Such bonds are weaker than normal chemical bonds (about 20 kJ mol⁻ but are stronger than the intermolecular forces between poly(ethene) chain These intermolecular forces are responsible for the toughness of PVC.

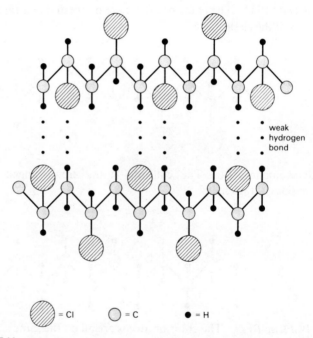

Figure 17.11

The problem of increasing the ease of flow in the melt of PVC in the productio of microgroove records has not been solved simply by reducing the relativ molecular mass. We must now try to reduce the intermolecular forces betwee the polymer chains.

A plasticizer is a compound having molecules which can get in betwee the polymer chains and separate them. Such molecules solvate charged par of the polymer molecule, for example the Cl atoms, in a manner analogous t that in which water molecules solvate an ion in solution. Some examples of plasticizers are:

tritolyl phosphate

$H_{17}C_8$—O—C—$(CH_2)_4$—C—O—C_8H_{17} dioctyl hexanedioate

di-2-ethylhexyl benzene-1,2-dicarboxylate

You may notice some resemblance between these molecules and those you are familiar with as detergents. In both cases the important feature of the structure is the presence of a polar and a non-polar group in the molecule.

When plasticizers were used to modify the PVC molecule there were two major drawbacks. The plasticizer reduces the attraction between the chains in the melted polymer in the record press, but i⁺ also reduces the attraction when the polymer solidifies; this means that the record may be too soft. Also, some plasticizer may separate out from the polymer in the press and cause staining of the finished record. Both these drawbacks occurred when plasticizers were tried, so that there was only one course of action left open to the company if it was to satisfy its customer; it had to modify the structure of the PVC molecule itself.

Solving the problem: stage 3 – co-polymers

Plasticizers were too effective for the purpose; they prised the PVC chains too far apart and gave a solid which was too soft. A less drastic way of separating the chains might be to build in a bulky side group at fairly regular intervals along the chain. This can be done by *co-polymerizing* chloroethene with another monomer (called the *co-monomer*) which contains a bulky group, and varying the amount of co-monomer until the right properties in the melt and solid forms are obtained.

What co-monomers are suitable? The choice of co-monomers with chloro-ethene is restricted because many readily available co-monomers are much more reactive than chloroethene. Take phenylethene, for example; this monomer, if co-polymerized with chloroethene, might introduce the bulky arene ring at intervals along the chain. However, phenylethene is so reactive that it polymerizes with itself rather than with chloroethene. As a result, we should obtain, not a co-polymer of chloroethene and phenylethene but a mixture of poly(phenylethene) and PVC!

Because of such factors, coupled with cost and availability, the choice of co-monomer narrowed down to either 1,1-dichloroethene, $CH_2=CCl_2$, or

$$O$$
$$\|$$

ethenyl ethanoate, $CH_2=CH—O—C—CH_3$. The choice was influenced by the fact that too marked a softening effect must not be produced; it was known that the effect of the co-monomer on the physical properties of PVC was directly related to the molecular volume of the co-monomer, that is, to its physical size. This latter fact focused attention on small co-monomer molecules. The final choice was made in favour of ethenyl ethanoate because it was appreciably cheaper than 1,1-dichloroethene. A series of co-polymers of chloro-ethene and ethenyl ethanoate were synthesized, and the ethenyl ethanoate content was varied and flow properties were tested until the most suitable composition was found. This proved to be a co-polymer, containing 16 per cent of ethenyl ethanoate, which had the desired flow characteristics and which was also tough and rigid enough in the solid state.

Large scale production

It was found that records made from the co-polymer wore out quickly in use. The reason was to be found in the process by which the polymer was manu-factured on the large scale.

Problems which are successfully solved in the laboratory do not always work out so well when we deal with tonnes rather than grams of material. There are many reasons why this should be so. For example, you will appreciate that it is easier to heat up and maintain at constant temperature a beakerful of liquid than a vessel containing hundreds of dm^3. Similarly, it is much simpler to add a gram of material to a liquid in a beaker and to stir thoroughly with a spatula than to accomplish the efficient mixing of tonnes of chemicals. The problems of heat and mass flow in large scale chemical apparatus are formidable but they must be solved if society is to be supplied with the vast quantities of chemical substances which it requires.

The design of plant is the province of engineers of many kinds; chemical, mechanical, electrical and electronic engineers work together in such a task.

Polymerization reactions are generally exothermic, the heat of poly-merization of chloroethene, for example, being $-960\,\text{kJ}\,\text{mol}^{-1}$. If the tem-perature of the reaction is to be controlled, this heat must be removed from the system and this is done by allowing the polymerization to take place in water. Chloroethene is a gas but it can easily be liquefied under pressure and introduced into a large volume of water where it is dispersed into droplets. If surface active agents such as a soap are added to the water in sufficient quantity the soap molecules form clusters called *micelles*, in which the non-polar tails point to the centre and the polar heads form the surface (see figure 17.12).

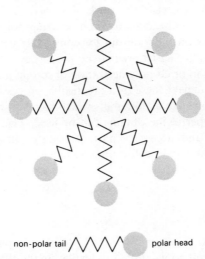

non-polar tail /\\/\\/\\ polar head

Figure 17.12
A cross section of a spherical micelle.

The monomer droplets find their way into the micelles and polymerize there, but the soap residues must be removed from the polymer particles at the end of the reaction. This process is called *emulsion polymerization* and was the method tried initially for the manufacture of the co-polymer. The stumbling-block was that some soap residues were present in the material that had been supplied to the manufacturer for trials in the early stages and it was these residues that gave rise to increased wear.

The problem was solved by a different production method. Soap was dispensed with and polymerization was allowed to take place in larger monomer droplets, initially stabilized by a different type of surface-active agent, which subsequently coalesced. This gave larger polymer particles than those obtained in the emulsion method. This process is known as *suspension polymerization* and it is now used to make the co-polymers needed in the record industry.

You have seen how, in this project, the company was unable to meet the requirements for records from its existing products; nevertheless, it was able to tailor the polymer molecule to give the desired combination of flow behaviour and physical properties by applying a fundamental knowledge of the relationship between polymer properties and molecular structure; lastly, it was able to improve the product by selecting the best production process. This is by no means the end of the story, for further development of these co-polymers is already in progress and they have also been found useful for other applications, for example in floor tiles.

The type of research which we have discussed, with a definite target in mind, is only one aspect of the work of a polymer chemist. Research is going on all the time to find out more about existing polymers and in addition the chemist is always on the look-out for completely new ones. The chemist may approach this by trying to predict the properties he (or she) should get from the polymerization or co-polymerization of certain monomers, and then seeing whether he can achieve these by a practicable process; alternatively, he may try and polymerize available monomers which have not been studied before to see whether the properties he obtains can be used. These various aspects form a complementary and necessary part of research in industry today, and the industrial chemist may be called upon to tackle any one of them.

17.2
DYES

Dyes are used to colour many things – foodstuffs, plastics, paper – but by far the most important application is to textiles. In the 1950s the textiles in common use were wool and cotton, but since then synthetic fibres have become readily available. Nylon and polyester are now widely used in textiles and the dyestuffs chemist has had to develop whole new ranges of dyes that can be used successfully with these synthetic fibres, and also with textiles of mixed composition. A large chemical company is likely to synthesize thousands of new compounds every year in the search for better or cheaper dyes. Without continuing research a chemical company could find that its products soon became uncompetitive.

This section is mainly concerned with one type of dye, the azo-dyes, but the task of dyeing textiles of different compositions is also briefly considered.

Azo-dyes

Azo-dyes are made from arylamines. You will recall from Topic 13 that amines are bases, and react with acids to form salts. However, one acid, nitrous acid, HNO_2, behaves differently.

Nitrous acid reacts with alkylamines to give a variety of products, including alkenes and alcohols; the proportion of the various products depends upon the alkyl group present in the amine. Because of the variety of products this reaction is not particularly useful as a method of synthesis.

Arylamines however are converted by nitrous acid to diazonium compounds, in a reaction known as *diazotization*. This reaction is an important distinction between aryl and alkyl amines.

An example of diazotization is the reaction of phenylamine with hydrochloric and nitrous acids to give benzenediazonium chloride:

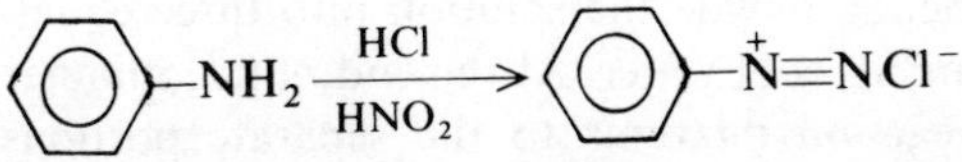

Diazonium compounds are noteworthy as important intermediates, that is, they are made for conversion into one of a number of other classes of compounds. Important in the present context is their reaction with arylamines, and with phenols, to give azo-dyes in a reaction known as *coupling*.

An example of coupling is the reaction between benzenediazonium chloride and naphthalen-2-ol (2-naphthol).

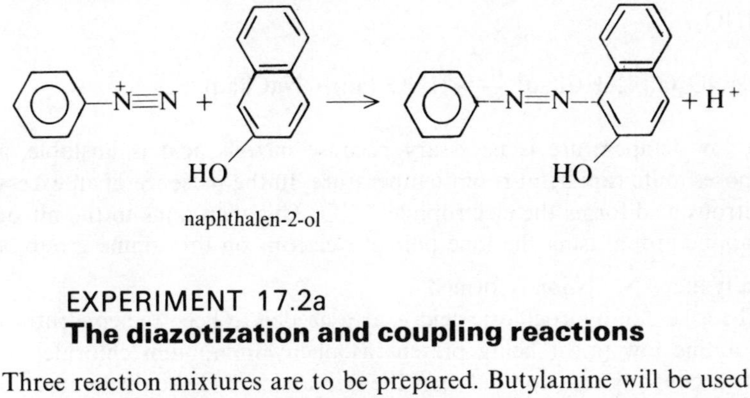

naphthalen-2-ol

EXPERIMENT 17.2a
The diazotization and coupling reactions

Three reaction mixtures are to be prepared. Butylamine will be used in one as an example of an alkylamine; phenylamine in the second as an example of an arylamine; the third mixture is a blank for comparison. Naphthalen-2-ol will be used as the phenol in the coupling reaction.

Procedure

1 *The diazotization reaction* Eye protection should be worn. To about 25 cm^3 of a crushed ice–water mixture at 5–10 °C in a 250-cm^3 beaker, add a solution of 0.5 cm^3 of butylamine dissolved in 10 cm^3 of 2M hydrochloric acid.

Prepare a similar solution using $0.5\,cm^3$ of phenylamine (aniline) in place of the butylamine.

Prepare a 'blank' solution using the inorganic reagents only and omitting any amine.

Also prepare a solution of $1.5\,g$ of sodium nitrite in $30\,cm^3$ of water and add $10\,cm^3$ to each of the three solutions you have just prepared.

Allow the three reaction mixtures to stand for five minutes (but no longer).

2 *The coupling reaction* In the meantime, prepare a solution of $3\,g$ of naphthalen-2-ol in $20\,cm^3$ of 2M sodium hydroxide. Warm, if necessary, to dissolve the phenol. Divide the solution into three equal portions, and dilute each with $50\,cm^3$ of cool water. At the end of five minutes, add small portions of the three reaction mixtures to the separate portions of naphthalen-2-ol solution. Does it look as if the amines have given distinctive reaction products? Finally, add all of the reaction mixtures to the naphthalen-2-ol solutions.

What does the 'blank' mixture tell you?

Interpretation

The mechanism of diazotization is complicated, but involves the following stages. Sodium nitrite reacts with hydrochloric acid to produce the weak acid nitrous acid, HNO_2.

$$NaNO_2(aq) + HCl(aq) \longrightarrow HNO_2(aq) + NaCl(aq)$$

A low temperature is necessary because nitrous acid is unstable, and decomposes quite rapidly at room temperature. In the presence of an excess of acid, nitrous acid forms the electrophile ^+NO. This then joins to the nitrogen of the amine group using the lone pair of electrons on this amine group, and eventually the $-\overset{+}{N}{\equiv}N$ ion is formed.

The excess of hydrochloric acid is also needed to keep the concentration of free amine low (most being present as phenylammonium chloride). This prevents a coupling reaction between the free amine and the diazonium compound. The diazonium compound will decompose if a low temperature is not maintained.

In the coupling reaction the diazonium compound is acting as an electrophile in an electrophilic substitution on the phenol.

The dyeing of different fabrics

A dye is a coloured compound which is capable of attaching itself firmly to fabrics. Once attached, it must be able to resist removal by water, soap, cleaning fluids, and other materials with which it might come in contact, and it must not be subject to atmospheric oxidation. Clearly, the best way for a dye to be attached to a fabric is by some form of chemical bonding to reactive groups on the molecules of the fabric. Because the molecules of different fabrics are of quite different types, it is not surprising to find that different fabrics require dyes of quite different composition.

EXPERIMENT 17.2b
To investigate the dyeing of different fabrics

You will be provided with a mixture of three dyes, Direct Red 23, Disperse Yellow 3, and Acid Blue 40 (their formulae are given in figure 17.13). Dissolve 0.05 g of the mixture in 200 cm^3 of hot water and add to the dyebath 25 cm^2 pieces of various fabrics. A good choice would be a piece of cotton, a piece of nylon, and a piece of cellulose ethanoate or polyester. Try to avoid fabrics that have a mixed composition and record the weave and surface texture of your fabrics so that you can identify them after dyeing.

Allow the pieces of fabric to boil gently in the dyebath for about 5–10 minutes, remove with a pair of tongs, and rinse under the tap.

What colours have your various fabrics been dyed?

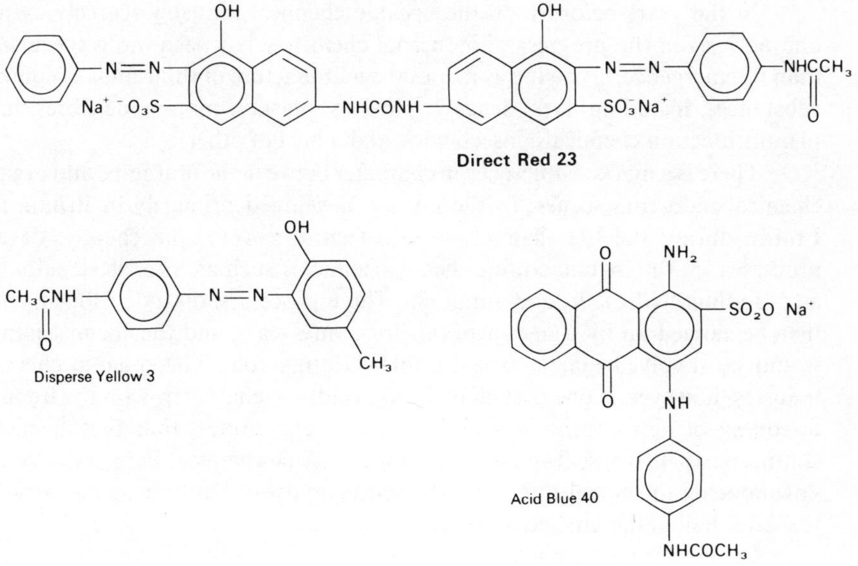

Direct Red 23

Disperse Yellow 3

Acid Blue 40

Figure 17.13
The molecular formulae of three dyestuffs.

Interpretation

Direct Red 23 dyes by hydrogen bonding to the compound making up a fabric; Acid Blue 40 dyes by ionic attraction of its sulphonic acid group to ionizable groups in a fabric; and Disperse Yellow 3 has a small molecule that will form a solid–solid solution with a fabric, using non-ionic forces of attraction.

Cotton consists essentially of cellulose, a polysaccharide based on glucose; nylon is a polyamide; cellulose ethanoate is chemically treated cellulose in which most of the hydroxyl groups have been ethanoylated; polyester contains only ester functional groups.

Which dyes have dyed which fabrics and by what method?

BACKGROUND READING 2
Dyestuffs: the origins of the modern organic chemical industry

The eighteenth century philosopher and statesman, Edmund Burke said, 'People will not look forward to posterity, who never look backward to their ancestors.' We should certainly not neglect the tremendous contributions to the present state of our science which were made by those individuals who effectively laid the foundations of the modern organic chemical industry.

In the years before 1850 the organic chemical industry scarcely existed, and nothing in the progress of industrial chemistry has been more spectacular than its emergence, involving as it does the manufacture of thousands of complex substances, including dyes, drugs, explosives, plastics, man-made fibres, fuels, plant protection chemicals, insecticides, and a host of others.

There is a marked difference in character between the inorganic and organic chemical industrial scenes. In the former, developed primarily in Britain and France during the first half of the nineteenth century, the chemist devised processes for the manufacture of heavy chemicals such as iron, steel, sulphuric acid, sodium hydroxide, and ammonia. These processes, once developed, could then be carried on by trained workers for many years, and the chemist himself would be involved mainly in a trouble-shooting role. The organic chemical industry, however, is one that changes so rapidly in character, with the frequent discovery of new compounds and new synthetic routes, that the chemist is continuously involved. The methods of the organic chemical industry were laid down during the establishment of the synthetic dyestuff and drug industries in the latter half of the nineteenth century.

Let us consider some aspects of the story of the dyestuff industry, because the principles of working which governed it are still basic to the philosophy of our modern organic chemical industry. We must begin with the discovery, in 1856, by an eighteen-year-old student, W. H. Perkin, of the first synthetic colouring material, a purple dye known as aniline purple or mauve. Perkin was a student of the German chemist, A. W. von Hofmann (then Professor of Chemistry at the Royal College of Chemistry in London). Hofmann suggested that the drug quinine might be synthesized from aryl amines derived from coal tar, and Perkin, on his own initiative, set out to attempt this. At that time the structural formulae of organic compounds had not been worked out, and chemists knew only the molecular formulae; for quinine, this was $C_{20}H_{24}N_2O_2$. Starting with an aryl amine, whose empirical formula was $C_{10}H_{13}N$, Perkin attempted his synthesis on the basis of the proposed reaction

$$2C_{10}H_{13}N + 3[O] \longrightarrow C_{20}H_{24}N_2O_2 + H_2O$$

He obtained, not the quinine he sought, but a dirty brown precipitate.

So he decided to investigate the oxidation of the simpler amine, phenylamine, C_6H_7N. After preparing the sulphate of phenylamine, he oxidized it with potassium dichromate(VI) and obtained a black precipitate which, after drying and extraction with ethanol, yielded a brilliant purple solution. This product proved to be an extremely good dyestuff and it was rapidly accepted by British and French dyers. It is interesting to note that the dye was as costly as platinum and that by the time theoretical knowledge had progressed far enough to elucidate its structure in 1888, the dye had fallen out of general use. It found its last major application in 1881 for the printing of lilac coloured 1d postage stamps.

The molecular structure of the dye, known as Perkin's mauve, is shown in figure 17.14; a colour plate showing the stamp is to be found in the *Book of data*.

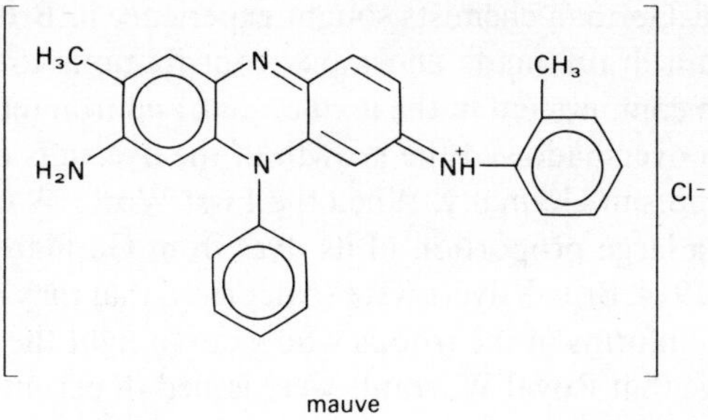

mauve

Figure 17.14

Hofmann forecast that Britain would become the chief dye manufacturing and exporting country in the World because of the ready availability of the starting material for dyestuffs, coal tar. As we shall see, this did not prove to be an accurate forecast. In the first twenty years of the dye industry, the inventive genius in synthesizing new phenylamine-based dyes lay almost wholly in Britain

Figure 17.15

W. H. Perkin, one of the founders of the synthetic dye industry, in a portrait painted by Sir A. S. Cope in 1906.

Copyright, the National Portrait Gallery, London.

and France. German chemists sought experience in Britain, and many involved in this 'brain drain' made enormous contributions to the field. However, the vast British commitment in the textile, coal, and iron industries of the Industrial Revolution overshadowed the growth of the dyestuffs industry which began to take firm roots in Germany. When the First World War broke out, Britain was importing a large proportion of its dyes from Germany. With the cessation of imports in 1914, British dyers were so deprived that they had insufficient dyestuffs to dye the uniforms of the troops who were to fight the Germans! So acute was the shortage that Royal Warrants were issued to permit trading with the enemy and dyes were purchased from Germany for a while by way of the Netherlands.

There are some modern instances of this sort of thing: the development of fundamental new techniques by the scientists and engineers of one country, and their subsequent exploitation by others. In this category we might include the computer, penicillin, and the hovercraft.

One of the most important contributions made by German scientists to the dyestuffs field was the discovery by Griess, in 1858, of the diazotization reaction.

You have seen an example of this reaction earlier in the Topic, in which phenylamine reacted with hydrochloric and nitrous acids to give the diazonium compound benzenediazonium chloride.

This reaction takes place with all primary aryl amines such as phenylamine. Benzenediazonium chloride, if isolated, crystallizes as ∠n explosive salt, but it is safe in solution, where it can be used as a valuable intermediate for further synthesis.

One important reaction of benzenediazonium chloride is its ability to couple with phenols and with aryl amines to give brightly coloured azo-compounds. Two examples are:

1 Coupling with phenol:

4-hydroxyazobenzene
(orange)

2 Coupling with N,N-dimethylphenylamine:

4-dimethylaminoazobenzene
(yellow)

The colour of compounds is due to their absorption of visible light but the compounds have the colour of the light *reflected* from the compound, not the light the compound absorbs. Thus a yellow colour is due to the absorption of blue light and a purple colour is due to the absorption of green light.

The absorption of light by a compound occurs when the compound can use the energy absorbed to change from its normal energy state to a higher energy state, from its 'ground' state to an 'excited' state. Such changes are associated with changes in structure and in the case of carbon compounds, the absorption of light energy is particularly associated with delocalized double bond systems. The delocalized double bond system of benzene has its longest wavelength absorption in the ultra-violet, at 260 nm, with maximum absorption at 204 nm; however, substituents which can form delocalized systems with the benzene ring can shift the wavelength of maximum absorption into the visible region of the spectrum. Thus the ion of 4-nitrophenol has its maximum absorption at 400 nm and is yellowish in colour.

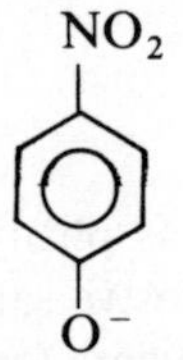

NO_2

O^-

Groups of atoms such as the benzene ring which, by absorbing light, have the potential to give rise to colour are known as *chromophores*, while substituents which shift the wavelength of absorption of a chromophore are known as *auxochromes*. The diazo group is an especially useful auxochrome because it forms a delocalized system linking two benzene rings.

Not all dyestuffs are azo-dyes with a diazo group, of course, but all do contain chromophoric groups of some kind. Here are some examples of azo-dyes in current use; the preparation of the sodium salt of methyl orange or of 'Dispersol' Fast Yellow G is one that you could undertake yourself.

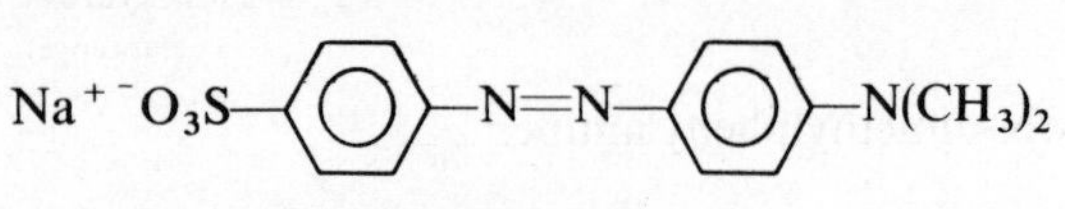

$Na^{+-}O_3S$—⟨O⟩—N=N—⟨O⟩—$N(CH_3)_2$

sodium salt of methyl orange
(compare with 4-dimethylaminoazobenzene,
example 2 on page 295)

Methyl orange is not a very good dye, but is a useful indicator.

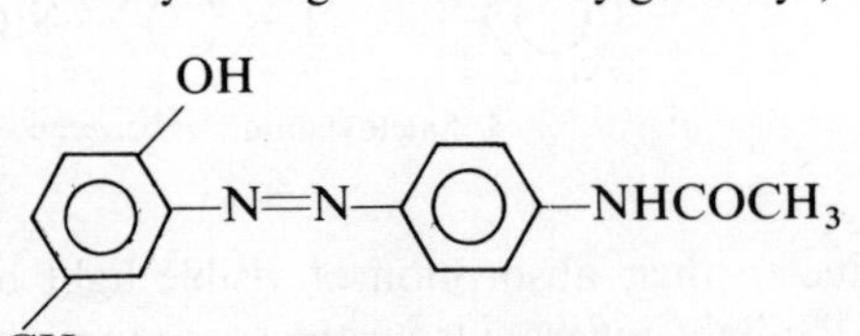

OH

⟨O⟩—N=N—⟨O⟩—$NHCOCH_3$

CH_3 'Dispersol' Fast Yellow G

The point to be borne in mind here is that there is a link between colour and constitution, between the structure of a molecule and the properties we desire. This seeking for, and exploitation of, the relationships between structure and properties is fundamental to the operation of the organic chemical industry.

17.3
DRUGS

The development of effective and safe drugs is a combined operation between chemists and doctors. While the first drugs were natural products of unknown molecular structure and variable quality, modern drugs are usually synthetic products whose production is based on a detailed knowledge of organic reactions and molecular structures. Nowadays the manufacture of drugs is subject to stringent controls, so that any material which fails to reach a specified level of purity has to be discarded. As a further check on safety, doctors continuously monitor the use of drugs, assessing their effectiveness and watching for side effects.

The first drug to be produced synthetically was aspirin, in 1899. Since then many thousands of compounds have been synthesized and tested to see if they might be effective drugs. Increasingly our knowledge of the chemistry of the body enables chemists to predict what molecular structures might be effective for particular illnesses.

EXPERIMENT 17.3
Preparations using 2-hydroxybenzoic acid

2-hydroxybenzoic acid (salicylic acid) can be converted by straightforward reactions into two products, both of which find application as medicines. Ethanoylation of the phenolic group produces aspirin

CO_2H

—OH + $(CH_3CO)_2O$ $\xrightarrow{\text{catalyst}}$ —O_2C—CH_3 + CH_3CO_2H

2-hydroxybenzoic   ethanoic   aspirin
acid   anhydride

and methylation of the carboxylic acid group produces oil of wintergreen.

CO_2H CO_2CH_3

—OH + CH_3OH $\xrightarrow{\text{catalyst}}$ —OH + H_2O

methanol   oil of wintergreen

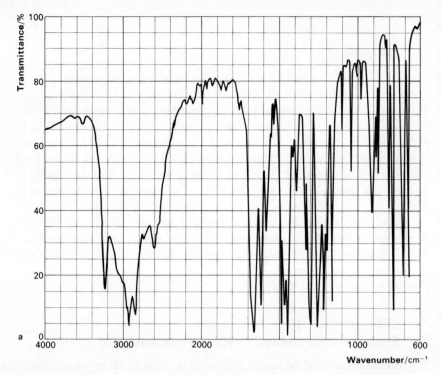

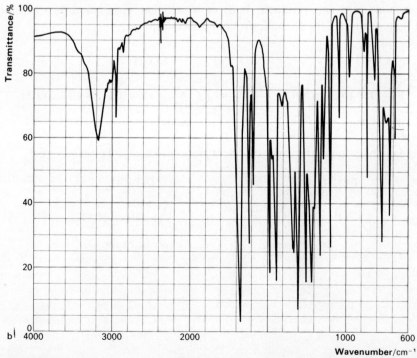

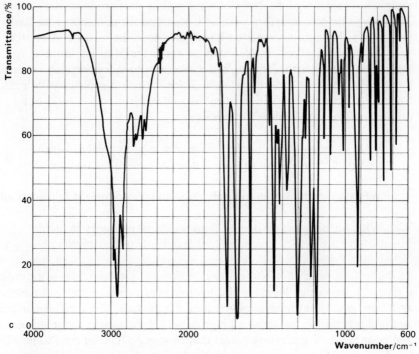

Figure 17.16
a The infra-red spectrum of 2-hydroxybenzoic acid.
b The infra-red spectrum of aspirin.
c The infra-red spectrum of oil of wintergreen.

Aspirin is described as having analgesic (pain-killing), anti-inflammatory, and antipyretic (fever-reducing) actions. Oil of wintergreen has the same properties, but is applied in liniments and ointments for the relief of pain in lumbago, sciatica, and rheumatism as the oil is readily absorbed through the skin.

The preparation of aspirin uses the same compounds as the industrial method of production.

Procedure

1 *Preparation of aspirin* Safety glasses should be worn. Add to a 50-cm^3 pear-shaped flask 2.0 g of 2-hydroxybenzoic acid and 4 cm^3 of ethanoic anhydride. (*TAKE CARE:* irritant, flammable.) To this mixture add 5 drops of 85% phosphoric(v) acid and swirl to mix. Fit the flask with a reflux condenser and heat the mixture on the steam bath for about 5 minutes. Without cooling the mixture, carefully add $2\,\text{cm}^3$ of water in one portion down the condenser. The excess ethanoic anhydride will hydrolyse and the contents of the flask will boil. When the vigorous reaction has ended, pour the mixture into $40\,\text{cm}^3$ of cold water in a 100-cm^3 beaker, cool to room temperature, stir and rub the sides of the beaker with a stirring rod if necessary to induce crystallization and, finally, allow the mixture to stand in an ice bath to complete crystallization. Collect the product by suction filtration and wash it with a little water. The product may be recrystallized from water.

What other compounds might have been used to prepare aspirin? Compare their cost with the cost of the compounds actually used.

Why do you think ethanoic anhydride and phosphoric(v) acid are used?

What further reaction do you think would be necessary to obtain 'soluble aspirin' from aspirin (2-ethanoylhydroxybenzoic acid)?

2 *Preparation of oil of wintergreen*   Safety glasses should be worn. Add to a 50-cm^3 pear-shaped flask $9\,\text{g}$ of 2-hydroxybenzoic acid, $15\,\text{cm}^3$ of methanol, and $2\,\text{cm}^3$ of concentrated sulphuric acid. Fit the flask with a reflux condenser and boil the mixture for about an hour. Cool the mixture to room temperature and pour it into a separating funnel that contains $30\,\text{cm}^3$ of cold water. Rinse the flask with $15\,\text{cm}^3$ of 1,1,1-trichloroethane and add this to the separating funnel. Mix the contents of the separating funnel, allow them to settle, and run the two layers into separate conical flasks. Return the (lower) 1,1,1-trichloroethane layer to the separating funnel and wash with $30\,\text{cm}^3$ of 0.5M aqueous sodium carbonate, releasing the pressure in the separating funnel frequently as there is likely to be considerable evolution of carbon dioxide. Dry the 1,1,1-trichloroethane extract with anhydrous sodium sulphate, filter, and remove the 1,1,1-trichloroethane by distillation (it has T_b $74\,°C$). Complete the distillation, collecting the distillate boiling above $220\,°C$ as methyl 2-hydroxybenzoate.

Note the characteristic odour of your product and compare it with a sample of oil of wintergreen. Why does methanol react with the carboxylic acid group and not the phenolic group?

What is the reason for adding 1,1,1-trichloroethane to the reaction mixture?

What is the reason for washing the 1,1,1-trichloroethane layer with sodium carbonate solution?

Examine the infra-red spectra of 2-hydroxybenzoic acid, aspirin, and oil of wintergreen (figure 17.16) and account for the major differences in the spectra.

BACKGROUND READING 3
Aspirin, from herbal remedy to modern drug

Look at the Bill of Mortality for London in the week 15–22 August, 1665, the year of the Great Plague, and notice how many of the causes of death could probably be treated successfully in modern London (figure 17.17). It was not that seventeenth century London lacked doctors and medicines but that the causes of illnesses were very imperfectly understood and many of the herbal and other remedies were ineffective or incorrectly administered.

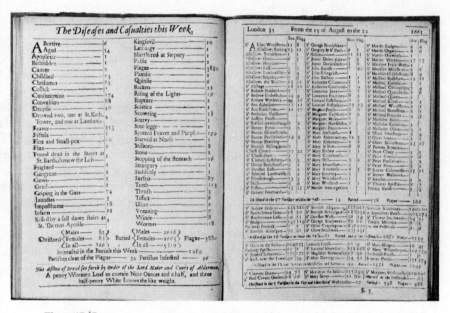

Figure 17.17
The Bill of Mortality for London, 15–22 August, 1665.
Photograph, The Museum of London.

Herbal remedies are included in the earliest medical writings from ancient Egypt and ancient China but the use of medicines was confused until quite recent times by the admixture of elements of magic and religion, together with theories that we now know were unsound. Nevertheless, there were many sound remedies, some better known to peasants than to the medical profession, and it has become common practice for the pharmaceutical industry to check all 'old wives' tales' for possible validity. For example, the dried roots of the *Rauwolfia* shrub had been used in Indian folk medicine for over two thousand years but were virtually unknown to Western medical science until 1952; now an alkaloid, reserpine, is extracted from the plant and used in the treatment of hypertension, making it one of the first drugs for the relief of high blood pressure.

The commonest drug of all, aspirin, has a typical history. The use of aspirin as a drug derives from the use of the bark and leaves of willow and poplar trees in a variety of ancient remedies. The active ingredient of the bark and leaves is salicin, the glucoside of 2-hydroxybenzoic acid (salicylic acid).

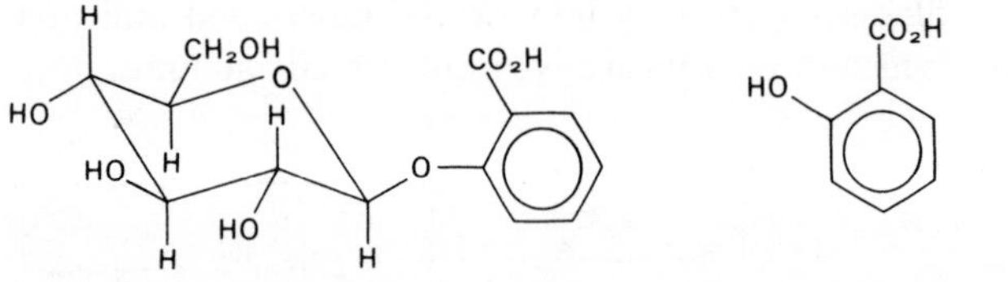

salicin   2-hydroxybenzoic acid

This was not known until 1826 although the bark and leaves had been in use for over 2000 years. From the writings of Hippocrates, the ancient Greek 'Father of Medicine', to those of the Renaissance, willow and poplar extracts had been recommended as remedies in a variety of illnesses – for eye diseases, for the removal of corns, as a diuretic, and in the treatment of sciatica and gout – although such treatments were mainly based on false hopes. In 1763, the Reverend Edmund Stone was the first to describe the value of a willow bark in treating fevers. Even he applied this treatment on the basis of false reasons:

'As this tree delights in a moist or wet soil, where agues (fevers)
chiefly abound, I could not help applying the general maxim, that many
remedies lie not far off from their causes.'

Natural extract of willow bark was replaced after 1874 by a synthetic process developed by the German chemist, Kolbe:

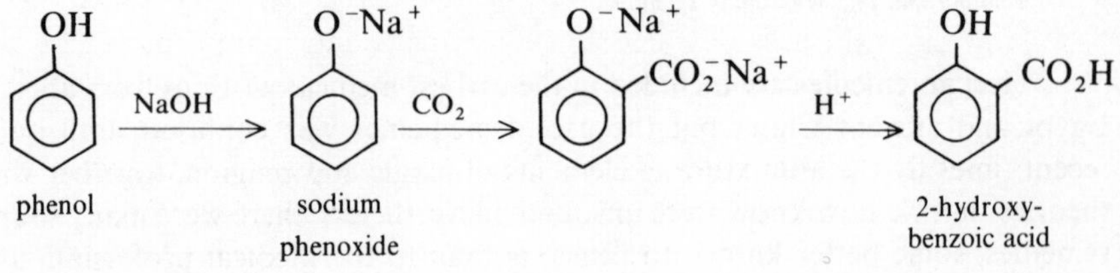

phenol   sodium   2-hydroxy-
                      phenoxide   benzoic acid

It was also about this time that the value of willow bark in treating rheumatism was reported by doctors. Although this was new to European medicine, it seems that the Hottentots in Africa had long been familiar with the remedy. However, prolonged treatment of rheumatism by 2-hydroxybenzoic acid has unpleasant side effects and the German chemist, F. Hoffmann, supplied his father with a variety of derivatives to try for his rheumatism. This led in a short time to the recognition of the merits of 2-ethanoylhydroxybenzoic acid (aspirin) and its marketing by the Bayer company in 1899.

The synthetic process developed by Hofmann is still used to manufacture aspirin today, and is based on the reaction between 2-hydroxybenzoic acid and ethanoic anhydride.

$$CH_2H \qquad\qquad\qquad\qquad CO_2H$$

$$\langle\bigcirc\rangle\!-\!OH + (CH_3CO)_2O \xrightarrow{\text{catalyst}} \langle\bigcirc\rangle\!-\!O_2C\!-\!CH_3 + CH_3CO_2H$$

2-hydroxybenzoic   ethanoic   aspirin   ethanoic
acid   anhydride   acid

The calcium salt is now marketed as soluble aspirin, and the sodium salt as effervescent (or 'fizzy') aspirin.

Nowadays, aspirin is one of the most widely used medicines, with 4000 million tablets produced each year in the United Kingdom. It is mainly taken for feverish colds, headaches, and rheumatism, but exciting new research is beginning to suggest that aspirin may have other important uses: in particular, recent work has shown that it is able to reduce the 'stickiness' of platelets in the blood stream. These platelets are normally freely circulating components of the blood but they can cause clots when they aggregate into sticky masses on the walls of veins and arteries. Clinical trials have shown that aspirin may be effective in treating diseases, like heart attacks and some strokes, which are caused by such clots. Since heart disease is the biggest single cause of death in the Western World this new research may be of great importance. Migraine, diabetes, and cataract are examples of other conditions against which some scientists think aspirin may be effective, and research is now being carried out to test these ideas.

However, aspirin is not without its hazards; some 200 people a year die of aspirin poisoning due to deliberate or accidental overdose, with children under five years forming a large proportion of those who die accidentally through eating the tablets. And since 1955, it has been increasingly recognized that aspirin causes internal bleeding. Consumption of aspirin causes a loss of blood of up to $6\,cm^3$ a day in 70 per cent of patients examined and may be the precipitating factor in 50 per cent of patients with gastroduodenal haemorrhage. It has been suggested, since safer drugs are available for headaches and feverish colds, that aspirin should only be available on prescription.

BACKGROUND READING 4
Drugs and medicines

The most remarkable applications of organic chemistry during the last 50 years have occurred in the treatment of disease and bodily disorder by chemical means. This has resulted in an increase in the health and longevity of the population, accompanied by relief from pain and physical suffering.

Most of the drugs in use up to 1935 were simple chemicals, or extracts from natural products, which often relieved the symptoms of the disease rather than attacking the source of the disease itself. Quinine had been used for the treatment of fever for three hundred years, aspirin was introduced to relieve pain during the nineteenth century, and in 1909 Ehrlich used Salvarsan, an organic arsenic derivative, for the first effective treatment of syphilis. Insulin was isolated from animal sources in 1922, offering relief to diabetics, but it was the discovery of Prontosil in 1935 which began a revolution in the treatment of disease. It was also the starting point for the growth of the pharmaceutical industry.

Antibiotics

Prontosil was the first sulphonamide drug found to be effective in the treatment of streptococcal infection: the realization that antibiotic substances could destroy disease-causing bacteria began a new era in medical therapy.

The essential feature of the sulphonamide molecule is the sulphonamide group —SO_2NH_2. This acts as a 'competitive inhibitor' in preferentially replacing the 4-aminobenzoic acid which is essential for bacterial growth. More than three thousand sulphonamides were prepared, but only a few were accepted for general use.

sulphanilamide   4-aminobenzoic acid

A dramatic extension of the antibiotic principle occurred with Fleming's observation of the restriction in the growth of bacteria caused by the mould

Penicillium notatum and the isolation of the active ingredients of penicillin by Florey and Chain in 1940. Many semi-synthetic penicillins have now been prepared, in which small chemical variations are made on the basic penicillin structure which is extracted from mould cultures. The structures of the three natural penicillins, G, F, and V are shown below.

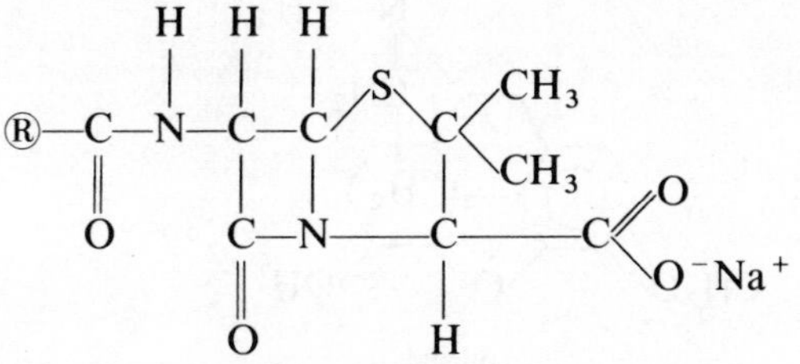

sodium salt of penicillin

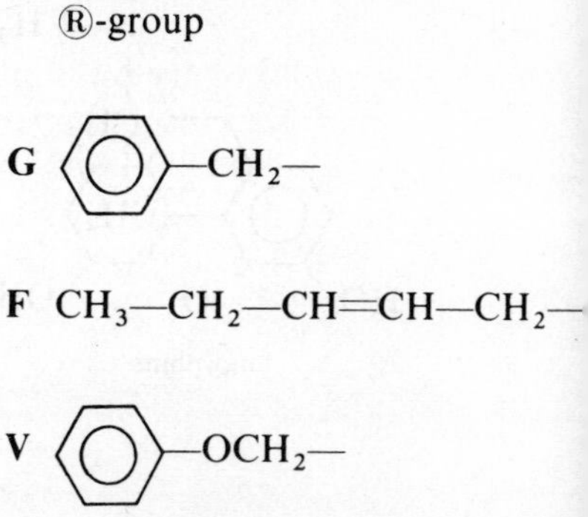

Ⓡ-group

G —CH_2—

F CH_3—CH_2—CH=CH—CH_2—

V —OCH_2—

Variations in the Ⓡ-group

The mechanism of the action of penicillin remains uncertain, but in spite of its remarkable efficacy there is a need for continued variation and replacement of any single antibiotic, since bacteria may eventually become resistant to it.

The tetracyclines form a third major group of anti-bacterial molecules and since these are effective against a greater variety of micro-organisms than penicillin they are known as 'broad spectrum' antibiotics.

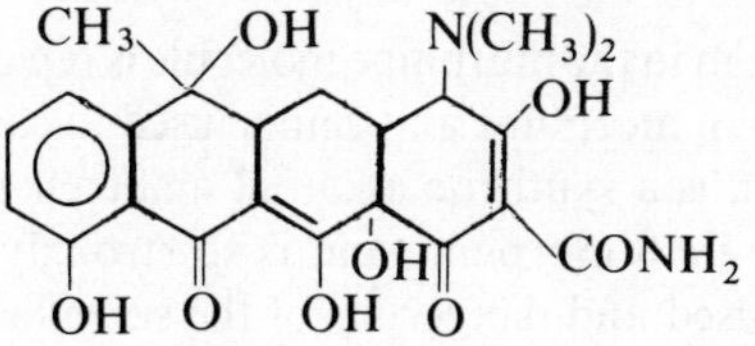

tetracycline

Analgesics

Some chemical molecules are able to relieve pain by modifying the pain signals as they approach the brain. Those which result in sleep or anaesthesia are known as 'anaesthetics' and some have been considered in Topic 9. Those which offer pain relief without sleep are known as analgesics and these have the capacity to depress the responses of the central nervous system without producing general anaesthesia. A wide spectrum of substances is available which range from the mild, widely used, aspirin to the powerful narcotic drugs of the morphine group which can lead to dependence, damage, and addiction.

All the powerful analgesics in clinical use are related to morphine, which is a nitrogenous base of the alkaloid group of naturally occurring substances; they have been available for many years from the opium obtained from poppy plants. The chief medical effects are all due to morphine: it has a remarkable ability to relieve pain, but repeated usage leads to physiological dependence or addiction.

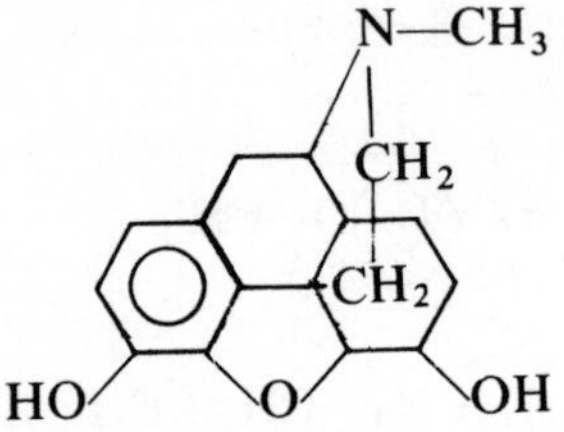

morphine

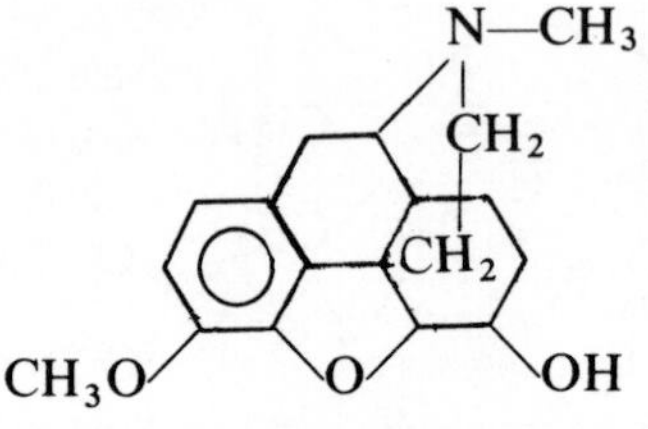

codeine
(methyl morphine)

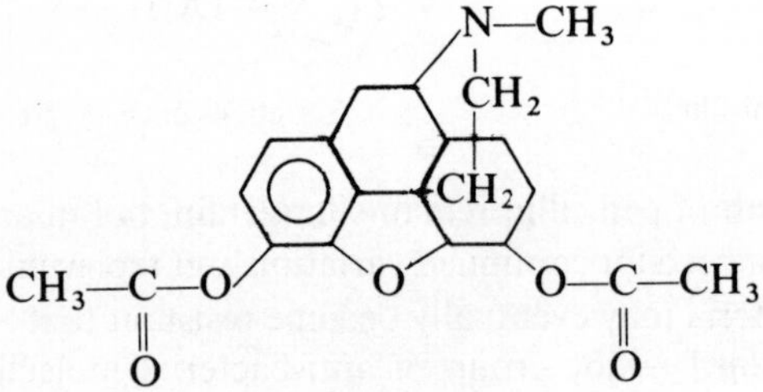

heroin
(diethanoyl morphine)

Codeine, in which a hydrogen atom in the morphine molecule is replaced by a methyl group, is much less potent than morphine and can be used to ease pain with little danger of addiction. Heroin is a synthetic alkaloid made chemically from morphine which is more potent than morphine and is so strongly habit-forming that it is the most widely abused and dangerous of the so called 'hard drugs'. Simpler alkaloid structures such as pethidine, although less powerful than morphine, can relieve pain with fewer side effects and are less likely to cause addiction.

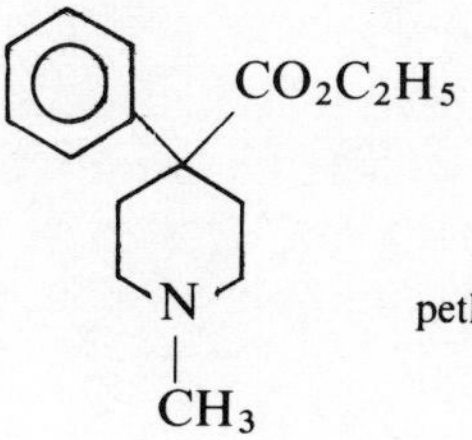

pethidine

Aspirin lies at the mild end of the spectrum of analgesics but, as we saw in Background reading 3, even this drug can have undesirable side effects.

Sedatives and stimulants

A wide range of substances can encourage sleep, reduce tension, limit sensation, or stimulate the imagination. The molecules which have been most frequently used as 'hypnotics' or 'sedatives', and which in small quantities reduce excitement, but in larger quantities induce drowsiness, are the barbiturates.

More than 50 variations of the barbituric acid structure are available by replacing the two hydrogen atoms at the base of the ring with differing side groups which result in a range of sedative substances that act on the body at different rates. In small quantities these drugs are safe and effective under medical supervision, but misuse results in addiction, and accidental death and suicide from barbiturate poisoning are not uncommon. As a result their use as sleeping pills has declined significantly.

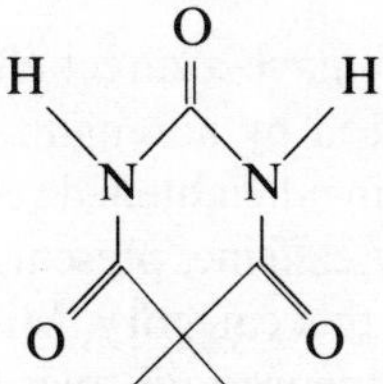

barbituric acid

The most famous of the sedative molecules is, of course, alcohol: the most infamous is thalidomide. This substance was introduced in the early sixties when it was thought to be safer than barbiturates. In fact, appalling abnormalities developed in children whose mothers had used the drug during pregnancy and over 500 deformed babies were born in the United Kingdom alone, before the drug was withdrawn. This experience has resulted in very much stricter controls being placed upon the testing and distribution of new drugs.

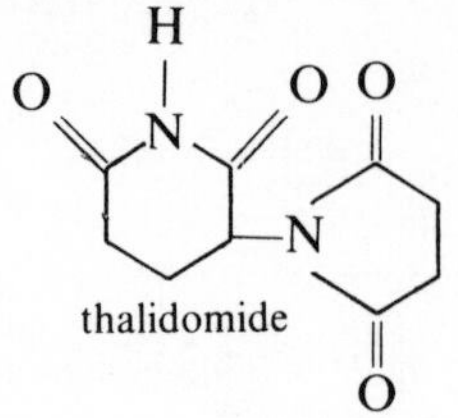

thalidomide

Many sedative drugs, unlike the barbiturates, are able to relieve tension and anxiety without inducing sleep or severely imparing mental or physical functions. Major sedatives, such as chlorpromazine (Largactil), were first introduced in 1953 for the treatment of severe mental disorders. Minor sedatives such as chlordiazepoxide (Librium) and diazepam (Valium) are now extensively used to relieve milder emotional distress and anxiety. But the taking of drugs to relieve normal anxiety, for instance before an examination, is a questionable action.

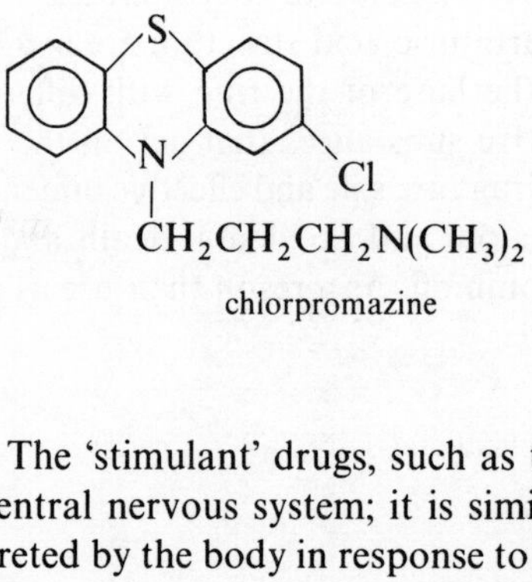

chlorpromazine

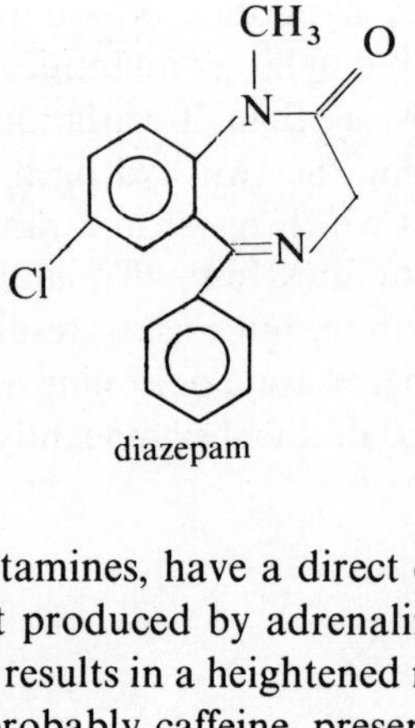

diazepam

The 'stimulant' drugs, such as the amphetamines, have a direct effect on the central nervous system; it is similar to that produced by adrenalin, which is secreted by the body in response to stress and results in a heightened response to danger. The most widely used stimulant is probably caffeine, present in tea, coffee, and 'cola', but nicotine may be another drug in this category. Misuse of stimulants is widespread, for the body develops dependence and tolerance of the drug; these lead to the need for an increased dosage to obtain satisfaction.

amphetamine

caffeine

nicotine

Hallucinogenic or 'psychedelic' drugs are little used in medicine, but can have a profound effect on the user's mood, memory, or perception. Marijuana (cannabis) has been known to Man for nearly 5000 years, but its active principle, tetrahydrocannabinol, has only recently been isolated and synthesized. Cannabis, however, is being used increasingly in medicine because of its property of stopping people from vomiting. It is used particularly for patients who have been given drugs to treat cancer; these drugs tend to make people vomit badly. Mescalin has been used for centuries in the religious ceremonies of the Mexican Indians and is obtained from the tops of the cactus plant, peyotl. Lysergic acid diethylamide (LSD), is a recent and extremely potent hallucinogen which is often unpredictable in its action and which can produce dangerous consequences.

lysergic acid diethylamide (LSD)

Drug dangers and dependence

For as long as Man has used chemical substances to relieve pain and cure illness he has used other chemical substances to alter his mood and produce feelings of well-being. The 'social' drugs such as alcohol and caffeine are usually only harmful when used to excess, but it should be remembered that alcoholism is a serious and growing problem in the modern world. And smoking, which involves the 'social' drug nicotine, is now regarded by the medical profession as harmful in any amount, even to non-smokers, by exposure to tobacco smoke. The 'soft' drugs, like cannabis and the amphetamines, may not lead to physical dependence, but an increase in availability has lead to an increase in abuse and has involved some users in serious crime. Addiction is probably inevitable with the 'hard' drugs of the morphine group, and heroin is the most dangerous of these substances: these narcotic drugs lead to complete dependence and often physical and mental damage. In order to obtain the desired effect, the addict finds it necessary to continually increase the amount of the drug until it reaches a level which is many times higher than might be administered for medical therapy. The result is the conditioning of the body to the substance and unpleasant withdrawal symptoms if the drug is not taken; these can only be diminished if removal is very gradual.

The possibilities of drug abuse should not obscure the enormous advantages to be gained from the careful medical application of chemical substances in the treatment of disease. All drugs carry the risk of possible side effects, for the greater the effect a drug has on one part of the body the more likely it is to affect another: these risks must be weighed against the advantages, which will vary with the patient and the nature of the illness. The potential of chemotherapy for the relief of disability and suffering is enormous, but it must be subject to proper medical control. No drug or medicine can now be marketed without the approval of the Committee on Safety of Medicines, which assesses the evidence in support of the safety and effectiveness of new drugs with scrupulous care and considerable expertise. Nevertheless some hazards may only be identified when a drug is in widespread use, and then the drug company can be forced to stop selling the drug.

To find new drugs which are more effective than those already discovered and which can meet the strict controls on safety, now results in only one new substance out of some 10 000 compounds investigated by the drug companies becoming commercially available for medical use.

17.4
THE IDENTIFICATION OF ORGANIC COMPOUNDS

In the last two decades the use of the mass spectrometer (Topic 4), the infra-red spectrometer, and the nuclear magnetic resonance spectrometer (Topic 7) have revolutionized our ways of obtaining knowledge of the composition and structure of organic compounds. Examination by these techniques will give precise information on the composition of a compound and the presence of various functional groups in the molecule. Nevertheless, older-established techniques are still useful and necessary.

Combustion analysis of organic compounds

One method for the determination of the elemental composition of organic compounds involves their complete combustion in pure, dry oxygen. An exact mass (0.1–0.3 g) of the compound is burned in a stream of oxygen diluted with helium gas and the combustion products are passed through a complex sequence of chemicals to ensure that the only gaseous products are carbon dioxide, water vapour, and nitrogen, mixed with helium (figure 17.18). The amount of carbon dioxide is used to calculate the carbon content, the amount of water vapour is used to calculate the hydrogen content, and the nitrogen gas any nitrogen content of the compound.

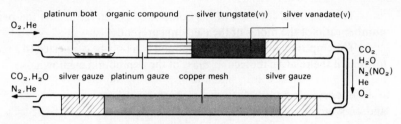

Figure 17.18
Apparatus for the quantitative combustion of organic compounds.

Volatile compounds of phosphorus, sulphur, and the halogens are all removed from the gas stream by reaction with chemicals that convert them to involatile substances, while the excess oxygen is removed by reaction with copper. The copper also serves to reduce any oxides of nitrogen to nitrogen gas. The mass of water vapour in the gas stream is determined by comparing the thermal conductivity of the gas stream before and after passing through magnesium chlorate(VII). Magnesium chlorate(VII) absorbs water vapour.

A similar procedure is used to determine the mass of carbon dioxide in the gas stream, using soda lime to absorb carbon dioxide, while the mass of nitrogen is determined by comparing the thermal conductivity of pure helium with the nitrogen–helium gas stream (under the same conditions).

To obtain a pure sample for analysis, or to separate a mixture before analysis, gas chromatography is commonly used when the compounds are sufficiently volatile and stable.

The results of combustion analysis are used to determine the empirical formula of organic compounds. The mass of carbon dioxide is converted into the mass of carbon in the sample of the compound, and then the mass of carbon is converted to the number of moles of carbon in the sample.

$$\text{g of carbon dioxide} \longrightarrow \text{g of carbon} \longrightarrow \text{moles of carbon}$$

The same procedure enables the number of moles of hydrogen and nitrogen in the sample to be determined.

To find out how much oxygen, if any, was present in the sample of the compound the original mass of the sample must be compared with the combined masses of carbon, hydrogen, and nitrogen (and other elements if present), as

determined by the combustion analysis. Any original mass of sample n
accounted for is attributed to an oxygen content and used to calculate th
number of moles of oxygen in the sample of the compound.

$$(g \text{ of sample}) - (g \text{ of C, H, N}) \longrightarrow g \text{ of oxygen} \longrightarrow \text{moles of oxygen}$$

The empirical formula of the compound is determined as the best who
number ratio of the moles of the elements present.

The empirical formula of a compound can only be converted to a molecula
formula if the relative molecular mass of the compound is known.

$$n \times \text{empirical formula} = \text{molecular formula}$$
$$\text{and} \quad n \times \text{empirical mass} \quad = \text{molecular mass}$$

The relative molecular mass can be obtained from a low resolution ma
spectrum, provided the spectrum contains a peak corresponding to positivel
charged ions which are complete molecules of the compound. Without a *parer
ion* peak a mass spectrum is not really sufficient for the determination of th
relative molecular mass of an unknown compound. An alternative techniqu
is the determination of the volume of gas formed in the vaporization of a samp
of known mass at a known pressure and a known temperature. The relationshi
$PV = nLkT$ is used to calculate the relative molecular mass.

Sample problem 1

0.205 g of the liquid, A, on complete combustion, produced 0.660 g of carbon
dioxide and 0.225 g water. (C = 12, H = 1.)
0.18 g of the liquid, A, produced 68 cm^3 of vapour at 100 °C and a pressure
of one atmosphere. ($Lk = 0.082 \text{ atm dm}^3 \text{ K}^{-1} \text{ mol}^{-1}$.)

What is the molecular formula of A?

Calculation

1 Determination of the empirical formula of A:

0.660 g of carbon dioxide contains $0.660 \times \dfrac{12}{44} \text{g} = 0.180 \text{ g of carbon}$

0.225 g of water   contains $0.225 \times \dfrac{2}{18} \text{g} = 0.025 \text{ g of hydrogen}$

0.180 g + 0.025 g $= 0.205$ g which is the mass of the original
sample, so A consists of carbon and hydrogen
only.

0.180 g of carbon   is $\dfrac{0.180}{12} = 0.015$ mole of carbon atoms

0.025 g of hydrogen   is $\dfrac{0.025}{1} = 0.025$ mole of hydrogen atoms

0.015 mole carbon atoms/0.025 mole hydrogen atoms $= 3C/5H$

so the empirical formula of A is (C_3H_5).

2 Determination of the relative molecular mass of A:

$$PV = nLkT$$

$$1 \times \frac{68}{1000} = \frac{0.18}{M_r} \times 0.082 \times 373$$

so $M_r = 81.0$

3 Determination of the molecular formula of A:

$n \times$ empirical mass $=$ molecular mass

$n \times (C_3H_5)$ $= 81.0$

$n \times 41$ $= 81.0$

         so $n = $ 2, to the nearest whole number

therefore the molecular formula is C_6H_{10}.

Sample problem 2

0.220 g of the liquid B, on complete combustion produced 0.472 g of carbon dioxide, 0.080 g of water, and 0.025 g of nitrogen. $(C = 12, H = 1, N = 14, O = 16.)$ 0.12 g of the liquid, B, produced 46 cm^3 of vapour at 290 °C and pressure of one atmosphere. $(Lk = 0.082 \text{ atm dm}^3 \text{ K}^{-1} \text{ mol}^{-1}.)$

What is the molecular formula of B?

Answer
You should find that the empirical formula is the same as the molecular formula and is $C_6H_5NO_2$, nitrobenzene.

The formation of derivatives to identify members of a homologous series

The determination of the molecular formula of an organic compound is rarely sufficient to identify the compound unambiguously. Chemical tests can be used to find out what functional groups are present, and the determination of the melting point, or boiling point, also gives useful information.

Individual members of a series, such as the series of aldehydes, can be identified by preparing a crystalline derivative and measuring its melting point. For carbonyl compounds the best derivative to make is that formed by reaction with 2,4-dinitrophenylhydrazine.

propanone and 2,4-dinitrophenyl-hydrazine

propanone 2,4-dinitro-phenylhydrazone

The reasons for using such a complicated reagent are as follows.

When preparing a derivative for identification purposes, the compound chosen should be capable of reacting with *all* the members of the particular functional group series, to give a derivative having the following properties.

1 It must be easily prepared by a reaction of high yield, involving little or no formation of by-products.

2 It must be a stable compound not decomposed at temperatures below or at its melting point.

3 It must be easily purified by recrystallization.

4 It should have a melting point in the range 50–250 °C, for convenience in determining the melting point.

These conditions rule out a number of possible compounds. In order to get melting points in a suitable range, compounds of fairly large molecular mass have to be made. As an example of this, the products formed by reactions between carbonyl compounds and hydrazine, NH_2NH_2, and its derivatives may be quoted. Hydrazine itself does not form products which are crystalline solids, but its derivative, phenylhydrazine, does so in many cases. Increasing the molecular mass still more by the introduction of the two nitro-groups to give 2,4-dinitrophenylhydrazine ensures the formation of a crystalline product.

EXPERIMENT 17.4a
The identification of a carbonyl compound

Safety glasses should be worn in this experiment.

Procedure

1 You are provided with a substance known to be either an aldehyde or a ketone. Try out the following test to decide which it is. Boil a little of the unknown substance with Fehling's solutions A and B and decide to which group it belongs.

2 Now take a few drops of the unknown aldehyde or ketone and dissolve them in the *minimum* quantity of methanol. Add this solution to about 5 cm^3 of Brady's reagent, which is a solution of 2,4-dinitrophenylhydrazine, shake the mixture, and allow it to stand. If no precipitate appears, carefully add 1–2 cm^3 of 2M sulphuric acid. Filter off the yellow precipitate using suction filtration, wash it (with the suction disconnected) with 1 cm^3 of methanol, and dry it by sucking air through it for a few minutes. Compare the melting point of your crystals with the values given in table 17.5 and see if you can identify the material. Now check the boiling point of the unknown substance and compare with those in the table.

Name	Formula	Boiling point/°C	Melting point of 2,4-dinitrophenylhydrazone/°C
Aldehydes			
Methanal	HCHO	−21	167
Ethanal	CH_3CHO	21	164, 146 (2 forms)
Propanal	CH_3CH_2CHO	49	156
Butanal	$CH_3CH_2CH_2CHO$	76	123
2-methylpropanal	$(CH_3)_2CHCHO$	64	187
Pentanal	$CH_3CH_2CH_2CH_2CHO$	103	98
Benzaldehyde	⬡—CHO	178	237
Ketones			
Propanone	CH_3COCH_3	56	128
Butanone	$CH_3CH_2COCH_3$	80	115
Pentan-2-one	$CH_3CH_2CH_2COCH_3$	102	141
Pentan-3-one	$CH_3CH_2COCH_2CH_3$	102	156
3-methylbutanone	$(CH_3)_2CHCOCH_3$	94	117
Hexan-2-one	$CH_3CH_2CH_2CH_2COCH_3$	128	107
Cyclohexanone	⬡=O	156	162
Phenylethanone	⬡—COCH$_3$	203	250, 237 (2 forms)

Table 17.5
Physical data for some aldehydes and ketones and their 2,4-dinitrophenylhydrazones.

3 Boiling points are most easily found by gently boiling 0.5 to 1.0 cm^3 of th liquid, in contact with one or two anti-bumping granules, in a 150×16 m test-tube, with a thermometer suspended just above the level of the liquid. Be results are obtained if the test-tube is immersed in a beaker containing dibut benzene-1,2-dicarboxylate. This should be heated until the thermometer show a constant reading, and the ring of liquid refluxing in the test-tube is about 1 c above the bulb of the thermometer.

State what you consider your unknown compound to be.

EXPERIMENT 17.4b
An investigation of three unknown organic compound

You will be provided with three organic compounds labelled A, B, and C whic contain carbon, hydrogen, and oxygen only.

The problem is to confirm their identity with as much certainty as possibl You should check, therefore, that all the data available are consistent with ar identity you propose and not 'jump to conclusions' using only part of the data

Procedure for compound A

1 The quantitative analysis of A gives C, 68.9 %; H, 4.9 %; O, 26.2 %. Use thes data to calculate the empirical formula of A.

2 The mass spectrum of A is given in figure 17.19a. Deduce the molecular mas of A from the mass spectrum, assuming that a parent ion peak is present, an determine the masses of the fragment ions from A. Use the molecular mass an empirical formula of A to calculate the molecular formula of A. What are the likel formulae of the seven fragment ions?

3 The infra-red spectrum of A is given in figure 17.19b. Use the correlatio chart in the *Book of data* to identify the functional group(s) and nature of th hydrocarbon group in A.

4 Carry out the following experiments with A. Wear safety glasses.

a Burn a small amount on a combustion spoon. What type of flame obtained? What can you deduce about A?

b Test the solubility of A in water by shaking a small amount with 5 cm^3 water. If it does not dissolve in cold water, see if A will dissolve in hot water. Wha can you deduce about A?

c Test a small amount of A with 5 cm^3 of 2M sodium carbonate solutio What can you deduce about A?

d Determine the melting point of A.

5 What is your conclusion about the identity of A?

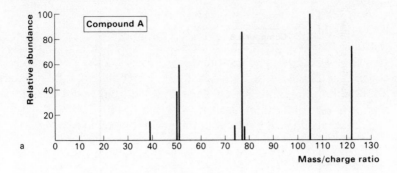

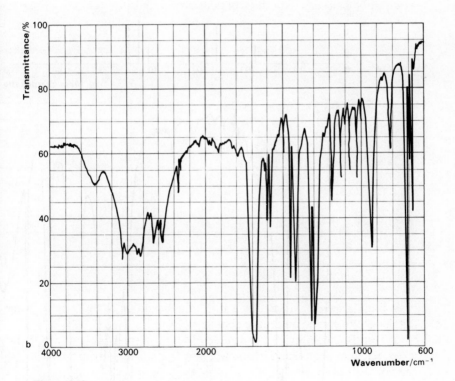

Figure 17.19

a The mass spectrum of compound A.

b The infra-red spectrum compound A.

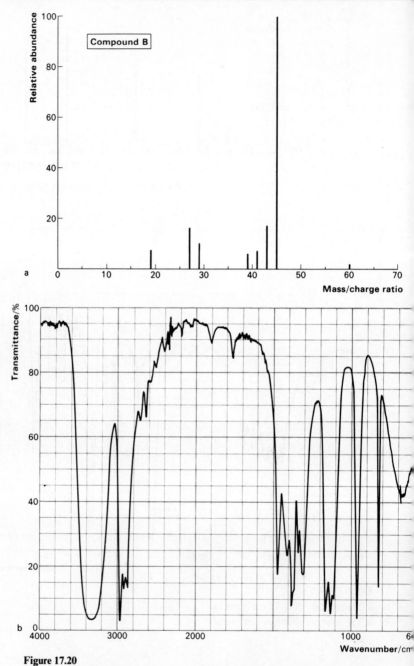

Figure 17.20
a The mass spectrum of compound B. **b** The infra-red spectrum of compound B.

Procedure for compound B

1–3 Follow the same procedure as for A. The analysis of B gives C, 60.0%; H, 13.3%; O, 26.7%; the mass spectrum and infra-red spectrum are given in figure 17.20.

4 Carry out the following experiments with B. Wear safety glasses.

a Burn a small volume on a combustion spoon. What type of flame is obtained? What can you deduce about B?

b Test the solubility of B by shaking 1–2 cm³ of B with 5 cm³ of water, hot and cold. What can you deduce about B?

c Warm a few drops of B with a mixture of aqueous potassium dichromate(VI) and 2M sulphuric acid. What can you deduce about B?

d Determine the approximate boiling point of B.

5 What is your conclusion about the identity of B?

Procedure for compound C

1–3 Follow the same procedure as for A and B. The analysis of C gives C, 66.7%; H, 11.1%; O, 22.2%, and the mass spectrum and infra-red spectrum are given in figure 17.21.

4 Carry out the following experiments with C. Wear safety glasses.

a Burn a small volume on a combustion spoon. What type of flame is obtained? What can you deduce about C?

b Test the solubility of C by shaking 1–2 cm³ of C with 5 cm³ of water, hot and cold. What can you deduce about C?

c To 5 cm³ of Brady's reagent add a few drops of C. If you have time, determine the melting point of the product. What can you deduce about C?

d Determine the approximate boiling point of C.

5 What is your conclusion about the identity of C?

17.5
SIMPLE SYNTHETIC ROUTES

A common problem of organic chemists is the synthesis of compounds of known structure. This can be important, for example, in the synthesis of drugs. When an active compound is discovered it is necessary to synthesize as many similar compounds as possible, as it is found that minor changes in structure can profoundly alter the effectiveness of a drug. By the synthesis of all possible

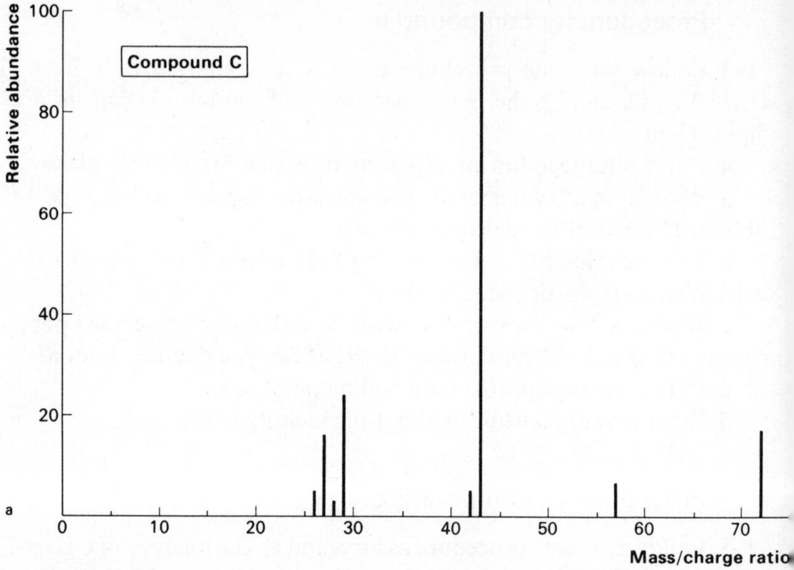

Figure 17.21
a The mass spectrum of compound C. **b** The infra-red spectrum of compound C.

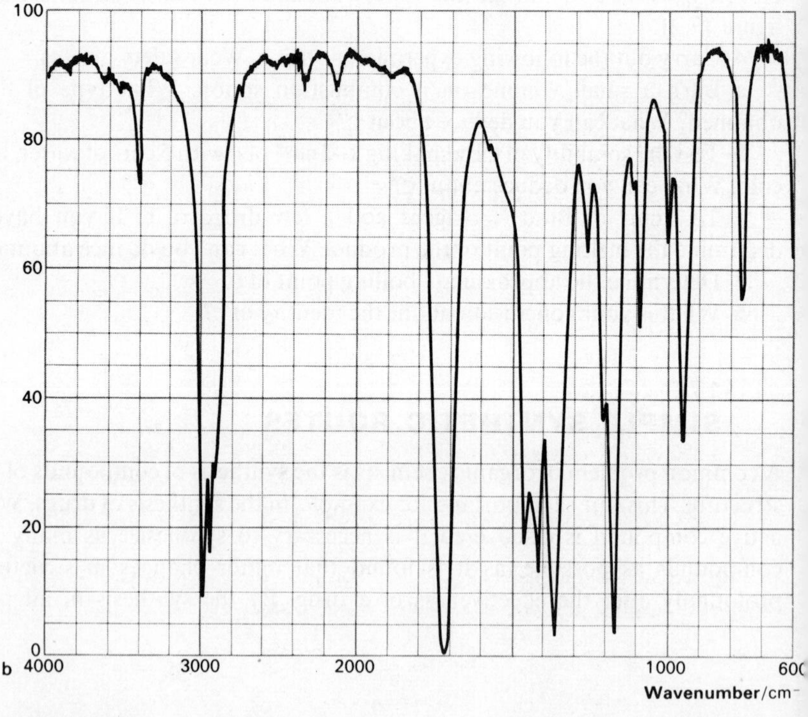

varieties of an active compound the most effective compound with the minimum of side effects can be made available to doctors. For example, a range of similar compounds has been found, all having useful properties as antibacterial drugs: these compounds are the sulphonamides.

H_2N—⟨O⟩—SO_2—NH_2 sulphanilamide

H_2N—⟨O⟩—SO_2—NH—C (ring with S, CH, N—CH)   sulphathiazole

H_2N—⟨O⟩—SO_2—NH—C (ring with N=CH, CH, N—CH)   sulphadiazine

Before you can tackle problems involving changing the functional group in compounds, it is necessary to organize your knowledge into a pattern that reveals which reactions are useful for producing a particular functional group. We shall first list the reactions that you have met in the previous Topics on organic chemistry that are useful for preparing functional groups. Secondly, we shall draw up a chart to illustrate the inter-relationship between the reactions. In organizing the reactions into a chart we need to remember that the most readily available starting materials in industry are the alkenes (from crude oil) and the carboxylic acids (from naturally occurring fats and oils).

Methods of producing a hydroxyl group

1 Carboxylic acids and esters are reduced by the nucleophile H^-, from $LiAlH_4$, to alcohols:

$$RCO_2H \xrightarrow{H^-} RCH_2OH$$

$$RCO_2CH_3 \xrightarrow{H^-} RCH_2OH + CH_3OH$$

2 Alkenes react with concentrated sulphuric acid in an electrophilic addition reaction and the hydrogen sulphates produced are hydrolysed by water to alcohols. In industry the reaction is carried out in a single stage.

$$RCH{=}CH_2 \xrightarrow{H_2O} RCHOHCH_3$$

3 Aldehydes and ketones are reduced by the nucleophile H^-, from $NaBH_4$, to alcohols:

$$RCHO \xrightarrow{\text{H}^-} RCH_2OH$$

$$RCOR \xrightarrow{\text{H}^-} RCHOHR$$

4 Halogenoalkanes react with aqueous alkali in a nucleophilic substitution, forming alcohols:

$$RBr \xrightarrow{\text{OH}^-} ROH$$

Methods of producing a halogenoalkane

5 Alkenes react with halogens and hydrogen halides in electrophilic addition reactions, yielding disubstituted and monosubstituted products:

$$R{-}CH{=}CH_2 \xrightarrow{\text{Br}_2} RCHBr{-}CH_2Br$$

$$R{-}CH{=}CH_2 \xrightarrow{\text{HBr}} RCHBr{-}CH_3$$

6 Alcohols will react with halide ions in a nucleophilic substitution reaction to product halogenoalkanes if the conditions are sufficiently acidic:

$$ROH \xrightarrow{\text{Br}^-} RBr$$

Methods of producing a nitrile

7 Halogenoalkanes react with cyanide ions in a nucleophilic substitution reaction:

$$RBr \xrightarrow{\text{CN}^-} RCN$$

8 Carboxylic acids will form ammonium salts which can be dehydrated in two stages to nitriles via acid amides:

$$RCO_2^-NH_4^+ \xrightarrow{-H_2O} RCONH_2 \xrightarrow{-H_2O} RCN$$

9 Carbonyl compounds react with sodium cyanide in a nucleophilic addition reaction, forming hydroxynitriles:

$$RCHO \xrightarrow{\text{CN}^-} RCHOHCN$$

Methods of producing an amine

10 Nitriles are reduced by $LiAlH_4$ to primary amines:

$$RCN \xrightarrow{\text{H}^-} RCH_2NH_2$$

11 Amides are reduced by $LiAlH_4$ to amines:

$$RCONH_2 \xrightarrow{\text{H}^-} RCH_2NH_2$$

12 Halogenoalkanes react with ammonia in a nucleophilic substitution to produce amines. As well as primary amines, secondary, tertiary, and quaternary amines are formed.

$$RBr \xrightarrow{\text{NH}_3} RNH_2$$

Methods of producing an alkene

13 Alcohols can be dehydrated by Al_2O_3 or by concentrated phosphoric(V) acid to alkenes:

$$RCH_2CH_2OH \xrightarrow{-H_2O} RCH{=}CH_2$$

14 Halogenoalkanes can be dehydrohalogenated by strong alcoholic alkali to alkenes:

$$RCH_2CH_2Br \xrightarrow{-HBr} RCH{=}CH_2$$

Methods of producing a carboxylic acid

15 Alcohols are oxidized by acidified sodium dichromate(VI) to carboxylic acids. Aldehydes are also oxidized to carboxylic acids but they will have been obtained by partial oxidation of alcohols.

$$RCH_2OH \xrightarrow{\text{oxidation}} RCO_2H$$

16 Nitriles are hydrolysed by strong aqueous acid or alkali to carboxylic acids:

$$RCN \xrightarrow{\text{H}_2O} RCO_2H$$

A method of producing aldehydes and ketones

17 Primary alcohols can be partially oxidized to give aldehydes, and secondary alcohols are oxidized to ketones:

$$RCH_2OH \xrightarrow{\text{oxidation}} RCHO$$

$$R{-}CHOH{-}R \xrightarrow{\text{oxidation}} RCOR$$

A method of producing acyl chlorides

18 Carboxylic acids treated with phosphorus pentachloride give acyl chlorides:

$$RCO_2H \xrightarrow{PCl_5} RCOCl$$

Methods of producing esters

19 Alcohols will react with acyl chlorides to produce esters, or with carboxylic acids if an acid catalyst is used. (Acid anhydrides can also be used.)

$$ROH \xrightarrow{R'COCl} R'CO_2R$$

$$ROH \xrightarrow{R'CO_2H} R'CO_2R$$

Methods of producing amides

20 Nitriles can be partially hydrolysed by water, if an acid or base catalyst is used, to form primary amides:

$$RCN \xrightarrow[H^+]{H_2O} RCONH_2$$

21 Acyl chlorides will react with ammonia and amines in a nucleophilic substitution, to form amides:

$$RCOCl \xrightarrow{R'NH_2} RCONHR'$$

22 Carboxylic acids can be neutralized by ammonia and then partially dehydrated to form primary amides:

$$RCO_2H \xrightarrow[-H_2O]{+NH_3} RCONH_2$$

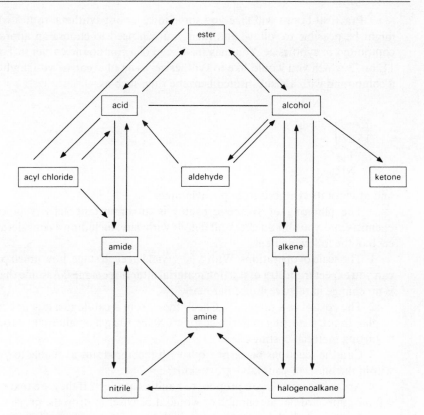

Figure 17.22
Synthetic routes.

EXPERIMENT 17.5
A problem in synthesis

If you have time you will be able to tackle this section of the course but the practical work is intended primarily for students who have carried out very few of the preparations suggested in the earlier Topics on organic chemistry.

In this final section on the organic chemistry of carbon, it is intended that you should attempt to plan and execute the synthesis of a compound of your own choice. A week is a reasonable length of time for you to spend on this work, so the synthesis should not include more than two or three separate steps, for example:

An alcohol $\xrightarrow{\text{dehydration}}$ an alkene $\xrightarrow{\text{halogenation}}$ a dihalogenoalkane

Practical books will give you some ideas about synthetic routes which it might be possible to follow, and then you will need to choose an appropriate compound to synthesize. You may have read of a compound earlier in Topics 9 11, or 13 which you would like to synthesize; as an alternative, you might select a compound with a disubstituted benzene ring, such as:

and attempt its synthesis from phenylamine.

The planning of synthetic routes is an important activity of organic chemists and you should check all details with care, including a consideration of each of the following items.

1 The scale of operations. With a 60% yield at each stage, how much product can you expect from 10 g of starting material after three stages? Assume that there is no change in relative molecular mass.

2 The cost of raw materials. Check these from a catalogue; it is not realistic to plan to use a starting material costing £20 for 100 g if an alternative route has a starting material costing £2.

3 Can the reactions be carried out with the apparatus available to you and within the limits of your laboratory skills?

4 Are the chemicals you propose to handle harmless? If they are toxic, would a fume cupboard be safe enough or would it be wiser to drop the project?

In spite of all the care you take in planning, there may be difficulties that you are unaware of and you MUST consult your teacher before starting.

We have included three examples of synthetic routes in case your own proposal proves to be impractical.

Reaction scheme One   Preparation of indigo

The dyestuff indigo has had a remarkable social and economic history. Indigo was well known in ancient India and its production was first described to Europeans by that intrepid traveller, Marco Polo, in his account of his travels between 1260 and about 1295:

'There is also plenty of good indigo, which is produced from a herb: they take this herb without the roots and put it in a big tub and add water and leave it till the herb is all rotted. Then they leave it in the sun, which is very hot and makes it evaporate and coagulate into a paste. Then it is chopped up into small pieces, as you have seen it.'

Figure 17.23a and **b**
Early records of the source of the natural dyestuff, indigo.
The Mansell Collection.

It remained an important commercial product in India and elsewhere (figure 17.23), extracted from plant leaves, until the purer and more reliable synthetic product from the German chemical industry destroyed the Indian trade in a short decade, and with much hardship.

For most of this century, even synthetic indigo has been replaced by better blue dyes but the fashion of the 1960s and 70s for blue jeans that faded in wear revitalized the manufacture of indigo. The popularity of indigo is due to the fact that as it fades, the colour remains bright.

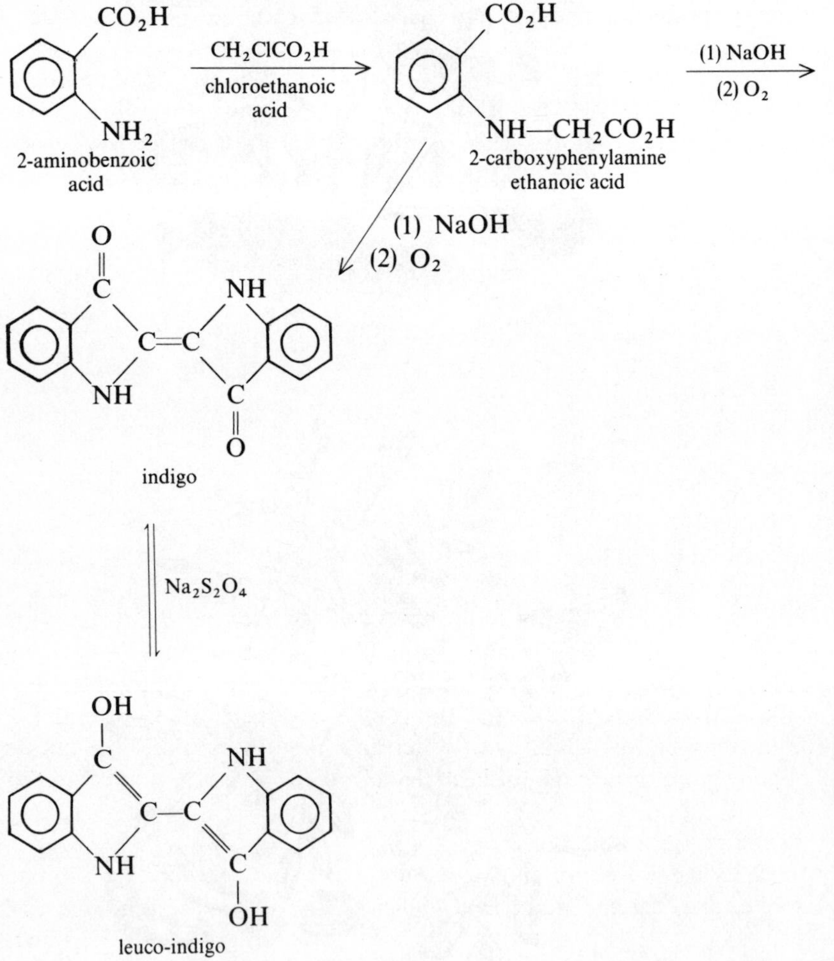

Chemicals required:
2-aminobenzoic acid (anthranilic acid)
Monochloroethanoic acid
Sodium dithionite, $Na_2S_2O_4$
Sodium carbonate (anhydrous)
Sodium hydroxide
Concentrated hydrochloric acid
Cotton cloth 10 cm × 10 cm

Preparation of the secondary amine
Wear eye protection and protective gloves, and carry out this preparation in a fume cupboard.

In a 250-cm^3 flask, mix 10 g of 2-aminobenzoic acid, 7 g of monochloro-ethanoic acid (*TAKE CARE* – it is corrosive and poisonous), 16 g of sodium carbonate (anhydrous), and 150 cm^3 of water. Reflux, using a microburner, for two or three hours.

Cool the refluxed mixture by standing the flask in cold water, then pour into a large beaker and add concentrated hydrochloric acid slowly and cautiously until the mixture is just acid to litmus. Allow to stand overnight, as complete crystallization of the product is slow, and collect the product by suction filtration. Dry the crude product in an oven at 100 °C and use directly for the next stage. (A pure sample can be obtained by recrystallization from hot water.)

Preparation of indigo
In a small evaporating basin, mix 5 g of 2-carboxyphenylamine ethanoic acid, 12 g of sodium hydroxide, and 5 cm^3 of water. Stir the mixture continuously with a thermometer protected by a glass or copper tube and heat to 250 °C. Wear eye protection and take care to avoid igniting the fumes evolved. The mass should melt and become orange in colour.

Allow the paste to cool, then dissolve the fused mass in 150 cm^3 of water in a large flask. On oxidation, indigo will be precipitated. This can be achieved by bubbling air through the solution, using a filter pump. Collect the indigo by suction filtration, decanting off the liquid as far as possible as indigo tends to clog the filter paper. Wash well with water and leave to dry. Test the filtrate by bubbling more air through to check that all the indigo has been precipitated.

Dyeing with indigo
Mix 0.2 g indigo, 6 cm^3 of 2M sodium hydroxide solution, 1 g sodium di-thionite, and 200 cm^3 water in a 400-cm^3 beaker, and bring to the boil. A clear yellow solution with a blue sheen on the surface should be obtained. Wearing glasses and protective gloves, immerse a piece of cotton cloth 10 cm × 10 cm in the solution for 5 minutes, then remove, rinse, and hang up to dry.

Reaction scheme Two   Preparation of benzocaine

Benzocaine is one of a group of similar compounds which are used as local anaesthetics:

H_2N—⟨O⟩—$CO_2CH_2CH_3$ benzocaine

H_2N—⟨O⟩—$CO_2CH_2CH_2N(C_2H_5)_2$ procaine

C_4H_9HN—⟨O⟩—$CO_2CH_2CH_2N(CH_3)_2$ amethocaine

These compounds were developed in order to find substitutes for cocaine which has dangerously addictive properties.

$$\langle O \rangle - CO_2 - \overset{\displaystyle CO_2CH_3}{\underset{\displaystyle }{(CH_3 - N}}$$

Cocaine is obtained from the leaves of the coca plant which grows on the high slopes of the Andes in Bolivia and Peru. Chewing the leaves of the coca plant reduces fatigue and increases endurance; the practice was known to the ancient Inca civilization and is still followed today by people who live in the Andes. The medical use of cocaine in Europe was pioneered by a Viennese physician, Dr Koller, who used cocaine as a local anaesthetic in 1884 for operations on the eye.

Before the danger of addiction to cocaine was recognized it was used quite casually as a stimulant. Conan Doyle showed Sherlock Holmes using it when he was relaxing from his work of detection. Here is an extract from Conan Doyle's story 'The Sign of Four'.

'Sherlock Holmes took his bottle from the corner of the mantelpiece, and his hypodermic syringe from its neat morocco case. With his long, white, nervous fingers he adjusted the delicate needle and rolled back his left shirt-cuff. For some little time his eyes rested thoughtfully upon the sinewy forearm and wrist, all dotted and scarred with innumerable puncture-marks. Finally, he thrust the sharp point home, pressed down the tiny piston, and sank back into the velvet-lined arm-chair with a long sigh of satisfaction.

'Three times a day for many months I had witnessed this performance, but custom had not reconciled my mind to it. On the contrary, from day to day I had become more irritable at the sight, and my conscience swelled nightly within me at the thought that I had lacked the courage to protest ...

'"Which is it to-day," I asked, "morphine or cocaine?" He raised his eyes languidly from the old black-letter volume which he had opened.

'"It is cocaine," he said, "a seven-per-cent solution. Would you care to try it?"

'"No, indeed," I answered, brusquely. "My constitution has not got over the Afghan campaign yet. I cannot afford to throw any extra strain upon it."'

Before the harmful effects of cocaine were recognized, coca extracts were used in a range of products and Coca-Cola was originally marketed with extracts from coca leaves and kola nuts. After much argument between the Coca-Cola company and the United States government the company was obliged to leave coca extract out of its drink, which is now an acceptable product.

In this experiment, benzocaine is prepared from 4-nitromethylbenzene, which is a readily available compound. Benzocaine is often the active ingredient in ointments used to alleviate pain caused by sunburn.

Chemicals required:
4-nitromethylbenzene
Sodium dichromate(VI) dihydrate
Concentrated sulphuric acid
1M sodium hydroxide
Concentrated hydrochloric acid
Ethanol
Calcium chloride (hydrated, if available)
Zinc dust
Ethoxyethane
Sodium sulphate anhydrous
Pentane

Oxidation of 4-nitromethylbenzene

Eye protection must be worn throughout this synthesis.

In a 500-cm^3 round-bottomed flask, dissolve 40 g (0.134 mol) of sodium dichromate(VI) dihydrate (*TAKE CARE:* avoid skin contact) in 100 cm of water Slowly and with stirring, add 50 cm^3 of concentrated sulphuric acid. Cool the reaction solution to less than 50 °C and add 13.6 g (0.1 mol) of 4-nitromethyl benzene (*TAKE CARE:* avoid skin contact). Add several anti-bumping granules and reflux the mixture for an hour.

Allow the mixture to cool for 10 minutes and pour the contents onto 200 cm^3 of ice in a 500-cm^3 beaker. Collect the resulting precipitate by suction filtration and wash the solid with two 50-cm^3 portions of water. This will take about 2 hours.

Transfer the filtered solid to a 500-cm^3 beaker and add 140 cm^3 of 1M sodium hydroxide solution to dissolve the 4-nitrobenzoic acid. Warm the result ing mixture on a hot-water bath for at least 10 minutes until any residual chromium(III) salts have coagulated as their insoluble hydroxides and then filter by suction, collecting the filtrate. This will take about $\frac{1}{2}$ hour.

Add 40 cm^3 of concentrated hydrochloric acid to 60 g of ice in a 500-cm beaker and slowly add the filtrate while stirring. Test the resulting solution with Full-range Indicator paper after mixing is complete, to ensure that the solution is strongly acidic. (4-nitrobenzoic acid is soluble in alkaline but not in acidic solutions.) Collect the resulting precipitate by suction filtration and wash the solid with two 20-cm^3 portions of cold water. Allow the precipitate to dry in a desiccator and then weigh it. This will take about $\frac{1}{2}$ hour.

Esterification of 4-nitrobenzoic acid

In this stage, the assumed quantity of product is 6.8 g of 4-nitrobenzoic acid. If you have more or less than this, then you will need to adjust the other reagent in proportion to your yield.

To 6.8 g (0.04 mol) of 4-nitrobenzoic acid, add 60 cm^3 of ethanol in 250-cm^3 round-bottomed flask with several anti-bumping granules. Attach reflux condenser and slowly pour 10 cm^3 of concentrated sulphuric acid down the condenser. Reflux the mixture for 1 hour and then allow the flask to cool for 10 minutes or so. Pour the solution into 100 cm^3 of 1M sodium hydroxide solution with 100 g of ice. Collect the precipitate by suction filtration, dry on steambath, and weigh the product. This will take about 2 hours.

Reduction of ethyl 4-nitrobenzoate

Again, the quantity of the reagents will need adjusting to the quantity of your product. The quantities here are based on 5.0 g (0.026 mol) of ethyl 4 nitrobenzoate.

Dissolve 2 g of calcium chloride in 25 cm^3 of water and mix this with

110 cm^3 of ethanol. Pour the resulting solution into a 250-cm^3 round-bottomed flask that contains 5.0 g of ethyl 4-nitrobenzoate with 50 g of zinc dust. Fit a reflux condenser and reflux for 1$\frac{1}{2}$ hours. Then allow the mixture to cool to room temperature. This will take about 2 hours.

The rest of the experiment involves the use of ethoxyethane (ether) as a solvent. Ethoxyethane is readily volatile and highly flammable so you MUST work well away from any naked flames and preferably use a fume cupboard.

Separate the unreacted zinc dust from the reaction mixture by suction filtration and wash the solid on the filter paper with two 30-cm^3 portions of ethoxyethane (ether). (*TAKE CARE:* it is highly flammable).

Take the combined filtrate of reaction mixture and ethoxyethane washings and shake with 300 cm^3 of saturated aqueous sodium chloride solution (about 17 g of sodium chloride will be needed) in a separating funnel. Ethoxyethane is volatile as well as flammable. Release any pressure from time to time by inverting the funnel and opening the tap. Remove and retain the ethoxyethane layer which contains the product. Extract further product from the aqueous layer by shaking twice with 30-cm^3 portions of ethoxyethane.

Wash the combined ethoxyethane extracts by shaking with 60 cm^3 of water, separate the layers, and dry the ethoxyethane extract with anhydrous sodium sulphate. Filter and collect the filtrate.

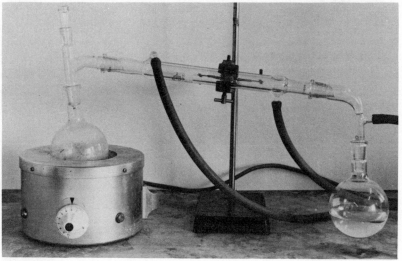

Figure 17.24
Apparatus for the distillation of ethoxyethane (ether).
Photograph, University of Bristol Faculty of Arts Photographic Unit.

In a fume cupboard set up the apparatus shown in the photograph (figure 17.24) and, well away from any naked flame, distil off the ethoxyethane (boiling point 35 °C) from the filtrate until the volume is reduced to 10 cm^3. Allow to cool and add 20 cm^3 of pentane (*TAKE CARE:* it is highly flammable) to the concentrated solution, to precipitate the product. If an oil separates, to obtain crystals chill and scratch the container with a glass rod. Dry and weigh your product, which should be an orange crystalline solid of melting point 90 °C. This last stage will take about 1$\frac{1}{2}$ hours.

Testing benzocaine

To test the effectiveness of benzocaine as a local anaesthetic, do not use your own product, which may contain hazardous impurities, but find a commercial product containing benzocaine (for an example, see figure 17.25). Apply a little of this product to a small area of skin. The forearm is convenient. Gently rub the treated area with your finger and compare with an untreated area of skin.

Be cautious in the use of products which contain benzocaine, as some people are sensitized by it so that repeated use can cause eczema.

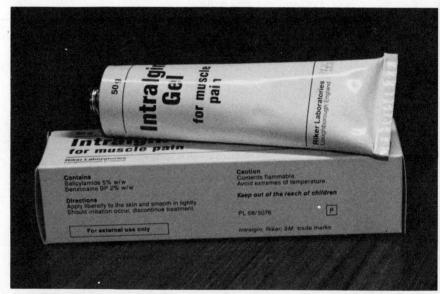

Figure 17.25
A commercial product containing benzocaine.
AVSU, Chelsea College. Photographed by permission of Bengué & Co. Ltd.

Reaction scheme Three   Preparation of a steroid

Steroids are powerful biochemical reagents of great significance to our health, and the ability of chemists to synthesize them has had, and will continue to have immense social significance. All of the sex hormones, male and female, are steroids, and the contraceptive pill has been developed from synthetic steroids.

The steroids are a group of organic compounds that have in common a four-ring carbon skeleton:

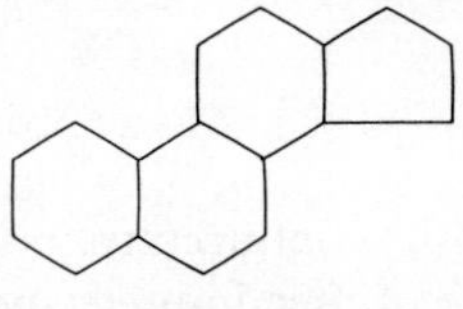

The most well known steroid is probably cholesterol:

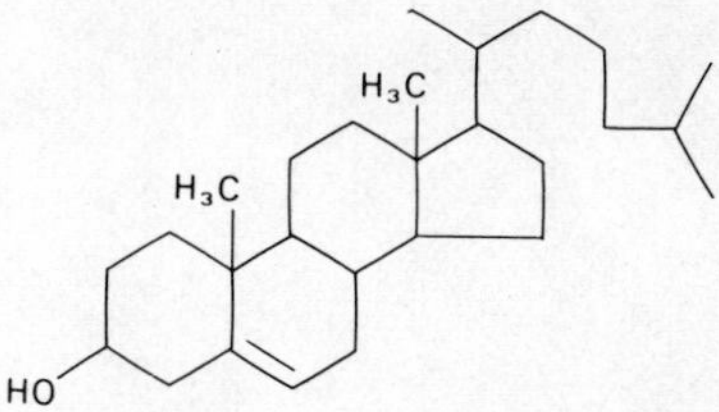

Cholesterol is made in the body from ethanoyl units and appears to be the steroid from which all other steroids in our bodies are derived.

This structural formula for cholesterol disguises the fact that with eight chiral centres cholesterol could be any one of 256 (that is, 2^8) stereoisomers. Cholesterol was first isolated in 1812; its molecular formula was established in 1888, the correct structural formula in 1932, and finally the stereochemical formula in 1955 (figure 17.26).

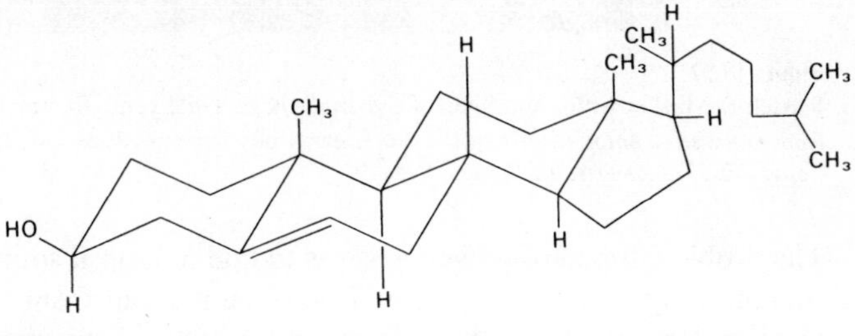

Figure 17.26
The stereochemical formula of cholesterol.

Contraceptive pills usually contain two steroids, based on two classes of female sex hormones, oestrogens and progestogens.

an oestrogen   a progestogen

The effect of taking the contraceptive pill is to raise the hormone levels in the body, creating a state of pseudo-pregnancy in which the normal monthly ovulation is suppressed (as it is in a true pregnancy).

Figure 17.27
Russell E. Marker with a specimen of *Cabeza de negro* (wild yam). *Reprinted with permission from LEHMANN, BOLIVAR and QUINTERO Journal of chemical education, 1973, 50, page 197. Copyright 1973 American Chemical Society.*

The problem for steroid chemists was to find a natural source from which an abundant supply of a steroid could be obtained and from which effective contraceptive steroids could be manufactured. The problem was brilliantly solved by the American chemist, Russell E. Marker, who found that the plant steroid diosgenin could be converted to progesterone and that a Mexican wild yam was an excellent source of diosgenin.

diosgenin   pregnenolone   progesterone

This experiment follows the last three steps of the Marker synthesis but we will use cholesterol (which is inexpensive) instead of pregnenolone as the starting material. The problem is the same in both cases: to convert an alcohol to a ketone and to carry out the isomerization of the double bond to a new position. The product, cholestene-3-one, is not, of course, a contraceptive.

cholesterol   dibromo derivative

dibromoketone   unsaturated ketone

cholestene-3-one

The bromination protects the double bond while the oxidation of the alcohol to the ketone is carried out. The double bond is then reinstated by treatment with zinc metal. Finally the isomerization of the double bond to a new position is effected by acid catalysis. The new position of the double bond is energetically favoured because of the formation of a delocalized system with the carbonyl double bond.

Chemicals required:

Cholesterol	Zinc dust
Bromine reagent in ethanoic acid–sodium ethanoate	Pyridine
Ethoxyethane	Sodium hydrogencarbonate
Ethanoic acid, glacial	Anhydrous sodium sulphate
Ice	Anhydrous ethanedioic acid
Sodium dichromate(VI) reagent in ethanoic acid	95% aqueous ethanol
Methanol	

Safety glasses and protective gloves must be worn for all stages of this synthesis.

This experiment requires at least a full morning's laboratory work. I possible it should be conducted as an all-day project. Chemicals and apparatus should be organized in advance of the time set aside for the experiment and, in particular, the two special reagent solutions should be prepared in advance.

Sufficient bromine reagent for five experiments is obtained by adding 0.5 g anhydrous sodium ethanoate and 2.05 cm^3 bromine to 60 cm^3 of pure (glacial ethanoic acid. Bromine is corrosive and poisonous; it must be handled with great care.

Sufficient sodium dichromate(VI) reagent for five experiments is obtained by adding 8 g of sodium dichromate(VI) dihydrate to 200 cm^3 of pure ethanoic acid. This reagent too must be handled with great care.

Hydrated sodium ethanoate and hydrated ethanedioic acid can be de-hydrated by gentle heating in crucibles.

Addition of bromine to the double bond

Dissolve 3.0 g cholesterol in 20 cm^3 ethoxyethane in a 50-cm^3 conical flask. Use a beaker of warm water to heat the mixture (*CAUTION:* no naked flames anywhere in the laboratory) until the crystals dissolve, and then cool the mixture back to room temperature. Add 12 cm^3 of the bromine reagent: the reaction mixture should solidify almost at once as the dibromo compound forms. Cool the reaction mixture and collect the solid dibromo compound by suction filtration, using a Hirsch funnel. Rinse out the conical flask and wash the solid dibromo compound, using a mixture of 21 cm^3 of pure ethanoic acid with 9 cm^3 of ethoxyethane that have been chilled in an ice-bath. Continue the suction filtration to remove as much liquid as possible from the damp dibromide.

Oxidation of the alcohol group to a keto group

Suspend the damp dibromide in 40 cm^3 pure ethanoic acid in a 100-cm^3 conical flask. Heat 40 cm^3 of the sodium dichromate(VI) reagent in a 100-cm^3 conical flask to 105 °C (*CAUTION:* flammable), using a thermometer to check the temperature carefully. Add the sodium dichromate(VI) reagent in one portion to the dibromo compound suspension. The reaction mixture should fall in temperature to between 55 °C and 58 °C and must be maintained within that temperature range for the next 3 to 5 minutes while the solids dissolve and the solution turns green-grey. Have one beaker of hot water and one beaker of cool water available to help adjust the temperature of the reaction mixture whenever necessary. When all the solids have dissolved, maintain the temperature for a further 2 minutes. Finally, allow the reaction mixture to stand for 20 minutes at room temperature.

Add 8 cm^3 of water to the reaction mixture and cool to 15 °C in an ice-bath. Small plate-like crystals of the dibromoketone should form. Collect the crystals by suction filtration, rinsing out the conical flask and washing the crystals with 10 cm^3 of ice-cold methanol. Transfer the crystals to a small beaker and stir with 25 cm^3 of ice-cold methanol to wash the crystals thoroughly. Collect the crystals by suction filtration. They should be almost colourless.

Debromination of the double bond

Dissolve the damp dibromoketone in 40 cm^3 of ethoxyethane (*CAUTION:* no naked flames) and 0.5 cm^3 of pure ethanoic acid in a 100-cm^3 conical flask. Use a beaker of chilled water to cool the mixture to between 15 °C and 20 °C. Add 0.8 g zinc dust in small portions over a period of 4 minutes, making sure the temperature does not rise above 20 °C. If the reaction is successful, small temperature rises should be observed. Shake the flask regularly to keep the zinc in suspension. Finally, allow the flask to stand for 10 minutes at room temperature.

Add 1.4 cm^3 of pyridine (*CAUTION:* harmful; strong odour) to the reaction mixture. The zinc ions in solution should then be precipitated as a white solid. Remove the solids by suction filtration, washing the solids on the filter paper with three 3-cm^3 portions of ethoxyethane. Retain the filtrate and washings (which may be slightly cloudy).

Transfer the filtrate and washings to a separating funnel and wash the mixture by shaking with a 15-cm^3 portion of water. When shaking the mixture, remember to release any pressure increase by inverting the funnel and opening the tap. Run off the lower water layer from the separating funnel and wash the organic layer twice more with further 15-cm^3 portions of water. Finally, wash the organic layer with 0.5 g of sodium hydrogencarbonate dissolved in 15 cm^3 of water. Test the organic layer with damp blue litmus paper to ensure that all the ethanoic acid has been removed by the washing process. Residual acid could cause premature isomerization of the double bond. Run the ethoxyethane layer into a 100-cm^3 conical flask and dry with anhydrous sodium sulphate.

Use a filter funnel and fluted filter paper to filter off the sodium sulphate and collect the filtrate in a dry 100-cm^3 conical flask that has been calibrated at the 20-cm^3 and 15-cm^3 levels. Add an anti-bumping granule and evaporate off ethoxyethane (*CAUTION:* no naked flames anywhere near) by standing the conical flask in a beaker of hot water in a fume cupboard. When the volume has been reduced to 20 cm^3, add 10 cm^3 of methanol and continue the evaporation until the volume is reduced to 15 cm^2. Cool the conical flask to room temperature and allow to stand for 30 minutes. Large ice-like crystals of the unsaturated ketone should form.

Finally, cool the solution in an ice-bath. Collect the crystals by suction filtration. Rinse out the conical flask and wash the crystals with 3 cm^3 of ice-cold methanol. The dry crystals should weigh about 1.5 g, a 50% yield, and have a melting point of 126–128 °C. If the debromination is unsuccessful, the product will melt at about 70 °C with decomposition to an orange–brown liquid.

Isomerization of the double bond by acid catalysis

To a 50-cm^3 pear-shaped flask fitted for reflux, add 1.0 g of the unsaturated ketone, 0.1 g of anhydrous ethanedioic acid, and 8 cm^3 of 95% aqueous ethanol (*TAKE CARE:* flammable). Reflux the mixture, using a microburner until the solids have dissolved (about 15 minutes) and then for a further 10 minutes. Allow the mixture to cool to room temperature, when crystals should appear, and finally chill in an ice-bath for 15 minutes.

Collect the needle-like crystals by suction filtration. Rinse out the pear-shaped flask and wash the crystals with 5 cm^3 of ice-cold methanol. The dry product should weigh about 0.9 g and have a melting point of 79–81 °C.

PROBLEMS

Revision questions

1 Remind yourself of the pattern of carbon chemistry by copying and completing the scheme below.

Covalent bonds may break in 2 different ways.

i   and **ii**

These bond-breaking processes give rise to 3 different types of reagent.

i   **ii**   and **iii**

The reactions which result are of 3 main types.

i   **ii**   and **iii**

Now find suitable examples to illustrate each of the statements you have made.

2 Classify the reactions given below, indicating clearly the reaction type and the reagent type in each case:

a $C_6H_{14} + Br_2 \xrightarrow[\text{light}]{\text{ultraviolet}} C_6H_{13}Br + HBr$

b $CH_3CH{=}CH_2 + Br_2 \longrightarrow CH_3CHBrCH_2Br$

c —$CH_3 + Br_2 \xrightarrow{Fe} Br$——$CH_3 + HBr$

d $3CH_3CH(OH)CH_3 + Cr_2O_7^{2-} + 8H^+$
$$\longrightarrow 3CH_3COCH_3 + 2Cr^{3+} + 7H_2O$$

e $CH_3CH_2CHO + HCN \longrightarrow CH_3CH_2CH(CN)OH$

f —$OH \xrightarrow{H_3PO_4}$ $+ H_2O$

g $CH_3COCl + CH_3OH \longrightarrow CH_3CO_2CH_3 + HCl$

3 Use your knowledge of the properties of organic compounds to compare and contrast the reactions of benzene and cyclohexene.
What similarities and differences in reactions would you predict for phenylamine (aniline) and cyclohexylamine?

4 A plan for a sequence of organic reactions is given below.

$$CH_3CH_2CH_2OH \longrightarrow CH_3CH_2CH_2Br \longrightarrow CH_3CH{=}CH_2$$
$$\text{A} \qquad\qquad\qquad \text{B} \qquad\qquad\qquad \text{C}$$

$$\downarrow$$

$$CH_3CH_2CH_2CN$$
$$\text{D}$$

$$\downarrow$$

$$CH_3CH_2CH_2CO_2CH_3 \longleftarrow CH_3CH_2CH_2CO_2H \longrightarrow CH_3CH_2CH_2COCl$$
$$\text{F} \qquad\qquad\qquad\qquad \text{E} \qquad\qquad\qquad\qquad \text{H}$$

$$\downarrow$$

$$CH_3CH_2CH_2CH_2OH$$
$$\text{G}$$

What reagents would you use and what type of a reaction is involved in each of the following conversions?

a A $\longrightarrow$ B   **b** B $\longrightarrow$ C   **c** B $\longrightarrow$ D
d E $\longrightarrow$ F   **e** E $\longrightarrow$ G   **f** E $\longrightarrow$ H

5 This question concerns the carbon compound M which has the structure:

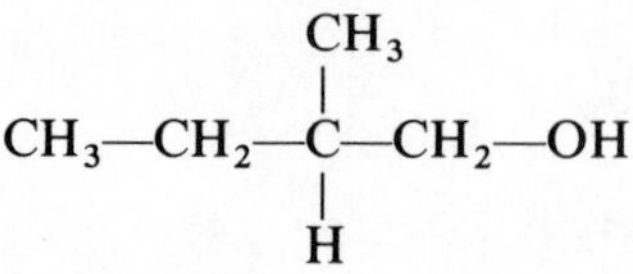

$$CH_3-CH_2-\overset{\overset{\displaystyle CH_3}{|}}{\underset{\underset{\displaystyle H}{|}}{C}}-CH_2-OH$$

a Name the substance M.
bi Draw a diagram to show the structure of substance N obtained by reacting M with a mixture of potassium iodide and phosphoric(v) acid.
ii What alternative reagent(s) could be added to M in order to obtain N?
iii Draw a diagram to show the structure of the product first formed when substance N reacts with ethanolic ammonia.
iv Suggest another substance that might also be formed if an EXCESS of N were added to ethanolic ammonia.
ci State how you would convert M into the substance P, of formula:

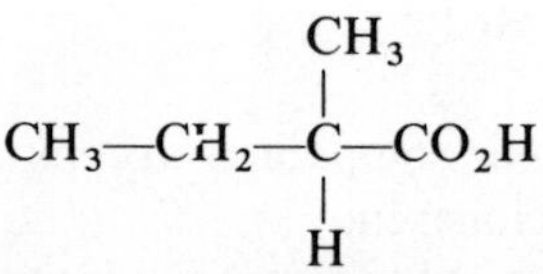

$$CH_3-CH_2-\overset{\overset{\displaystyle CH_3}{|}}{\underset{\underset{\displaystyle H}{|}}{C}}-CO_2H$$

ii What reagent(s) would you use to re-convert P into M?
di M exists as a pair of optically active isomers. Indicate by means of suitable diagrams the structures of the two forms.
ii What feature of the molecule of M gives rise to these isomers?
iii Indicate a method whereby a mixture of optical isomers might be separated.

6 Some of the products which can be obtained from 2-hydroxypropanoic acid (lactic acid) by a single step synthesis are shown opposite. The reagents for some of the reactions are also shown.
a Suggest suitable reagents for the reactions which form:
   B; C; F; I.
b Classify the type of reaction leading to:
   A; B; C; I.
c Which one of the reactions gives an ionic compound?
d With what reagent would you treat 2-hydroxypropanoic acid if you wished to form $CH_3\overset{\underset{\displaystyle |}{\underset{\displaystyle OH}{}}}{C}HCO_2Na$ rather than F?

e What substance would first be formed when 2-hydroxypropanoic acid reacted with ammonia? What further operation would be necessary in order to obtain E?

$$CH_2\!=\!CHCO_2H \quad (B)$$

$$\underset{O}{\overset{O}{CH_3CCO_2H}} \quad (C)$$

$$\underset{\underset{Cl}{|}}{CH_3CHCOCl} \quad (D)$$

$$PCl_5$$

$$\underset{\underset{OH}{|}}{CH_3CHCO_2C_2H_5} \quad (A) \qquad ethanol \qquad \underset{\underset{OH}{|}}{CH_3CH\!-\!CO_2H} \qquad ammonia \qquad \underset{\underset{OH}{|}}{CH_3CHCONH_2} \quad (E)$$

$$heat\ alone$$

$$\underset{\underset{OCOCH_3}{|}}{CH_3CHCO_2H} \quad (I) \qquad \underset{\underset{O\!-\!CO\!-\!CHCH_3}{|}}{CH_3CH\!-\!CO\!-\!O} \quad (G) \qquad \underset{\underset{ONa}{|}}{CH_3CHCO_2Na} \quad (F)$$

f How can you account for the formation of substance G?

g Give one test which would enable you to confirm that substance C was a ketone rather than an aldehyde.

h Substance B is found to react with hydrogen bromide.

i Write structural formulae for the two possible products.

ii Write a mechanism for the formation of the product you think is most likely to be formed.

Synthesis

There now follows a group of questions concerned with the conversion of one carbon compound into another. A good deal of guidance is given to help you solve the problems. Work through these carefully and try to remember the way in which they are solved. You will then be better able to solve problems in which little or no guidance is given. All reactions involved in this group of questions have been met in earlier Topics or are given in the questions.

7 This problem is to find a way of converting 1-bromobutane into butanoic acid.

a What are the structural formulae of these compounds?

b Is there the same number of carbon atoms in the product as in the starting material?
   (*Note:* if the number of carbon atoms had to be changed, special reactions would be necessary.)

c To what series of compounds does the product belong?

d This series of compounds can be prepared by an oxidation reaction. What would be a suitable compound to treat in this manner?

e Can this compound be made from 1-bromobutane, and if so, how?

f Write down the equations, using full structural formulae, for the changes you have suggested.

8 The problem is to make butanoic acid from propan-1-ol.

a What are the structural formulae of these compounds?

b Is there the same number of carbon atoms in the product as in the starting material?
   (*Note:* if more carbon atoms have to be introduced, a reaction involving the cyanide ion would increase the number by one.)

c What type of reaction must the desired conversion therefore involve?

d Carboxylic acid functional groups can be made by oxidation of —CH_2OH groups or by hydrolysis of —CN groups. Does this suggest a suitable compound from which butanoic acid might be made, bearing in mind your answer to **c**? If so, what is its structural formula?

e Think of a way by which the compound can be made from something easily obtained from propan-1-ol. Write down your suggested route, using structural formulae.

9 Some 1,2-dibromobutane has to be made from butan-1-ol.

a What class or series of compounds does butan-1-ol belong to?

b Write down any reactions of this class of compounds that you know. Use words to describe these, rather than equations.

c Write down the structural formula of the desired product.

d What sort of reaction can place two bromine atoms on neighbouring carbon atoms?

e What compound must you have in order to make 1,2-dibromobutane by this sort of reaction? Write its name and its structural formula.

f You should be able to make the compound you have named in **e** from butan-1-ol by one of the reactions you mentioned in **b**. Write the equations, using structural formulae, both for making this intermediate compound from butan-1-ol, and for converting it into the desired product.

The remaining questions also involve the conversion of one compound into another, but do not give any guidance on how this should be done. If you cannot think of a suitable method, refer to the summary charts you have constructed to find reactions that would achieve the result.

10 How would you carry out the following changes?

a Ethyl ethanoate from ethanol using no other carbon compound.
b Two isomeric esters of formula $CH_3CO_2C_4H_9$ starting from ethanol, butan-1-ol, and butan-2-ol.
c 2-methylphenyl benzoate from 2-methylphenol and benzoic acid.
d Propylamine from propan-1-ol.
e Propan-2-ol from propan-1-ol.

11 Suggest how the following conversions could be brought about.

a $CH_3CH_2Br \longrightarrow CH_3CH_2CH_2NH_2$

b $CH_3CH_2CH_2NH_2 \longrightarrow CH_3CH_2CH_2NHCOCH_3$

c ⟨O⟩—NH_2 ⟶ ⟨O⟩—$N{=}N$—⟨O⟩—NH_2

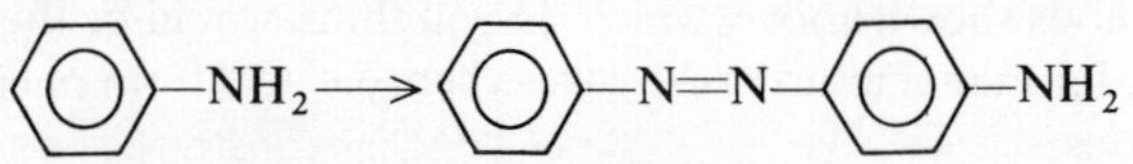

d $CH_3CH{=}CH_2 \longrightarrow CH_3COCH_3$

e $CH_2{=}CH_2 \longrightarrow H{-}C{\equiv}C{-}H$

12 The production of aspirin and Disprin from phenol, ⟨O⟩—OH, by two alternative methods is shown below. During the synthesis, substances **B** and **C** must be purified before being used for the next stage.

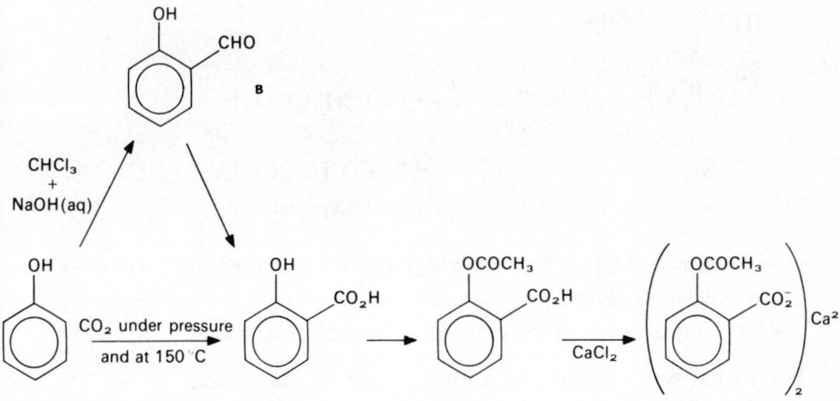

a Name one reagent for carrying out the following stages, and state the type of reaction involved.

i B ⟶ C
ii C ⟶ D

b Describe tests (one for each group) to identify the two groups attached to the benzene ring in compound **B**, and give the results of the tests.

i —OH
ii —CHO

c Name one other useful class of substances (other than drugs) which is obtained from phenol.

d How could the solid c be purified?

e Of the two methods shown above, which do you think would be the more economic for manufacturing 2-hydroxybenzoic acid from phenol? Give your reasons.

f Disprin is said to be more effective than aspirin for relieving pain because it is more soluble in water. Why is it more soluble?

13 This question is concerned with the synthesis and properties of polymers. Two reaction schemes are shown for producing compounds X and Y from readily available materials:

$$CH_2CH_2CH_2CH_2 \xrightarrow{(i)} CN(CH_2)_4CN \xrightarrow{(ii)} NH_2(CH_2)_6NH_2$$

with Br and Br substituents; Compound X

$$\text{OH} \quad \text{OH}$$

phenol $\xrightarrow{(iii)}$ cyclohexanol $\xrightarrow[\text{nitric acid}]{\text{conc.}}$ $HO_2C(CH_2)_4CO_2H$

$$\downarrow (iv)$$

$$ClOC(CH_2)_4COCl$$
Compound Y

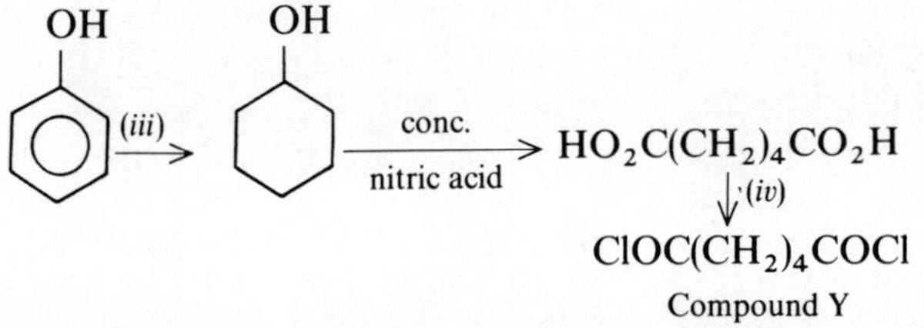

a Name the principal reagents which could be used to carry out each of the reactions (*i*) to (*iv*).

Compounds X and Y react together by polymerization.

b By what type of polymerization reaction do they react?

c Name the polymer produced.

d Indicate by means of an equation how a molecule of X and a molecule of Y react together.

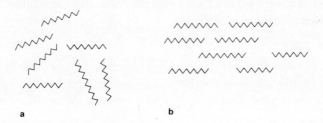

Figure 17.28
a Non-crystalline form. **b** Crystalline form.

Figures 17.28a and b represent polymer chains packed together under different conditions.
e How can the non-crystalline form be converted into the crystalline form?
f How will the two forms differ
i in elasticity?
ii in tensile strength?
iii when viewed between crossed polaroids?

14 Describe, with examples, the types of chemical reaction that lead to the formation of polymers.
How are the properties of polymers affected by their composition and structure?

15 During the second half of the twentieth century, many objects which used to be made of other materials came to be made of plastics. Give examples of such objects, explaining why the use of plastics has superseded the other material.

To what extent do you think that plastics will become increasingly important in the next 30 years? Justify your answer.

16 Diazonium salts such as $\bigcirc\!\!-N^+\!\!\equiv\!\!NCl^-$ are important starting materials for a wide variety of synthetic procedures which include the production of azo dyes.

a What experimental conditions are necessary to make an aqueous solution of the diazonium ion from phenylamine?
b Suggest reasons for the following:
i Solid diazonium salts are often explosive (and therefore rarely isolated).
ii Stable solutions of salts are only obtained from primary amines containing a benzene ring.

c The diazonium ion reacts with phenol in alkaline solution to form an az
dye.

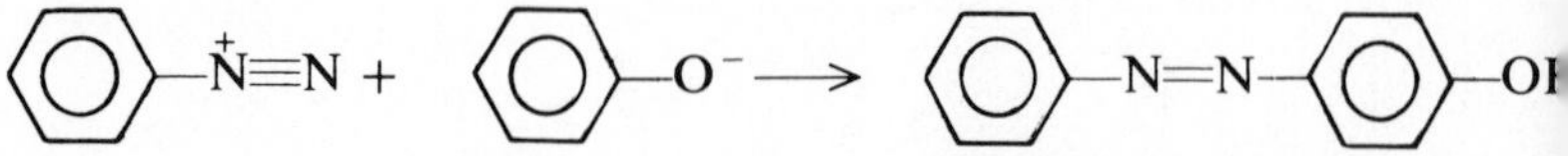

(4-hydroxyphenyl) azobenzene

i Why is the phenate ion shown in the equation rather than phenol?
ii What kind of reagent is the diazonium ion in this reaction?
iii What kind of attack has taken place on the phenate ion?
d The diazonium ion can be reduced by sodium sulphite, Na_2SO_3, to
phenylhydrazine, $\bigcirc$—$NHNH_2$. Describe an important use of
compounds of this type in organic chemistry.
e Benzene rings are usually attacked by electrophilic reagents, but if an
attached group is sufficiently electron-withdrawing, electrophilic attack
may be discouraged to the point where nucleophilic attack is preferred.
i Give two reasons why you might expect diazonium salts to react with
nucleophilic reagents.
ii Which atom is normally substituted in arene reactions? Which atom, or
group of atoms, would you expect to be substituted when diazonium
salts react? Give reasons.
iii Account for the formation of phenol when solutions of diazonium salts
are boiled.
iv What do you think might be formed if a solution of a diazonium salt
were warmed with potassium iodide solution?

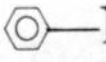

17 Limonene, which occurs in lemons and oranges, in peppermint oils, and
oil of turpentine, has the structure shown below and belongs to the class
of natural products known as terpenes.

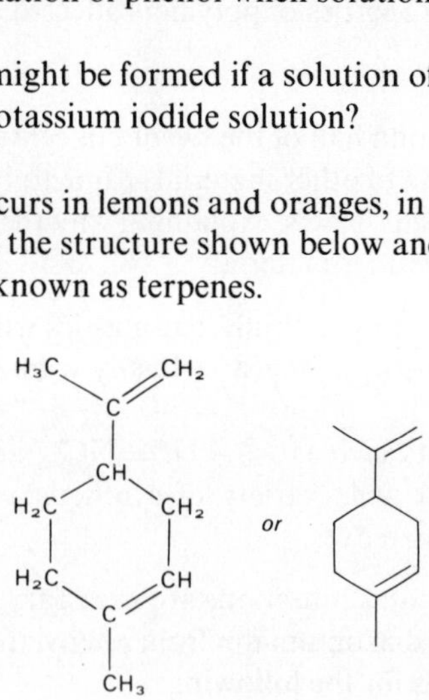

a How many moles of bromine molecules would react with one mole of
limonene?

b What would be obtained if limonene were reacted with hydrogen in the presence of a nickel catalyst?

c State how limonene could be converted into

d State how the halide of **c** could be converted into the terpineol

e What product, if any, would be likely to be obtained if the terpineol of **d** were reacted

i with a mild oxidizing agent?

ii with a dehydrating agent?

18 The structures of two female sex hormones, oestrone and oestradiol, are given below. They belong to the class of natural products known as steroids.

oestrone   oestradiol

a Suggest one physical and one chemical method of distinguishing between the two structures.

b What reagent could you use to convert oestrone into oestradiol?

c How would you attempt to remove the double bonds in the ring structure of oestradiol?

d List as many reactions as you can in which the two —OH groups in oestradiol behave:
 i in a similar way;
ii differently.
 e It is thought that the biosynthesis of steroid structures derives from ethanoic acid. What kind of technique could be used to determine whether the individual carbon atoms in the steroid structures originate in the methyl group or the carboxyl group of ethanoic acid?

19 The tropane group of naturally occurring nitrogenous bases, known as alkaloids, includes atropine and cocaine. Cocaine can be obtained from the leaves of *Erythroxylon coca* (coca leaves) and is used as a local anaesthetic, and atropine is extracted from *Atropa belladonna* (deadly nightshade).

The first total synthesis of a tropine alkaloid was accomplished by Willstätter (1901–1903) using a systematic fourteen-step process, of which the last four stages are given below:

$$
\begin{array}{ccc}
CH_2\!\!-\!\!-\!\!-CH\!\!-\!\!-\!\!-CH_2 & & CH_2\!\!-\!\!-\!\!-CH\!\!-\!\!-\!\!-CH_2 \\
\vert \quad\quad NCH_3 \quad CH & \longrightarrow & \vert \quad\quad NCH_3 \quad CHBr \\
CH_2\!\!-\!\!-\!\!-CH\!\!-\!\!-\!\!-CH & & CH_2\!\!-\!\!-\!\!-CH\!\!-\!\!-\!\!-CH_2 \\
\text{tropidine I} & & \text{II}
\end{array}
$$

$$
\begin{array}{ccc}
CH_2\!\!-\!\!-\!\!-CH\!\!-\!\!-\!\!-CH_2 & & CH_2\!\!-\!\!-\!\!-CH\!\!-\!\!-\!\!-CH_2 \\
\longrightarrow \vert \quad\quad NCH_3 \quad CHOH & \longrightarrow & \vert \quad\quad NCH_3 \quad C\!\!=\!\!O \\
CH_2\!\!-\!\!-\!\!-CH\!\!-\!\!-\!\!-CH_2 & & CH_2\!\!-\!\!-\!\!-CH\!\!-\!\!-\!\!-CH_2 \\
\text{III} & & \text{tropinone IV}
\end{array}
$$

$$
\begin{array}{c}
CH_2\!\!-\!\!-\!\!-CH\!\!-\!\!-\!\!-CH_2 \\
\longrightarrow \vert \quad\quad NCH_3 \quad CHOH \\
CH_2\!\!-\!\!-\!\!-CH\!\!-\!\!-\!\!-CH_2 \\
\text{tropine V}
\end{array}
$$

 a Why do you think the last two steps are included when structure III is apparently the same as structure V?
 b Suggest possible reagents for each of the four steps.

c The structures of cocaine and atropine are given below:

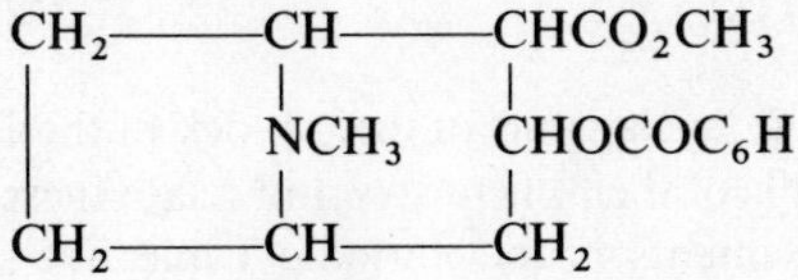

$$CH_2 \text{———} CH \text{———} CHCO_2CH_3$$
$$| \qquad\qquad |$$
$$NCH_3 \qquad CHOCOC_6H_5$$
$$| \qquad\qquad |$$
$$CH_2 \text{———} CH \text{———} CH_2$$

cocaine

$$CH_2 \text{———} CH \text{———} CH_2 \qquad CH_2OH$$
$$| \qquad\qquad | \qquad\qquad |$$
$$NCH_3 \qquad CHOCOCHC_6H_5$$
$$| \qquad\qquad |$$
$$CH_2 \text{———} CH \text{———} CH_2$$

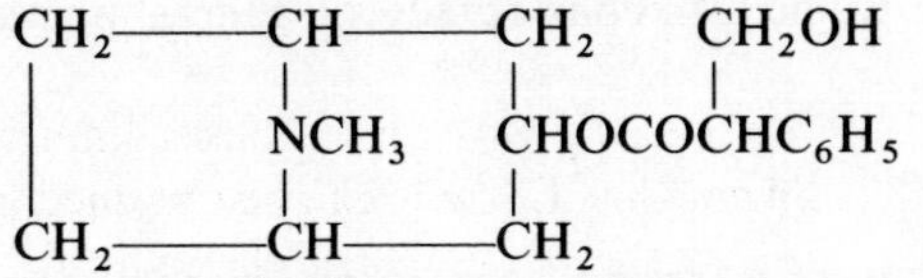

atropine

i By what kind of functional group does atropine differ from tropine and what type of operation would be necessary to convert atropine into tropine (structure v)?

ii What three operations would be necessary to convert cocaine into tropinone (structure ɪv)? (*Note.* —CO$_2$H may be removed from a molecule by a decarboxylation reaction.)

Questions 10, 12, 17, and 18 in this Topic are reproduced by permission of the University of London University Entrance and School Examinations Council.

The Periodic Table 5: the elements of groups III, IV, V, and VI

In this Topic we shall study some of the elements of the p-block of the Periodic Table. We shall examine some details of their chemistry and relate these details to the positions occupied by the elements in the Periodic Table. We shall be using some of the methods and ideas that we have met in previous Topics. Before beginning this study, however, we need to consider several general points about the p-block as a whole.

18.1
SOME GENERAL FEATURES OF THE CHEMISTRY OF THE p-BLOCK ELEMENTS

Expansion of the octet

When an element such as nitrogen forms covalent bonds it normally does so by the sharing of three of the five electrons in the outermost quantum level of the atom. In the ammonia molecule, for example, three electrons are used to make covalent bonds and the remaining two constitute a 'lone pair':

$$\text{H} \overset{\times\times}{\underset{\underset{\text{H}}{\times\bullet}}{\overset{\bullet}{\bullet}}} \text{N} \overset{}{\overset{\times}{}} \text{H}$$

In the nitrate ion, the three electrons are again used for covalent bonding and the lone pair is used to form a dative bond, thus:

The resulting ion is a structure with delocalized electrons so that all N—O bond lengths are equal:

In both the ammonia molecule and the nitrate ion, there are eight electrons in the outermost quantum level of the nitrogen atom. This group of eight electrons corresponds to the filling of the 2s and the 2p sub-levels of the nitrogen atom. It will be referred to as an *octet* of electrons.

A possible way in which nitrogen could form five ordinary, single covalent bonds would be for one of the electrons in the 2s level to be promoted to the next available level, 3s, so giving five unpaired electrons. This does not happen because the energy required for the promotion is about 2900 kJ mol^{-1} (see figure 18.1) and the energy obtained by forming two more covalent bonds, about 1260 kJ, is not enough.

In the phosphorus atom a 3d level is available, only 1370 kJ above the 3s (as shown in figure 18.2), so that promotion can occur, and a phosphorus atom can form five single covalent bonds. When this happens *expansion of the octet* is said to have taken place, because a phosphorus atom which is forming five single covalent bonds has *ten* electrons in its outermost quantum level.

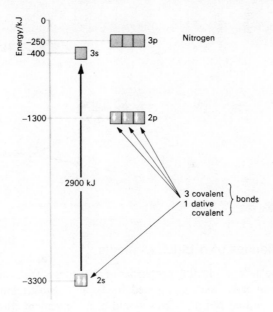

Figure 18.1

Expansion of the octet cannot happen in elements of the second period of the Periodic Table (lithium to neon) but it often does so in elements of later periods. It is possible, for example, for phosphorus to form phosphorus pentachloride, PCl_5, as well as phosphorus trichloride, PCl_3, but nitrogen only forms nitrogen trichloride, NCl_3.

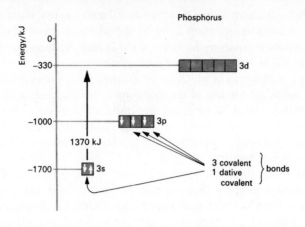

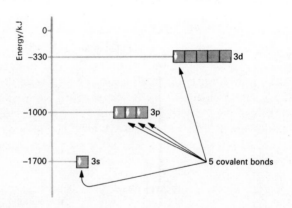

Figure 18.2

Chlorides of p-block elements

One of the most interesting properties of chlorides is their tendency to hydrolyse. This has already been mentioned in Topic 5, which includes details of the structures of the chlorides. This would be a convenient stage at which to revise the account which was given there.

Many chlorides of non-metals are readily hydrolysed. The usual pattern is:

non-metal chloride + water
$\longrightarrow$ oxide or oxoacid of non-metal + hydrochloric acid

Examples of this are summarized in the following equations:

$$PCl_3(l) + 3H_2O(l) \longrightarrow H_3PO_3(aq) + 3HCl(aq)$$

$$SiCl_4(l) + 2H_2O(l) \longrightarrow SiO_2(hydrated)(s) + 4HCl(aq)$$

It is thought that the mechanism of this hydrolysis involves a step where a lone pair of electrons on the oxygen atom of the water molecule attacks the non-metal atom:

$$H-\overset{\displaystyle |}{\underset{\displaystyle H}{\ddot{O}}}\!: \;-\!-\!\!\rightarrow \overset{\displaystyle Cl \quad Cl}{\underset{\displaystyle Cl \quad\;\; Cl}{Si}}$$

In the resulting structure, the lone pair of electrons is accommodated in the electronic structure of the non-metal atom. This is made easier if the non-metal atom has empty quantum sub-shells conveniently available. In the case of elements in the third and later periods of the Periodic Table, that is, sodium to argon, potassium to krypton, and so on, such empty sub-levels are available, for example the 3d levels. In the case of elements of the second period (lithium to neon), however, the next available levels, the 3s, are of much higher energy than the 2p levels which are being used for bonding. Consequently the chlorides of the elements of the second period are much more resistant to hydrolysis by this mechanism than those of elements of later periods unless, as in the case of halides of Group III elements, a lone pair can be accommodated in the same sub-level that is being used for bonding. The hydrolysis of boron trichloride, BCl_3, mentioned in Topic 5, takes place in this way.

Similar arguments apply to the hydrolysis of hydrides, and of compounds in which identical non-metal atoms are bonded together. For example,

$$H-\overset{\displaystyle \overset{H}{|}\;\;\overset{H}{|}}{\underset{\displaystyle \underset{H}{|}\;\;\underset{H}{|}}{C-C}}-H \text{ is stable towards hydrolysis}$$

$$\text{whereas } H-\overset{\displaystyle \overset{H}{|}\;\;\overset{H}{|}}{\underset{\displaystyle \underset{H}{|}\;\;\underset{H}{|}}{Si-Si}}-H \text{ hydrolyses readily.}$$

Oxides of the elements

Among the most important features of the chemistry of oxides are their acid–base properties. These can be demonstrated in a number of ways:

i Some oxides react with water to give either acidic or alkaline solutions:

e.g.

BASIC OXIDE $\quad CaO(s) + H_2O(l) \longrightarrow Ca^{2+}(aq) + 2OH^-(aq) \quad$ alkaline

ACIDIC OXIDE $\quad P_4O_6(s) + 6H_2O(l) \longrightarrow 4H_3PO_3(aq) \quad\quad$ acidic

ii Some oxides and the corresponding hydroxides react with acids to give salts:

e.g. $\quad$ BASIC OXIDE $\quad CaO(s) + 2HNO_3(aq) \longrightarrow Ca(NO_3)_2(aq) + H_2O(l)$

iii Some oxides react with alkalis such as sodium hydroxide to give salts:

e.g. $\quad$ ACIDIC OXIDE $\quad CO_2(g) + 2NaOH(aq) \longrightarrow Na_2CO_3(aq) + H_2O(l)$

iv Some oxides react with both acids and alkalis. Such oxides are said to be *amphoteric*. So are the corresponding hydroxides.

e.g. $\quad$ AMPHOTERIC OXIDE $\quad ZnO(s) + 2HCl(aq) \longrightarrow ZnCl_2(aq) + H_2O(l)$

$$ZnO(s) + 2NaOH(aq) \longrightarrow$$
$$Na_2ZnO_2(aq) + H_2O(l)$$

v A few oxides are neutral; they have no identifiable acidic or basic properties. Dinitrogen oxide, $N_2O(g)$, is an example.

The p-block of the Periodic Table shows three main trends of acid–base behaviour in the oxides of the elements.

1 In going from left to right of the Periodic Table the oxides become more acidic and less basic.

2 In going down a Group of the Periodic Table the oxides become more basic and less acidic.

3 If an element forms more than one oxide, the one having the element with its highest oxidation number will be the most acidic.

These various points are summarized in figure 18.3.

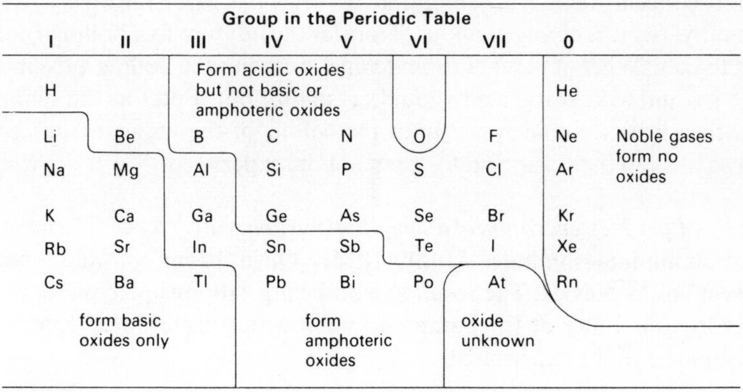

Figure 18.3
The Periodic Table and the classification of oxides. *After* MOODY, B. J. Comparative inorganic chemistry. *Ed. Arnold.*

18.2
GROUP III: BORON AND ALUMINIUM

The commonest compound of boron is disodium tetraborate-10-water, $Na_2B_4O_7 \cdot 10H_2O$, otherwise known as borax. The oxidation number of boron in borax is $+3$, as it is in most other boron compounds. Borax is a rather unconventional salt of boric acid, H_3BO_3. The molecular structure of boric acid is

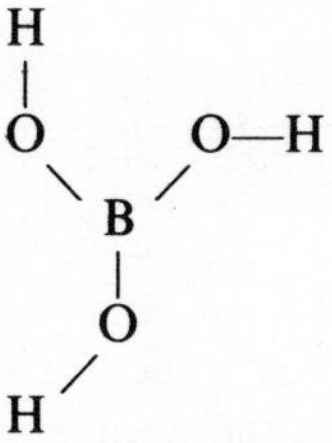

In the solid state, boric acid molecules are held together by hydrogen bonding.

The aluminium equivalent of boric acid is aluminium hydroxide, $Al(OH)_3$. Both boron and aluminium form oxides of formula type X_2O_3.

EXPERIMENT 18.2
Boric acid and aluminium hydroxide

Part 1 Preparation of a sample of boric acid Dissolve 2.5 g of borax in 25 cm^3 of hot water in a beaker. Add a few drops of methyl red solution and then add concentrated hydrochloric acid (*TAKE CARE*) drop by drop until the methyl red just changes colour. Transfer the mixture to a boiling-tube and allow it to cool. When it reaches room temperature put the boiling-tube into a mixture of ice and salt. Boric acid should crystallize out. Filter at the pump and wash with a little ice-cold water until the colour of the indicator has been washed away. Keep the boric acid for part 3 of the experiment.

Part 2 Preparation of a sample of aluminium hydroxide Dissolve about 0.5 g of aluminium sulphate, $Al_2(SO_4)_3 \cdot 16H_2O$, in $10\,cm^3$ of cold water and add $3\,cm^3$ of M NaOH. The result should be a gelatinous precipitate of aluminium hydroxide. Filter at the pump and wash with a little water. Keep the product for part 3 of the experiment.

Part 3 Comparison of the acid–base properties of boric acid and aluminium hydroxide

i Transfer about half of your solid aluminium hydroxide to each of two test-tubes. Add M HCl to the first and M NaOH to the second. Put a cork in each test-tube and shake them well, holding your thumb over the cork.

> Bearing in mind that aluminium hydroxide is not soluble in water, what do the results tell you about the acid–base behaviour of aluminium hydroxide?

ii Make up a set of four test-tubes, labelled A–D, as follows:

A 0.1 g of boric acid $+$ 5 cm^3 of pure water
B 5 cm^3 of pure water
C 0.1 g of boric acid $+$ 5 cm^3 of pure water
D 5 cm^3 of pure water

To each of the four test-tubes add 2–3 drops of Full-range Indicator solution. To A and B in turn add 0.1M HCl drop by drop, counting the number of drops required to produce a red colour. To C and D in turn, add 0.1M NaOH drop by drop, counting the number of drops required to produce a blue colour.

> Do the results suggest that boric acid is acting as an acid and/or a base? Compare your results with the general discussion of the acid–base properties of oxides and hydroxides given in section 18.1.

The anhydrous chlorides of boron and aluminium, BCl₃ and Al₂Cl₆

If possible, examine the appearance of the two chlorides. Boron trichloride is a gas which may be condensed to a clear, colourless liquid, whereas aluminium chloride is a solid. Boron trichloride hydrolyses vigorously with water and and aluminium chloride also hydrolyses, though it is possible to prepare aluminium chloride as a hydrated solid. These reactions were mentioned in the context of the behaviour of a wider range of chlorides, in Topic 5.

Personal work

Write an account of the properties of the oxide, hydroxide, and chloride of aluminium, and of the corresponding compounds of boron. Your account should include equations wherever possible.

BACKGROUND READING 1
The extraction of aluminium

According to its electrode potential, aluminium should be more reactive than zinc, though less reactive than magnesium.

$$Al^{3+}(aq)|Al(s); \qquad E^{\ominus} = -1.66\,V$$

$$Zn^{2+}(aq)|Zn(s); \qquad E^{\ominus} = -0.76\,V$$

$$Mg^{2+}(aq)|Mg(s); \qquad E^{\ominus} = -2.37\,V$$

Aluminium usually seems much less reactive than this and has many familiar everyday manufacturing uses where lightness, strength, and an attractive, non-tarnishing appearance are valuable assets. This is due to the formation of a very thin layer of oxide on its surface. This layer, though only about 10 nm thick, has a cohesive, impervious structure which protects the metal underneath from chemical attack whilst being too thin to be seen. Although iron is a much less reactive metal

$$Fe^{2+}(aq)|Fe(s); \qquad E^{\ominus} = -0.44\,V$$

hydrated iron oxide is a porous material which gives no protection at all to the surface underneath. Thus, iron corrodes much more readily than aluminium.

A more detailed account of the corrosion of iron is given in the Special Study *Metals as materials*.

Although aluminium is the third most abundant element in the Earth's crust (after oxygen and silicon) its extraction from clay and many other complex and widespread minerals is not commercially viable. The World's aluminium is extracted by electrolysis from the mineral bauxite. Bauxite consists of aluminium oxide in a hydrated form (either the trihydrate, gibbsite, $Al_2O_3 \cdot 3H_2O$, or the monohydrate, boehmite, $Al_2O_3 \cdot H_2O$) together with varying amounts of iron(III) oxide and silicon(IV) oxide as impurities.

Aluminium is often required in a pure form and its purification after extraction is difficult. It is necessary, therefore, to purify the bauxite before it is electrolysed. The process of purification depends to a large extent on the amphoteric nature of aluminium oxide ('alumina') and its related hydroxide. The stages of purification are as follows:

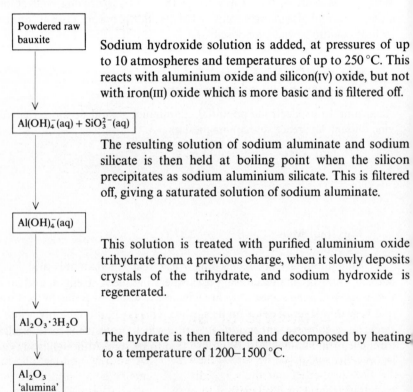

Powdered raw bauxite

Sodium hydroxide solution is added, at pressures of up to 10 atmospheres and temperatures of up to 250 °C. This reacts with aluminium oxide and silicon(IV) oxide, but not with iron(III) oxide which is more basic and is filtered off.

$Al(OH)_4^-(aq) + SiO_3^{2-}(aq)$

The resulting solution of sodium aluminate and sodium silicate is then held at boiling point when the silicon precipitates as sodium aluminium silicate. This is filtered off, giving a saturated solution of sodium aluminate.

$Al(OH)_4^-(aq)$

This solution is treated with purified aluminium oxide trihydrate from a previous charge, when it slowly deposits crystals of the trihydrate, and sodium hydroxide is regenerated.

$Al_2O_3 \cdot 3H_2O$

The hydrate is then filtered and decomposed by heating to a temperature of 1200–1500 °C.

Al_2O_3 'alumina'

Aluminium oxide has a very high melting point (2345 °C). In order to melt it, so that it can be electrolysed, it is mixed with the mineral cryolite, Na_3AlF_6, and the mixture is maintained at about 950 °C.

A direct current, which may exceed 100 000 amperes, and which maintains the electrolyte at the necessary temperature, is passed through the molten mixture by means of large carbon anodes; the cathode is formed by the carbon lining of the bath containing the mixture (see figure 18.4). As the aluminium oxide is decomposed, liquid aluminium collects at the bottom of the bath, where it is drawn off at intervals. The carbon anodes, which are steadily burned away by reaction with liberated oxygen, are renewed all the time from above, and fresh aluminium oxide is fed into the bath so that the process is continuous.

This electrolytic process consumes about 14 kilowatt-hours of electricity and 0.6 kg of anode carbon for each 1 kg of aluminium made.

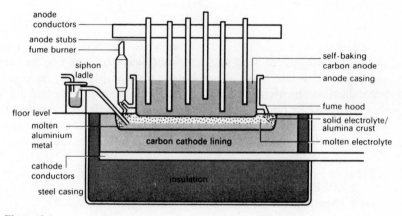

Figure 18.4
The reduction cell in which aluminium oxide ('alumina'), Al_2O_3, is dissolved in molten cryolite, Na_3AlF_6. Aluminium metal from the alumina is deposited on the cathode at the base of the cell.
Aluminium Federation.

18.3
THE ELEMENTS OF GROUP IV

One feature of the chemistry of the elements of Group IV is particularly striking: carbon and silicon at the top of the Group are entirely non-metallic in their chemical properties but the succeeding elements show a progressive increase in metallic character until, with lead, we have an element whose properties are predominantly metallic. A second feature of the group is that each of the elements has two main oxidation numbers in its compounds, $+2$ and $+4$. The $+2$ oxidation number, seldom encountered in carbon and silicon, becomes more prominent as the group is descended. On the other hand, the $+4$ oxidation number is of greatest stability at the top of the group.

The following table shows the formulae of some of the more important oxides and chlorides of the elements of Group IV:

	Oxidation number +2	Oxidation number +4
Carbon	CO	CO_2
		CCl_4
Silicon	SiO	SiO_2
		$SiCl_4$
Germanium	GeO	GeO_2
		$GeCl_4$
Tin	SnO	SnO_2
	$SnCl_2$	$SnCl_4$
Lead	PbO	PbO_2
	$PbCl_2$	$PbCl_4$

Use the *Book of data* to find out the physical properties of these compounds and from these deduce what you can about their structure. For the oxides, find out if they are acidic, basic, or amphoteric and compare your findings with the generalizations mentioned in section 18.1. For the chlorides, find out whether they hydrolyse or not and compare your findings with the remarks also made in section 18.1. You will need to consult works of reference for this information, discuss it with your teacher, and summarize the results in tabular form.

EXPERIMENT 18.3a
Preparation of some oxides of tin and lead

Tin(IV) oxide SnO_2

C A U T I O N: Concentrated nitric acid is highly corrosive. Be particularly careful to avoid spillages.

Safety glasses must be worn.

Procedure

Tin(IV) oxide can be obtained by warming tin with concentrated nitric acid

$$Sn(s) + 4HNO_3(aq) \longrightarrow SnO_2(s) + 4NO_2(g) + 2H_2O(l)$$

Because nitrogen dioxide is evolved, this preparation should be done in a fume cupboard.

Put about 0.5 g of tin in an evaporating basin and add about 2 cm³ o concentrated nitric acid. Warm very gently until there is a steady evolution o brown fumes of nitrogen dioxide and then remove the flame. Add further drop of concentrated nitric acid until no more nitrogen dioxide is given off. The yellow solid which remains is a hydrated form of tin(IV) oxide, $SnO_2 \cdot xH_2O$, where the value of x is uncertain. Add some water to the tin(IV) oxide and filter the mixture, using a Buchner funnel and flask. Wash the product well with water to remove excess nitric acid. Keep a sample of the product for use in experiment 18.3b.

Lead(IV) oxide PbO_2

CAUTION: Concentrated nitric acid is highly corrosive. Be particularly careful to avoid spillages.

Safety glasses must be worn.

Procedure

Concentrated nitric acid will only oxidize lead to oxidation number $+2$. Therefore, it is necessary to oxidize the lead further by choosing a stronger oxidizing agent. Use the electrode potentials given in the *Book of data* to verify this statement and show that chlorate(I) ions, ClO^-, would be expected to be suitable for this purpose.

Dilute some nitric acid by adding about $2\,cm^3$ of concentrated nitric acid to $2\,cm^3$ of water. Add the diluted nitric acid to $1\,g$ of lead in a boiling-tube. Heat the boiling-tube in a beaker of boiling water to accelerate the reaction as necessary. When all the lead has reacted, cool the contents of the boiling-tube to room temperature and add 2M sodium hydroxide solution drop by drop to neutralize the excess nitric acid, continuing to add until there is a faint white precipitate. The lead now has oxidation number $+2$ and is present as a solution of lead nitrate, $Pb(NO_3)_2$.

Add $10\,cm^3$ of 2M sodium hydroxide and $6\,cm^3$ of 2M sodium chlorate(I), (sodium hypochlorite in 15% solution) (*TAKE CARE:* continue wearing safety glasses). Heat the mixture in a beaker of boiling water. The colour of the precipitate changes from white to dark brown. Remove the boiling-tube from the boiling water and allow the precipitate to settle. Filter the lead(IV) oxide precipitate, using a Buchner funnel and flask, and wash the residue well with water.

Dry your sample of lead(IV) oxide and keep it for use in experiments 18.3b and 18.3c.

EXPERIMENT 18.3b
Comparison of tin(IV) oxide and lead(IV) oxide

Procedure

1 Warm a very small quantity of each oxide with M hydrochloric acid in a test-tube. When no further action can be observed, allow the mixtures to cool. The crystals which form in one of the two cases contain the metal with oxidation number + 2 as its chloride.

2 Add a very small quantity of each oxide to portions of a solution of potassium iodide, acidified with M hydrochloric acid. If either of the oxides is capable of oxidizing iodide ions, iodine will be produced.

Questions

1 What do the results of these two experiments reveal about the redox properties of the two oxides?

2 What feature of the methods used for the preparation of the oxides confirms the answer to question 1?

EXPERIMENT 18.3c
An estimation of the percentage purity of lead(IV) oxide

Procedure

The principle of this estimation is to oxidize iodide ions to iodine, using lead(IV) oxide, and to titrate the iodine with sodium thiosulphate solution.

Weigh accurately about 0.3 g of the dry lead(IV) oxide prepared in experiment 18.3a. Transfer this sample to a rubber-stoppered conical flask and add about 1 g of solid potassium iodide followed by 50 cm^3 of 2M hydrochloric acid. Shake the mixture in the stoppered flask until no more iodine is produced. Transfer the mixture to a 100-cm^3 standard volumetric flask and rinse out the conical flask with several small portions of water, adding the rinsings to the volumetric flask. Make up the solution to 100 cm^3 with water and mix the contents of the flask thoroughly. Allow the contents of the flask to settle.

Titrate 10-cm^3 portions of the solution with 0.05M sodium thiosulphate solution, using starch as an end-point indicator as described in Topic 5.

Calculation

1 Write the equation for the oxidation of iodide ions to iodine by lead(IV) oxide in acid solution.

2 Write the equation for the reaction between iodine and sodium thiosulphate.

3 Calculate the percentage purity of the lead(IV) oxide, using a method based on that given in Topic 5 (section 5.3).

BACKGROUND READING 2
The Chemist and micro-electronics

There have been many interesting developments in the chemistry of the solid state, but the materials that have had the greatest effect on electronics are the elemental semiconductors. They are used to make the transistors and diodes that carry out the main circuit functions of rectification, switching, and amplification. These are required to convert, for example, the weak signal received at the radio aerial to the current finally used to drive the loudspeaker.

You have all heard of the transistor. It was discovered in the Bell Telephone Laboratories in New York in 1947, and was originally made in pure germanium. At first, it was much more difficult to produce silicon that was pure enough. In practice, devices made in silicon have advantages in their electronic performance. Some unique features of silicon chemistry made the integrated circuit possible. As a result, silicon-based devices have now almost entirely replaced those made of germanium.

The need for electronic systems of the utmost reliability and minimum weight in ballistic missiles spurred American industry to invest large sums of money in the development of microelectronics. The system first used in the nuclear missile *Minuteman* relied on 'integration'. This is the name of the process of making miniature or integrated circuits complete with a number of transistors, diodes, resistors, capacitors, and interconnections, all in the surface of one piece of solid semiconductor (figure 18.5).

Integration involves a variety of chemical operations. To understand the various steps, we must start with an account of the elementary physics of the solid state and the nature of semiconduction.

Electrons in solids

You will recall that the narrow lines in the emission spectrum of the atoms of an element heated to a high temperature can be explained by the transfer of electrons between a limited number of fixed energy levels. The study of these allowed or 'quantized' energy levels begun by Rydberg and Bohr, together with the work of Rutherford, Moseley, Pauli, and many others, provides a model for the atom which allows two electrons of opposite spin to occupy each level. When two atoms combine, the electrons of both atoms share a new molecular orbital which has two energy levels. One is higher and one is lower than in

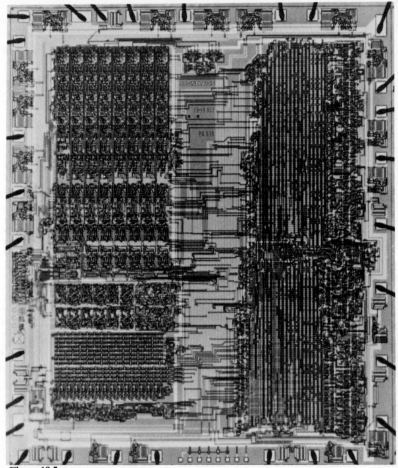

Figure 18.5

A typical microcomputer integrated circuit designed to distribute the 8 bit data words of a microprocessor to the outside world. Magnification × 5.
Photograph, Mullard Ltd.

the original isolated atoms. Both electrons fall into the lower level with a corresponding gain in energy, the bond energy (figure 18.6a).

In a solid such as diamond, the carbon atoms in any one crystal are bonded together in three dimensions to form one large molecule. All the valency electrons (namely, those involved in the formation of bonds) are shared between all the bonds in the crystal. The two simple levels of the hydrogen atom become two bands packed with as many closely spaced energy levels as there are atoms in the crystal.

The lower band containing the valency electrons is naturally called the *valence band*. In an insulator such as diamond, the upper band remains empty unless an electron absorbs enough energy, for example from an energetic quantum of ultra-violet radiation, to allow it to jump the energy gap between the two bands (figure 18.6b).

An applied electric field is unable to transfer an electron from the valence band to the upper band; with the valence band filled there are no empty levels to allow electrons to move through the crystal. This is the band structure of an insulator.

In a metal, the upper band and the valence band overlap and electrons may exchange freely between the common levels of the two bands. The electrons in the upper, partly filled, band are mobile and give the metal its conductivity, so that we call this the *conduction band* (figure 18.6c). This overlap between the valence band and the conduction band is a fundamental characteristic of the metallic state.

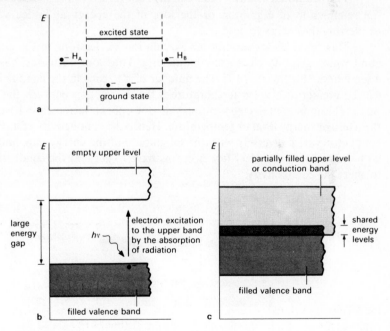

Figure 18.6
a Electron energy levels for two isolated hydrogen atoms, H_A and H_B, and for the hydrogen molecule formed from them.
b Energy level diagram showing electronic energy levels in a solid insulator.
c Energy level diagram showing electronic energy levels in a typical conductor.

Semiconductors

What happens if we have a material with no overlap between the conduction and valence bands but a rather narrow energy gap between them? If we recall that an energetic quantum can transfer an electron from the valence band to the upper conduction band in an insulator, we can expect the same sort of thing to happen, but with quanta of lower energy.

As we lower the energy of a quantum, Planck's quantum relationship $E = hv$ tells us that the frequency falls and the wavelength increases. In nature there is, in fact, an almost continuous range of 'insulators' with energy gaps of all sizes. Thus we can expect to find materials that will begin to absorb radiation at wavelengths down through the visible and into the infra-red.

Although at room temperature the average thermal energy of the atom in a solid is quite low, the energy of an individual atom is not always equal to the average. There is a spread or distribution in the energies of the atoms, so that a few have sufficient energy for their electrons to jump the gap into the conduction band.

The promoted electrons can equally well return to the valence band with the re-emission of a photon or the loss of the energy as lattice vibrations converting the energy to heat.

This provides an equilibrium population of electrons in the conduction band which gives an electrical conductivity. Thus, our 'insulator' has become a semiconductor (figure 18.7). The number of electrons in the conduction band will be greater at a given temperature if the energy gap between the bands is small. The number of electrons pushed into the upper level increases if we increase the average energy level or temperature. Hence the conductivity of a pure semiconductor varies inversely with the magnitude of the energy gap, and directly with the temperature. Very few electrons have to get into the conduction band to affect the conductivity.

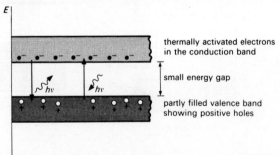

Figure 18.7
Energy level diagram of electronic energy levels in a semiconductor. This shows electron promotion in the conduction band by low energy quanta and the creation of positive holes in the valence band.

Holes

There is one important feature of semiconduction that we have omitted so far. The electrons transferred to the conduction band have left an equal number of gaps in the population of the valence band so that it is no longer full (figure 18.7). Each gap or 'hole' in the valence band can be filled by an electron, but this leaves another hole. By repeating this step the hole can move anywhere in the crystal. The hole behaves as a mobile positive charge; with an applied electric field it moves and provides electrical conduction. Some of the properties of holes are a little baffling at first, but a moving hole in a sea of electrons is equivalent to an electron moving in the opposite direction.

Think of a bubble in a spirit level; when it is tilted in the gravitational field we say that the bubble rises, but this is exactly equivalent to the oil falling. A good analogy for a semiconductor is provided by the model in figure 18.8. With the lower half full and the upper part empty the gravitational field can provide no motion, hence no conduction. Transfer a few 'electrons' out of the bottom half into the top and water can flow in both halves, equivalent to conduction in both bands.

The mechanism of semiconduction described so far is dependent only on the structure of the material. It is called the *intrinsic conductivity*.

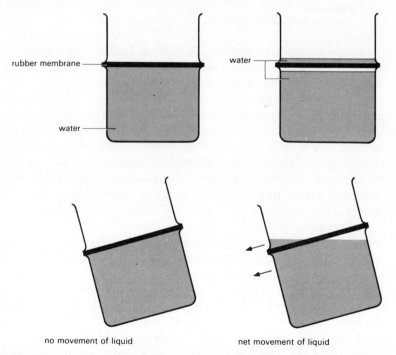

Figure 18.8
A simple hydraulic analogue of electron and hole conduction in a semiconductor.

Impurities in semiconductors

The crystal lattice of silicon is exactly like that of the diamond. Each atom has four identical nearest neighbours at the same distance in the tetrahedral geometry, familiar to us from the stereochemistry of carbon. The three dimensional crystal produced by connecting these tetrahedral units can be constructed from identical cubical units, giving the crystal its cubic symmetry (figure 18.9).

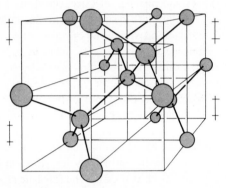

Figure 18.9
The crystal structure of diamond.

Now consider an impurity atom from Group V in the Periodic Table with five valency electrons; provided the atom is small enough to fit into the lattice, it can complete the four bonds but has one electron spare.

The single isolated energy level of the excess electron is inside the energy gap but is usually very much closer to the conduction band than the valence band. The separation of the impurity level and the conduction band is usually so small that there is a very high probability of the impurity electron gaining enough thermal energy to pass into the conduction band. The net result is that the Group V impurity atom remains as a positively charged atom in the lattice (without a gap in the valence band so that it is *not* a hole, it is a *fixed* charge) and an additional electron appears in the conduction band (figure 18.10).

If we put in a Group III impurity atom with three valency electrons, there is one bond incomplete, so that there is a hole but it is initially uncharged. This would be a fixed hole and would contribute nothing to conductivity. However, like the spare electron in the Group V atom, it provides another localized energy level just above the top of the valence band. There is then a very high probability that an electron from the filled valence band will jump up to the impurity level; this will give it a fixed negative charge and will leave a positive hole in the valence band free to move throughout the lattice (figure 18.11). Although it may not fit the normal usage we speak of the impurity atoms as being 'ionized'.

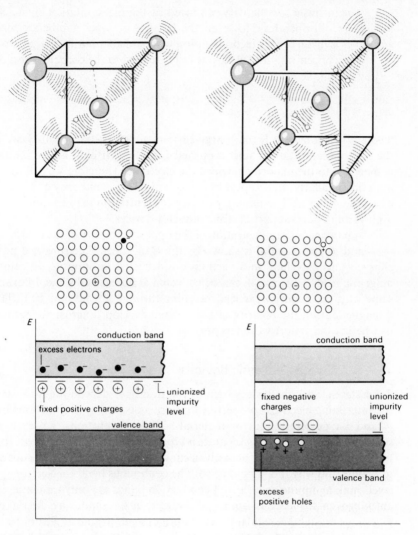

Figure 18.10
The donation of excess electrons to the
conduction band by Group V
impurity in a Group IV semiconductor.

Figure 18.11
The production of positive holes by
a Group III impurity in a Group
IV semiconductor.

A Group V impurity atom gives an electron to the conduction band and
increases the number of *negative* charge carriers. Hence it is called an *n*-type
impurity or donor. Similarly, a Group III impurity element takes electrons out
of the valence band to give *positive* charge carriers, and is called an acceptor,
or *p*-type impurity.

The intrinsic conductivity produced by thermal excitation giving electron hole pairs still operates. However, the balance of equal numbers of electron and holes is disturbed by recombination with added impurity charge carriers.

The concentration of electrons $[n]$ is related to the concentration of hole $[p]$ by the simple equation

$$k = [n] [p]$$

where k is a constant. If the charge carriers added by the impurity are n-type then the concentration of holes is pushed down and virtually all the conductivity is then due to the majority carriers, the electrons. Therefore we call this n-type material. Similarly, an acceptor impurity will give us a majority of positive hole or p-type material. The minority carrier concentration may be quite small, but it plays an important part in semiconductor devices.

A convincing demonstration of the opposite polarity of the charge carriers in n- and p-type material is shown by the Hall Effect. Electrons and positively charged ions travelling in a vacuum are deflected in opposite directions in a magnetic field; their charge can be collected at an earthed plate. In exactly the same way, the majority carriers travelling through a solid may be deflected in a magnetic field; the polarity of the voltage developed across the sample tells us whether the majority carriers are n or p.

Solid state electronic devices

Solid state electronic devices such as the transistor depend for their action on the interesting electrical properties that are associated with junctions between p- and n-type semiconductor materials. The p–n junctions, however, cannot simply be made by bringing separate p-type and n-type semiconductors together because the junctions must actually occur within a single crystal. It is not difficult to understand why this should be so. The continuous bands of electronic energy levels making up the valence and conduction bands are only formed if we have an unbroken sequence of atoms, all bonded in the same way. At the edge of any single crystal this regular array of atoms comes to an end. So do the valence and conduction bands. In a polycrystalline solid the boundaries between one crystal and the next present major barriers to the movements of electrons and holes which are essential to semiconductor action.

A typical transistor consists of a single crystal in which n–p–n or p–n–p junctions are present. In other words, within a single crystal lattice both p- and n-type impurity atoms are present in regions of the crystal, so as to give rise to the sequence n–p–n or p–n–p. We shall return to this single crystal aspect later.

Semiconductor materials

The commonest semiconductor material is silicon, although germanium devices do still exist. Let us see how these elements are obtained from natural sources, purified, and manufactured into transistors before we highlight the differences that give us silicon integrated circuits.

Germanium There is no major mineral source of germanium, but its oxide GeO_2 occurs in flue dusts from zinc and lead smelting processes. It is also found in flue dusts from gasworks using certain varieties of coal. The first stage in producing the metal is to purify the oxide. To do this, the oxide is dissolved in hydrochloric acid, in the presence of nitric acid, so as to form the volatile tetrachloride $GeCl_4$. The tetrachloride is distilled off and the oxide regenerated by pouring it into an excess of water. These processes can be summarized as

$$GeO_2(s) + 4HCl(aq) \rightleftharpoons GeCl_4(l) + 2H_2O(l)$$

A single distillation is not enough to obtain GeO_2 of high purity so the whole cycle is repeated several times. Finally, the purified oxide is reduced to the metal by heating it in a graphite boat in a stream of hydrogen.

$$GeO_2(s) + 2H_2(g) \longrightarrow Ge(s) + 2H_2O(g)$$

The germanium produced at this stage is not sufficiently pure for use in semiconductors. When we have seen how silicon is obtained, we will deal with the purification of both elements together.

Silicon As mentioned in the Background reading 'Minerals and their structure' (*Students' book I*, page 216), the enormous quantity of silicon present in the Earth's crust is mainly in the form of silicates (from which it is extremely difficult to extract) and silicon dioxide, SiO_2. Partly because of the very high bond energy of the Si—O bond ($466\,kJ\,mol^{-1}$), it is only possible to reduce this stable oxide directly with a powerful reducing agent such as metallic sodium:

$$SiO_2(s) + 4Na(l) \longrightarrow Si(s) + 2Na_2O(l)$$

If a less energetic reducing agent, such as magnesium, is used, a compound rather than the free element is formed.

$$SiO_2(s) + 4Mg(l) \longrightarrow Mg_2Si(s) + 2MgO(s)$$

The magnesium silicide can then be used to make silane, SiH_4.

$$Mg_2Si(s) + 4HCl(aq) \longrightarrow SiH_4(g) + 2MgCl_2(aq)$$

This can be decomposed on a heated surface to give silicon.

$$SiH_4(g) \longrightarrow Si(s) + 2H_2(g)$$

This route is in fact useless for the preparation of silicon of semiconductor grade. It is, however, of historical interest as it was the reaction first used to show the existence of silane.

The best routes to silicon are those which involve the reduction of its chlorides by hydrogen. The oxide can be chlorinated directly, in the presence of a reducing agent, thus:

$$SiO_2(s) + 2C(s) + 2Cl_2(g) \longrightarrow 2CO(g) + SiCl_4(g)$$

followed by

$$SiCl_4(g) + 2H_2(g) \longrightarrow Si(s) + 4HCl(g)$$

In practice, trichlorosilane $SiHCl_3$ is easier to reduce than $SiCl_4$. It is currently used in the preferred commercial route. Trichlorosilane is prepared by the following reactions:

$$FeO(s) + 5C(s) + 2SiO_2(s) \longrightarrow FeSi_2(s) + 5CO(g)$$
$$FeSi_2(s) + 8HCl(g) \longrightarrow FeCl_2(s) + 2SiHCl_3(g) + 3H_2(g)$$

The chlorides of silicon are reactive chemicals and must be prevented from attacking the containers, pipes, and valves used to handle them. The only materials which may be used in the apparatus for reducing them are quartz, silver, stainless steel, and the inert plastic polytetrafluoroethene (PTFE).

The purification of germanium and silicon Solid state electronic devices require very pure starting materials for their manufacture. The addition of only one part in 10^9 of an electrically active impurity can, for example, halve the resistivity of germanium. Compare this with the natural concentrations of trace impurities in foodstuffs which are considered to be harmless: one part of arsenic in 10^6 is quite acceptable!

The process by which very high purity germanium and silicon are obtained is known as *zone refining*. In general, impurities are more soluble in the melt than in the solid and this is the property which is exploited in the method.

A solid bar of the material to be purified is passed slowly through a source of heat so that a liquid zone is created which travels along the length of the bar. The sections of the bar solidifying behind the travelling liquid zone will always be purer than the liquid until a state of affairs is reached where the impurities are equally distributed between liquid and solid. The zone cannot then purify the material any further. What we have just described is a single zone 'pass'. We can now start a second liquid zone passing along the bar, which once again will sweep impurities along with it leaving purer solid behind.

During zone refining, germanium can be held in a graphite crucible but silicon presents a very difficult problem. It will react with almost any material with which it comes into contact, therefore nothing must touch it during the refining process.

In practice the silicon rod is supported vertically and a single floating zone is passed down it. This zone can be maintained suspended by surface tension between the solid portions of the rod, something of a conjuring trick! Only one molten zone at a time can be passed along the rod so that the refining process is a slow one. Figure 18.12 illustrates this floating zone refining and in figure 18.13 we see a photograph of the process in operation.

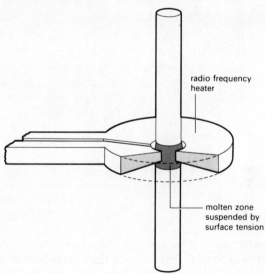

radio frequency
heater

molten zone
suspended by
surface tension

Figure 18.12
Floating zone refining for silicon.

As we saw earlier, in order to make sure that valence and conduction bands are continuous, we need the most perfect single crystals we can get.

To grow large single crystals, a seeding technique is used. The molten element is brought into contact with an already formed single crystal so that crystallization begins at the crystal–melt interface. In one technique, known as crystal pulling, the seeding crystal is attached to the end of a shaft which is

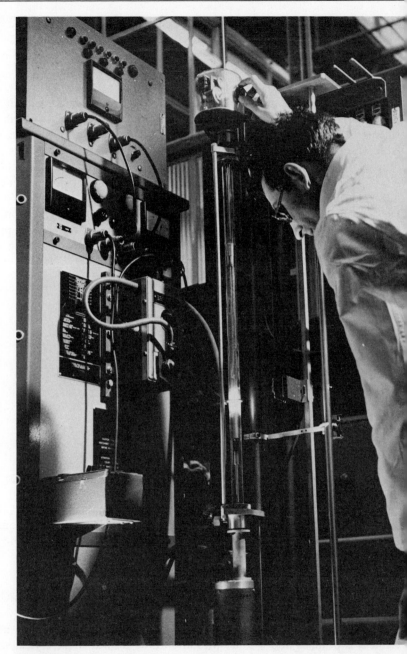

Figure 18.13
A laboratory scale floating zone system used in the purification of single crystal silicon rod
Photograph, Mullard Ltd.

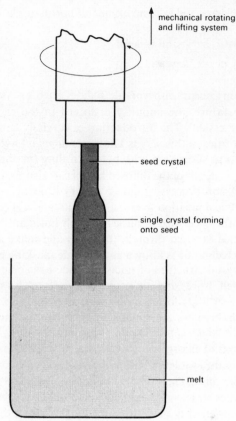

↑ mechanical rotating
and lifting system

— seed crystal

— single crystal forming
onto seed

— melt

Figure 18.14
The production of a single crystal by the crystal pulling technique.

slowly rotated and lifted out of the melt as crystallization proceeds. Figure 18.14
illustrates this procedure.

The crucible holding the melt has to be made from pure quartz. Although
some silicon may react with the quartz (silicon dioxide, SiO_2) to produce an
unstable monoxide of formula SiO, the main impurity introduced into the crystal
is oxygen.

Making transistors

Germanium transistors had to be made practically one at a time, for example
by melting metallic indium contacts to give *p*-type alloy junctions on either side
of an *n*-type chip, and by similar metallurgical processes. We can take advantage
of the chemical reactivity of silicon, however, to introduce *p*- or *n*-type impurities
into the surface of a single crystal slice. For example, silicon will reduce borate

or phosphate glasses and take up elemental boron or phosphorus.

$$3Si + 2B_2O_3 \longrightarrow 3SiO_2 + 4B$$
$$3Si + 2P_2O_3 \longrightarrow 3SiO_2 + 4P$$

Similar reactions occur with volatile halides such as $AsCl_3$ and PCl_3. At the reaction temperature, the impurity dissolved in the crystal surface can then diffuse into the crystal. The introduction of carefully controlled amounts of p or n-type impurities in this way is known as *doping*. Layers of p and n material can be built up in succession and, by controlling the time and temperature of the diffusion, the depth of the diffused layers can also be accurately controlled.

This diffusion process would be of very little advantage if we could only make one large p–n junction over the whole area of a slice; however, if we cover the surface of the slice with a mask in which holes are cut, we can then dope the areas of slice exposed through the holes and make a set of transistors on one slice. This technique is known as the oxide masking process and is possible because the crystal structure of silicon dioxide has some similarities to that of the element itself. If we carefully oxidize the surface of a silicon slice, a strongly adherent layer of SiO_2 forms on it, so that doping agents cannot get at the silicon beneath. Now we must make an array of small holes in the silica mask exposing silicon which can be doped to form transistors.

A thin layer of an organic material, the photoresist, is spread over the oxide and exposed to ultra-violet radiation through a negative which has the required pattern of holes on it. The organic material polymerizes where the ultra-violet radiation reaches it, leaving unpolymerized material where we wish the holes to be. The photoresist is then 'developed' by washing the slice in a solvent to remove unpolymerized material. We now have a tough rubbery layer covering the slice with holes exposing silica. These are just where we want them to be.

Now we must get rid of the silica in the holes so as to expose the silicon beneath, ready for doping in a suitable vapour. This is done by dipping the slice in a solution of hydrogen fluoride which results in the reaction

$$SiO_2(s) + 6HF(aq) \longrightarrow H_2SiF_6(aq) + 2H_2O(l)$$

The polymer can now also be stripped off the slice by means of a suitable solvent. Treatment in a furnace deposits the doping element in the holes; here it diffuses into the silicon, producing an array of n-type regions. By repeating the whole process we can put down a p layer. Thus, a large number of transistors can be mass-produced in the surface of the silicon slice.

The sequence of events is shown in figure 18.15. Finally, the individual transistors are separated by scratching the back of the slice with a diamond so that it will split under light pressure.

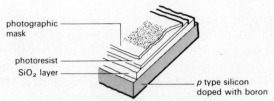

photographic mask

photoresist

SiO₂ layer

p type silicon doped with boron

First stages in the production of transistors on a slice of single crystal silicon. The silicon surface has been oxidized, the photoresist material is spread, and the photographic mask is in place.

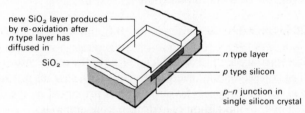

SiO₂

p type silicon

silicon surface exposed ready for diffusion process

A later stage, following exposure, development of the photoresist, etching by hydrofluoric acid, and the removal of the photoresist with a solvent.

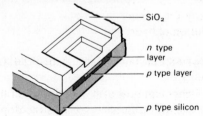

new SiO₂ layer produced by re-oxidation after n type layer has diffused in

SiO₂

n type layer

p type silicon

p–n junction in single silicon crystal

Diffusion now takes place; a new layer of n type material diffuses into the p type silicon to form a p–n junction, and a fresh oxide layer is added.

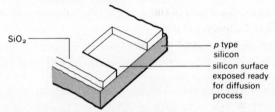

SiO₂

n type layer

p type layer

p type silicon

A second diffusion process has been completed and a p type layer has bèen diffused in through an oxide window to form the complete p–n–p junction transistor.

Figure 18.15
Making transistors in a silicon slice.

The integrated circuit

To make a complete working circuit we add conducting layers of evaporate aluminium to link up the transistors or junction diodes corresponding to t wiring in conventional electronic circuits. The adherent thermal oxide on t silicon surface now provides the electrical separation between the conductii pattern and the planar devices in the surface of the slice. Holes etched in t oxide allow contacts to be made between the conductor pattern and the termina of the individual devices. Some devices use a thin oxide layer to make a capa tative contact inside a transistor, the field effect transistor or FET. These a particularly useful in computer memory circuits.

The choice of particular sequences of active layers and oxides gives ri to numerous different device technologies. As well as the thermal oxide, sor of these use chemically deposited silicon dioxide or silicon nitride as an inte mediate insulator or a protective coating. The oxide can be deposited, usii the simple gas phase reaction:

$$SiH_4 + 2O_2 \longrightarrow SiO_2 + 2H_2O$$

but the reaction temperature is inconveniently low. Using CO_2 or NO to provi the oxygen, the dissociation reaction

$$2CO_2 \rightleftharpoons 2CO + O_2$$

keeps the partial pressure of oxygen low until the temperature is raised.

The corresponding reaction of silane with gaseous nitrogen to deposit t nitride is energetically less favourable. It is promoted by the energy of reacti provided by the elimination of hydrogen from ammonia:

$$3SiH_4 + 4NH_3 \longrightarrow Si_3N_4 + 12H_2$$

Where next?

The complex patterns required for the photolithographic processes start as lar scale drawings that are reduced photographically until the actual dimensio of some of the areas in the pattern are only a few micrometres acro (1 micrometre $= 10^{-6}$ m). Some conception of the actual size of a finished circu on a chip can be gained from the photograph of an early integrated circuit figure 18.16. Modern circuits can now have thousands of devices in a chip this size. The main trend is towards ever finer geometry to pack more devic onto each chip. For example, an electron beam may be used instead of lig to polymerize the photoresist, in order to make masks with dimensions less tha a micrometer (figure 18.17).

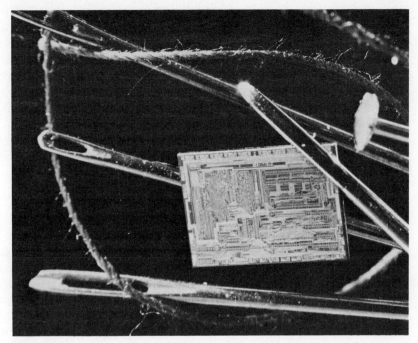

Figure 18.16
An Integrated circuit containing thousands of components fits the eye of a needle. The 'rope'
is ordinary sewing cotton.
Photograph, Mullard Ltd.

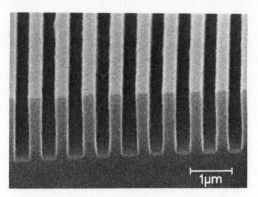

Figure 18.17
An electron microscope picture of a photoresist exposed in an electron beam pattern
generator. This shows the deep channels produced in a film that is only 2 μm thick.
Some impression of the size comes from realizing that ten of these pictures, or one
hundred individual channels, would just fit across the diameter of a human hair.
Photograph, Philips Research Laboratories.

Chemical processing using liquid etches runs into a number of practical problems in small scale geometry; a variety of gas phase reactions using a low pressure electrical discharge to generate reactive ions and free radicals is now in common use. The chemistry of these plasma-activated reactions is complex but they already have important applications in production. Other physico chemical processes, such as the use of powerful ion beams to sputter away unwanted material or to implant doping atoms just where they are wanted, are increasingly used. For really high speed applications we can expect the III–V compounds which are analogues of the Group IV elements to come into use. Here organo-metallic compounds provide the chemical deposition processes, for example:

$$Ga(CH_3)_3 + AsH_3 \longrightarrow GaAs + 3CH_4$$

These compounds are particularly important where the emission and detection of visible light are needed. They form the basis of the light-emitting diodes used in display, and the semiconductor lasers for the video disc and the audio compact disc. But that is another story!

Whatever the nature of the materials and processes used in electronics tomorrow, electronic devices will become ever more commonplace. The technological revolution of microelectronics depends upon chemists who can go beyond the confines of their original discipline and work in close collaboration with physicists and electronic engineers.

BACKGROUND READING 3
Lead pollution

Lead is one of the most troublesome pollutants of the environment. All lead compounds are poisonous, damaging the nervous system and other organs. Children are particularly prone to lead poisoning. They may suffer from hyper activity – a state of frenzied restlessness – and their development can be retarded. Lead poisoning is usually chronic, that is slow, as constant small doses of lead build up in the body, from which they are eliminated only slowly. Acute, sudden lead poisoning from one large dose is rare, though it does occur.

Lead pollution is widespread. It affects the remotest areas, as can be seen from analysis of ice layers in Greenland. The ice is deposited in annual layers which can be dated, rather like tree rings. Lead pollution has increased sharply since 1950. The surge is due almost entirely to the use of lead in petrol, allowing it to enter the air in the exhaust gases of cars and so to be distributed worldwide through wind, rain, and snow. Once in the environment, lead does not readily disperse, for there are no natural processes to eliminate it, other than slow washing away by rainfall and other forces of erosion. So lead builds up in soil and dust in ever increasing amounts.

Lead is not one of the most abundant elements in the Earth's crust. Other elements in Group IV – silicon, and carbon – are much more common, as are many in other groups. Nevertheless there is a fair amount present, mostly as galena (lead sulphide, PbS), and each year about 25 000 tonnes reach the soil, the air, and living organisms through natural processes. But far more, currently about 450 000 tonnes per year, enters the environment through human activities.

There are other heavy metals in the environment which are hazardous, notably mercury and cadmium. But the sheer volume of lead makes it a far more serious pollutant than those.

Despite the recent increase in pollution, lead poisoning is nothing new. In particular the ancient Romans suffered from it, since their towns were supplied with water by way of lead-lined aqueducts and storage tanks, and they also used kitchen and drinking vessels made of lead. Historians have even blamed the decline of the Roman empire on mass lead poisoning, as well as suggesting it as an explanation of the famously crazy behaviour of some of the Roman emperors. Although the Romans had little chemical knowledge, some did suspect that lead might be harmful. In fact vessels made of pewter, an alloy mostly of lead and tin, continued in domestic use until the nineteenth century. Since then pewter has been made without lead.

In modern times humans absorb lead in various ways. Food may contain small amounts of lead from the solder used in cans or from lead-glazed pottery, or deposited from polluted air. Old houses often have lead water pipes, and sometimes lead-lined storage tanks which are quite as dangerous as Roman ones. Such houses may also have paintwork containing lead (which is not used in most modern paint). Lead from paint can get into house dust and, more importantly, can fatally poison young children who gnaw old painted items. Factories processing or using lead can give off harmful fumes and dust. Lead is much used, for example in car batteries, roofing, and solder.

The most notorious modern source of lead pollution is petrol, which is now the largest source of airborne lead in Britain and most developed countries. Organic lead compounds are added to petrol to increase its octane rating, that is to allow low-grade, relatively cheap hydrocarbons to give a similarly good performance in car engines to that of hydrocarbons of higher octane rating (see the Background reading 'octane number of petrol hydrocarbons', *Students' book I*, pages 298–301). However, airborne lead is seldom the chief source of lead taken into the body, except in a few urban areas.

Lead from food

Some lead may enter food naturally through the weathering of lead-rich minerals, and hence into the soil and plants. There are few regions where this is a significant

problem. Most lead in food comes from human activities, in three main ways
First, the lead solder used in making cans may contaminate their content
Second, lead plumbing systems may pollute tap water used in cooking. Third
lead in air and dust, coming from car exhausts and factories, may contaminat
soil, crops, and food: this can seriously affect fruit and vegetables grown i
polluted areas. In comparison with these, lead-glazed pottery is a relatively un
important source and lead glazes are less used now than formerly.

A British study of dietary intake of lead shows that the amount consume
daily by an adult is usually in the range 70 to 150 µg (micrograms) a day, tendin
towards the lower figure.

About 15% of this comes from soldered can seams: ordinary solder is a
alloy of lead and tin. Canned foods contribute much more than canne
beverages, partly because of the lead-dissolving acidity of certain foods such
as canned fruit, and partly because of the design of the cans. Conventional foo
cans have soldered bases and side seams. Increasingly, drink cans are being mad
with one-piece, seamless bodies. (All can tops are crimped on over a plasti
sealing ring.) The average concentration of lead in canned drinks is less than
$20\,\mu g\,kg^{-1}$. Levels in canned food are often more than ten times as high.

British regulations concerning foods intended for babies and youn
children specify a maximum lead level of $200\,\mu g\,kg^{-1}$. Manufacturers of canne
baby foods have had to use expensive pure tin solder, so that the food contain
only such lead as was already present in the ingredients.

The Metal Box company, Britain's largest can manufacturer, has studie
ways of reducing lead contamination of foods of all kinds. No practical wa
was found of covering the solder with a protective lacquer coating: althoug
tinplate is routinely lacquered before forming the can, it is hard to coat an interio
seam. Tin solder was too expensive. The solution has been to weld seams o
to use seamless, beverage-type cans. In 1982 Metal Box announced that it woul
phase out lead solder entirely by 1985.

A less common reason for high lead levels in food is contamination o
vegetables and fruits while they are actually growing. Rarely, this is caused by
the plant taking up lead compounds naturally present in the soil. More usuall
it is due to airborne lead contaminating the place where the plant is growing
Lead enters the plant in two ways at once. It falls directly on to it: this ha
a particularly strong effect on the broad leaves of vegetables such as cabbage
and lettuce. And lead also falls on to and builds up in the soil so that it is take
up through the plant's roots into leaves and fruits. Unacceptably high lead level
have been found in Britain both in produce grown on the urban allotment
which are popular with town dwellers and in the blackberries which people ofte
thoughtlessly pick from roadside hedges. The heavier the traffic on the road
the more poisonous the berries. It is advisable to pick blackberries and othe
wild fruits only from places safely away from busy roads.

Lead from tap water

There is seldom much lead in mains water. It gets into tap water from domestic lead piping. Lead was formerly the chief material for house water pipes because it was easy to shape and solder. It was also sometimes used to line water storage tanks to protect their iron or steel walls from corrosion. Modern houses have iron or copper pipes, or more recently plastic ones for cold water only. Lead solder is still used in some metal pipe joints but adds little to lead contamination since most of it is on the outside of the pipe. Most modern pipe joints are solderless. Also in modern plumbing systems, drinking water is drawn directly from the main, not from a tank.

The pH of the water supply affects its 'plumbosolvency' – the extent to which it dissolves lead (Latin *plumbum*, lead, as also in the chemical symbol Pb). Soft water tends to be slightly acidic and therefore plumbosolvent, while hard water is faintly alkaline and much less plumbosolvent. However, some hard water is slightly plumbosolvent because of other substances in it. It is not yet fully known which substances are chiefly responsible.

The amount of lead in tap water also depends on the length of lead piping it travels through and the amount of time it spends in contact with lead. Thus in Scotland, where the water is mostly very soft and where there are many older houses with lead piping and sometimes with drinking water drawn from lead-lined tanks, high concentrations of lead are often found, and this is clearly reflected in raised blood lead levels of the inhabitants. It is recommended that each morning the cold tap should be allowed to run for a minute or so to flush out water which has stood in the pipes all night, accumulating lead.

The only satisfactory way to eliminate lead from tap water is to replace lead pipes or tanks. After this the high blood levels of the house's inhabitants return to normal in about six months. But replacement is expensive: in 1981 the British National Water Council estimated the average cost at £600 per dwelling. Of this £200 would fall to the water authority for the pipe connecting the house to the main, and £400 to the householder for his or her own pipework. Replacing a lead-lined tank would cost an extra £150. In 1982 the Government extended the home improvement grant scheme to cover such work.

Lead from air

For most people uptake of lead from air is much less than that from food or water. In Britain the average concentration of lead in the air of rural and most urban areas is below $0.5\,\mu g\,m^{-3}$, a safely low level.

Nine-tenths of all airborne lead comes from the exhausts of cars using leaded petrol. Lead, in the form of tetraethyl lead, $(C_2H_5)_4Pb$, and similar compounds, has been added to petrol since the mid 1930s. Although some other substances can be used instead to give fuel satisfactory burning properties, these are uneconomically expensive. In 1978 in the United Kingdom 18 million tonnes of petrol were used, to which 10 300 tonnes of lead as organic lead compounds were added. About 7000 tonnes of this lead were re-emitted along with the other pollutants of car exhaust, mainly in the form of lead halides which in the atmosphere are gradually converted into oxides, carbonates, and sulphates.

Airborne lead levels can be dangerously high near busy roads. A survey carried out on the centre reservation of a motorway in 1975 showed an annual mean concentration of over $12 \mu g\,m^{-3}$, though more recently this has fallen to $9 \mu g\,m^{-3}$ as the amount of lead in petrol has been reduced. Near a busy urban road the concentration is usually in the range 2 to $6 \mu g\,m^{-3}$. This is high enough to cause concern about its effect on children. There has been powerful lobbying for the total removal of lead from petrol.

Car manufacturers and the oil industry have been opposed to the introduction of lead-free petrol. It is more expensive to produce; car engines have to be specially designed to use it; and even then they use more fuel. Nevertheless the United States has for several years required all new cars to run on unleaded fuel. The EEC countries are committed to phase lead out by 1988.

Lead in paint and dust

Before the insidious effects of lead were known it was much used in paint and even in cosmetics. Various lead compounds give durable, opaque white, red or yellow colours. Most countries have now banned leaded paint for domestic use and banned the importation of lead-painted goods, especially toys. However, much old paintwork remains, both in and out of doors.

Small children, who tend to suck and chew anything that they pick up, including broken-off pieces of domestic woodwork or plaster, are particularly at risk of acute poisoning. In England and Wales from 1968 to 1976, 13 children are known to have died from lead poisoning. Four of these deaths were attributed specifically to lead paint.

More usually, old lead paintwork deteriorates and flakes, so that lead gets into dust, where it is joined by lead from other sources such as the air. Lead levels in town dust are typically in the range 500 to $5000 \mu g\,g^{-1}$, and higher outdoors than in. The difference is due partly to car exhausts and partly to the fact that the paint for the yellow lines on roads contains about 1% of lead by mass from the use of lead chromate(VI) as pigment.

It has been estimated that a two-year-old child takes in about 0.1 g a day of soil and dust from licking fingers and dirty objects. Thus in towns quite a lot of lead can be ingested in this way, enough to cause longterm though not acute effects.

Some British local authorities have analysed dust in school playgrounds and classrooms and have found levels over $1000 \mu g\,g^{-1}$. The highest figures have been associated with old paintwork in poor condition, but the nearness of roads has also been found to have an influence. In 1981 the Greater London Council reduced the upper allowable limit of lead in school dust from $5000\,\mu g\,g^{-1}$ to $500\,\mu g\,g^{-1}$. The regulation is rather hard to enforce since British inspectors do not have a portable instrument for on-the-spot analysis. In the United States they use portable X-ray fluorescence analysers. Schools can be made safe in the short term by vacuum cleaning and washing, and in the long term by removing all old paint.

Lead fishing weights

Lead is used for anglers' fishing weights because it is heavy, relatively cheap, and also soft, so that a split lead shot can be squeezed shut around a line to fix it in place. Most anglers do this with their teeth, a habit which might be dangerous, in particular to the young.

But the chief sufferers are waterfowl, in Britain especially swans, which accidentally eat discarded weights along with the water plants which they scoop up from river beds. In England alone about 3000 swans die from lead poisoning every year, solely in this way. They perish slowly and painfully from liver and kidney damage, suffering also from paralysis which causes their necks to droop in a sadly characteristic way.

Appeals have been made to anglers not to discard lead weights, but with the best intentions many are lost by accident. So far no acceptable substitute for split lead shot has been found. A putty containing tungsten, a metal even heavier than lead, works quite well but is expensive. Steel weights also cost more than lead: the metal is cheaper but manufacturing costs are higher. They are awkward to use because the hard metal cannot be pressed on to the line, and the weight must be attached in some other way.

18.4
GROUP V: THE PROPERTIES OF SOME NITROGEN COMPOUNDS

In its compounds, nitrogen shows oxidation numbers ranging from $+5$ in nitric acid and nitrates to -3 in ammonia and ammonium compounds. In some of the following experiments you will need to look at the chart of electrode potentials (figure 18.18):

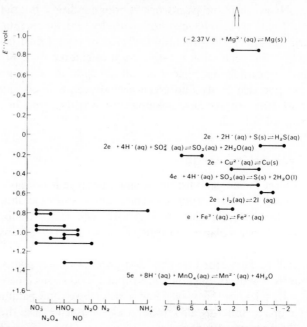

Figure 18.18
Electrode potentials at pH = 0.

Notes on the formulae shown on the horizontal axis of figure 18.18

NO_3^- is the nitrate ion. It is present in solutions of nitric acid and in nitrates. Nitric acid is a strong acid and may be regarded as being completely ionized in aqueous solution.

N_2O_4 is a dimer of NO_2. When nitrogen has this oxidation number in an oxide, an equilibrium mixture of NO_2 and N_2O_4 is present. It is the nitrogen dioxide, NO_2, which has the brown colour. This oxide is acidic and reacts with alkalis to give mixtures of nitrate, NO_3^-, and nitrite, NO_2^- ions.

HNO_2 is the formula for nitrous acid. At pH = 0, nitrous acid is shown as HNO_2 and not as NO_2^- because, unlike nitric acid, nitrous acid is a weak acid.

NO is the formula for nitrogen monoxide. It is a colourless, neutral gas which reacts immediately on contact with the oxygen of the air to give nitrogen dioxide, NO_2.

N_2O is the formula for dinitrogen oxide, a colourless, neutral gas.

NH_4^+ is the formula for the ammonium ion. Since these electrode potentials are given at pH $= 0$, ammonia itself is not shown.

EXPERIMENT 18.4a
Some reactions of nitric acid, HNO₃, in concentrated solution

CAUTION: Concentrated nitric acid is corrosive. Be particularly careful to avoid spillages.

Safety glasses must be worn.

Concentrated nitric acid is a powerful oxidizing agent. Using figure 18.18, predict the outcome of reactions between concentrated nitric acid and

1 copper
2 sulphur dioxide solution
3 potassium iodide solution

In each case confirm your prediction experimentally. Record your prediction, your experimental method, and your result, together with an ionic equation, in your notebook.

EXPERIMENT 18.4b
Some reactions of nitrous acid, HNO₂, in dilute solution

Nitrous acid can be made as required by acidifying sodium nitrite. To about 0.5 g of sodium nitrite (*TAKE CARE:* it is very poisonous) add 10 cm³ of M hydrochloric acid. This solution contains nitrous acid and sodium chloride.

Using figure 18.18, predict the outcome of reactions between nitrous acid and

1 potassium iodide solution
2 potassium manganate(VII) solution.

Confirm your predictions experimentally.

Nitrous acid will *disproportionate* on being warmed; that is to say, it reacts in such a way that the oxidation number of the nitrogen in one product is greater than $+3$ and in the other product is less than $+3$. What possible products of this disproportionation can you suggest? Warm a sample of nitrous acid, and satisfy yourself that disproportionation does in fact take place.

Record the predictions, methods, and results of this experiment in your notebook, and write equations for the reactions wherever possible. In the disproportionation, several products are possible so that a single equation cannot be written.

EXPERIMENT 18.4c
Some reactions of ammonia solution

Ammonia reacts with water reversibly, thus:

$$NH_3(aq) + H_2O(l) \rightleftharpoons NH_4^+(aq) + OH^-(aq)$$

The equilibrium constant for this reaction is about $2 \times 10^{-5}\,mol\,dm^{-3}$, which means that ammonia is a weak base. Ammonia solution, however, contains some hydroxide ions and is therefore alkaline. Ammonia can also act as a ligand, as was explained in Topic 16.

Bearing these things in mind, investigate the action of M ammonia solution, $NH_3(aq)$, on 0.1M solutions of

1 lead(II) ions, $Pb^{2+}(aq)$
3 copper(II) ions, $Cu^{2+}(aq)$
3 zinc ions, $Zn^{2+}(aq)$

and try to explain the observations that you make.

Investigate the action of

1 M ammonia solution, $NH_3(aq)$, and
2 a mixture of equal volumes of M ammonia solution and M solution of ammonium ions, $NH_4^+(aq)$

on 0.1M solution of magnesium ions, $Mg^{2+}(aq)$. Try to account for the observations that you make.

18.5
GROUP VI: THE PROPERTIES OF SOME SULPHUR COMPOUNDS

If we apply the same considerations to oxygen and sulphur as we applied to nitrogen and phosphorus earlier in this Topic, we can see that electrons in oxygen cannot be promoted from the second to the third energy shell for the formation of compounds. In sulphur, however, we would expect that electrons could be promoted within the third energy shell from 3s and 3p to 3d.

This promotion makes it possible for sulphur to form up to six covalent bonds, as shown in figure 18.19. Formation of two covalent bonds occurs in hydrogen sulphide (oxidation number -2) and sulphur dichloride ($+2$); four bonds are formed in the sulphite ion and sulphur dioxide ($+4$), and six bonds in the sulphate ion and sulphur trioxide ($+6$).

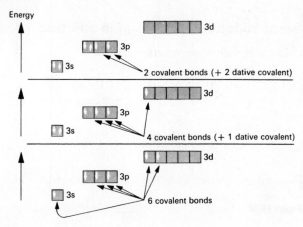

Figure 18.19

Sulphur can have oxidation numbers ranging from $+6$ to -2 as shown in the following table:

$+6$ sulphate,	SO_4^{2-}	sulphur trioxide, SO_3
$+5$ dithionate,	$S_2O_6^{2-}$	
$+4$ sulphite,	SO_3^{2-}	sulphur dioxide, SO_2
$+3$ dithionite,	$S_2O_4^{2-}$	
$+2$ thiosulphate,	$S_2O_3^{2-}$	sulphur dichloride, SCl_2
0 sulphur	S_8	
-2 sulphide,	S^{2-}	hydrogen sulphide, H_2S

The electrons are delocalized in sulphate and sulphite ions in a similar manner to those in nitrate and nitrite ions (figure 18.20).

EXPERIMENT 18.5a
Disproportionation in some sulphur compounds

Using the charts of electrode potentials (figures 18.21 and 18.22) predict what disproportionation of sulphur could occur

1 with thiosulphate ions in acid solution
2 with sulphur itself in alkaline solution.

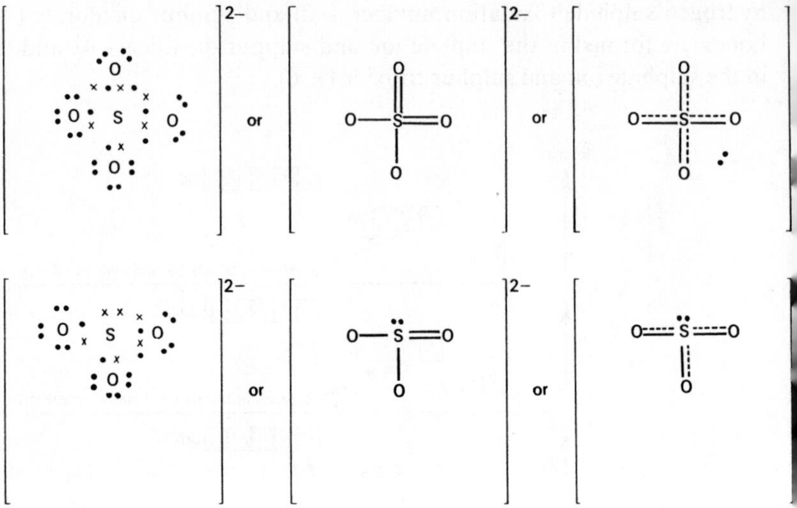

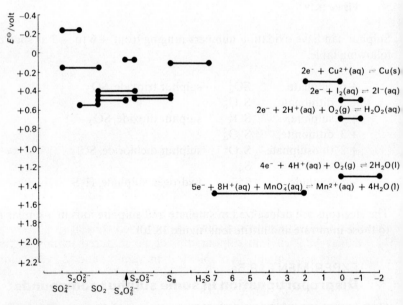

Figure 18.20

Figure 18.21
Electrode potentials at pH = 0.

Test your predictions experimentally. In **2** you might use a method based on reaction **1** to decide whether there has been a disproportionation.

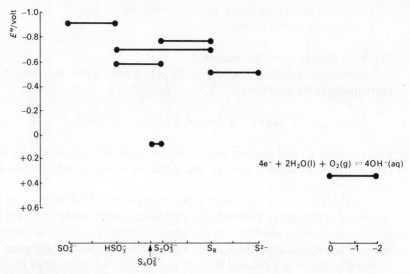

Figure 18.22
Electrode potentials at pH = 14.

EXPERIMENT 18.5b
Some reactions of sulphite ions, SO_3^{2-}(aq), and of sulphide ions, S^{2-}(aq)

Sulphurous acid, H_2SO_3, is a weak acid formed when the gas sulphur dioxide is dissolved in water. This situation gives rise to a series of equilibria:

$$SO_2(g) + H_2O(l) \rightleftharpoons H_2SO_3(aq) \rightleftharpoons H^+(aq) + HSO_3^-(aq) \rightleftharpoons$$

$$2H^+(aq) + SO_3^{2-}(aq)$$

The addition of dilute hydrochloric acid to a sulphite, therefore, displaces all these equilibria to the left, producing sulphur dioxide.

TAKE CARE: sulphur dioxide has a strong, choking smell and is especially dangerous to anyone suffering from a respiratory complaint.

Satisfy yourself that sulphur dioxide is produced in this reaction, by warming very small quantities of solid sodium sulphite and M hydrochloric acid in a fume cupboard.

The presence of sulphur dioxide can be verified by holding a piece of filter paper soaked in acidified potassium manganate(VII) at the mouth of the test-tube in which the gas is being made; the colour of the manganate(VII) ion is taken away.

Using the electrode potential chart at pH = 0, predict what should happen when sodium sulphite solution is added to

1 acidified potassium manganate(VII) solution
2 potassium iodide solution.

Test your predictions experimentally.

Hydrogen sulphide is a very weak acid which, when dissolved in water participates in the equilibria:

$$H_2S(aq) \rightleftharpoons H^+(aq) + HS^-(aq) \rightleftharpoons 2H^+(aq) + S^{2-}(aq)$$

Explain what should happen when sodium sulphide is acidified with M HCl (The situation is rather similar to the equilibria involving sulphur dioxide and water mentioned above.)

If this reaction is attempted experimentally it MUST be done in a fume cupboard. Hydrogen sulphide has an unpleasant smell ('rotten eggs') but besides this it is extremely poisonous. In dangerous concentrations it anaesthetizes the sense of smell, and thus the danger may not be realized. Its presence can be detected by holding a piece of filter paper soaked in a solution of lead(II) ions (lead(II) nitrate or lead(II) ethanoate solution) at the mouth of the test-tube in which the gas is being made. Hydrogen sulphide reacts with the lead ions to form lead sulphide, a dark brown or black insoluble material.

Small quantities of hydrogen sulphide are present in the atmosphere. They have been formed by the decomposition of some organic matter (notably sulphur-containing proteins) and by some industrial processes. Artists' oil colours are generally made with white lead carbonate as a base. Exposure to atmospheric hydrogen sulphide, even in the small quantities in which it is normally present, causes the slow conversion of white lead carbonate to black lead sulphide. After many years, this results in a general darkening of an entire painting. It is possible to restore the former colour by careful treatment of the painting. The varnish covering the pigments must first be removed; then a dilute solution of hydrogen peroxide is applied. This oxidizes the black lead sulphide to white lead sulphate, and brings back the former colours.

If you have tested for hydrogen sulphide using a filter paper as described above, try putting a few drops of hydrogen peroxide on the lead sulphide that you have made. Study the chart of electrode potentials, figure 18.21, to see the steps involved in this reaction.

The manufacture of sulphuric acid

Sulphuric acid is an industrial chemical of really major importance. It is essential for the production of many basic requirements – steel, oil, fertilizers, fibres, paints and detergents – and is used at some stage in the manufacture of practically all consumer goods. Without sulphuric acid, standards of living would revert to those of the Middle Ages and World food production would be only a small fraction of its present level.

The way in which sulphuric acid is used in the United Kingdom is shown in figure 18.23.

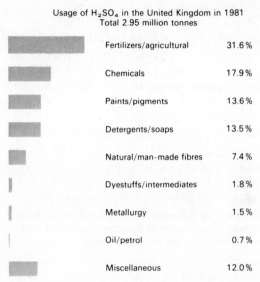

Usage of H_2SO_4 in the United Kingdom in 1981
Total 2.95 million tonnes

Fertilizers/agricultural	31.6%
Chemicals	17.9%
Paints/pigments	13.6%
Detergents/soaps	13.5%
Natural/man-made fibres	7.4%
Dyestuffs/intermediates	1.8%
Metallurgy	1.5%
Oil/petrol	0.7%
Miscellaneous	12.0%

Figure 18.23
(*Based on information obtained from The National Sulphuric Acid Association Ltd.*)

The level of sulphuric acid consumption has long been considered to be a good indicator of a country's prosperity and the substantial reduction in consumption in the United Kingdom between 1973 and 1975 and from 1979 onwards reflected accurately the general economic depression at that time (see figure 18.24).

The contact process

This derives from the work of Peregrine Phillips, a vinegar manufacturer who in 1831 patented a system for oxidizing sulphur dioxide to sulphur trioxide by air, using a platinum catalyst. Commercial development of his ideas was slow because of the high cost of the process, low demand for the concentrated acid it produced, and ignorance of the physical principles which govern exothermic reactions.

By 1901, however, the application of Le Châtelier's principle was understood and contact plants became a practical proposition. At the same time demand from the dyestuffs industry for 96–98% acid and oleum (a solution of

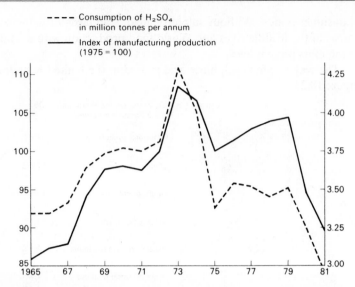

Figure 18.24
A comparison of the consumption of H_2SO_4 (−−−) with the index of manufacturing production (———) in the United Kingdom from 1965 to 1981.
Imperial Chemical Industries plc, Mond Division.

sulphur trioxide in sulphuric acid) was increasing. The resulting plants demonstrated two important points:

1 an excess of air was needed in the air/sulphur dioxide mixture;

2 the catalyst beds and reacting gases needed to be cooled, not heated, once the reaction started.

Modern contact plants use the same principles except that vanadium(v) oxide, promoted by potassium sulphate on a silica support, has entirely replaced platinum as the catalyst, on grounds of cost and resistance to impurities in the gas stream. With the aid of this solid catalyst, sulphur dioxide is converted to sulphur trioxide gas according to the following exothermic reversible reaction:

$$2SO_2 + O_2 \rightleftharpoons 2SO_3; \qquad \Delta H = -197\,\text{kJ}\,\text{mol}^{-1}$$

The result is an equilibrium mixture of sulphur trioxide, sulphur dioxide, and oxygen.

From the Equilibrium Law and a knowledge of catalyst operation, it is possible to deduce the effects of changing reaction conditions on two important aspects of the operation: the sulphur trioxide yield and the reaction rate. These effects are shown in the following table.

Change made to reaction condition	Effect on yield	Effect on reaction rate
Increase of temperature	Decrease	Increase
Increase of pressure	Increase	Small increase
Increase in excess oxygen concentration	Increase	Small increase
Removal of SO_3 from reaction zone	Increase	Small increase
Use of suitable catalyst	No change	Great increase

In practice, a compromise temperature in the range 670–870 K is chosen so that a reasonable yield of SO_3 is obtained at an acceptable reaction rate. With one exception, all the conditions listed which maximize the yield are taken advantage of in modern sulphuric acid plants. The exception is increased pressure, for although potential benefits can accrue from this, satisfactory yields ($>99.5\%$) can be obtained at pressures only slightly above atmospheric.

At present the technology for high pressure operation is little developed and high pressure plants are not competitive, but this may be an area of significant future development.

The absorption of sulphur trioxide

Sulphur trioxide gas cannot be directly absorbed in water because the gas first comes into contact with the water vapour over the liquid and reacts with it to produce a stable mist of separate, minute droplets of sulphuric acid. This mist would pass right through the absorber and be vented to the atmosphere as a visible plume. The gas is therefore absorbed in 98 % sulphuric acid, which has a very low water vapour pressure. Absorption is carried out in ceramic packed towers, and sufficient water is then added to the circulating acid to maintain the required concentration. If acid at a concentration greater than 99.5 % is used, the increased vapour pressure of sulphur trioxide prevents complete absorption, and again a visible plume results.

$$SO_3 + H_2O \rightleftharpoons H_2SO_4; \qquad \Delta H = -130 \, \text{kJ mol}^{-1}$$
$$\text{(in 98 % } H_2SO_4)$$

The absorption reaction, unlike the contact oxidation reaction, releases its energy at a low temperature (333–373 K) and cannot be used to raise steam directly.

The energy is usually dissipated in large air or water coolers. However,

Figure 18.25
Photograph, Imperial Chemical Industries plc, Mond Division.

since this represents a total loss of one-third of the energy produced in th
sulphuric acid process, many plants (see figure 18.25) are now installing equip
ment to recover part of this energy as hot water for space heating or as boile
feed water.

Although one might expect sulphuric acid plants to be responsible fc
considerable atmospheric pollution, this is not the case. A large double
absorption sulphuric acid plant loses to the atmosphere less than 1% of the SO
emitted by a medium-sized coal-burning power station. Moreover, the who
of the UK sulphuric acid industry produces less atmospheric pollution tha
one small power station or domestic coal-burning in a small town.

SUMMARY

At the end of this Topic you should:
1 know the meaning of the expression 'expansion of the octet';
2 understand the significance, in relation to the Periodic Table, of acidic,

basic, and amphoteric oxides and of the hydrolysis of chlorides;

3 understand the application of chemical principles to the industrial extraction of aluminium;

4 know the reactions of a number of compounds of boron, aluminium, tin, lead, nitrogen, and sulphur that have been mentioned in the text, and be able to relate them to the position of the element in the Periodic Table, and to the relevant electrode potentials;

5 understand the application of kinetic and equilibrium concepts to the manufacture of sulphuric acid;

6 be aware of the use of silicon and germanium as semiconductor materials.

PROBLEMS

* indicates that the *Book of data* is needed.

*1 Use the table of electrode potentials in the *Book of data*, and information given in this Topic, to decide what reactions, if any, are possible between each of the following pairs of substances:

a tin(II) chloride solution and potassium dichromate(VI) solution
b tin(II) chloride solution and iodine solution
c boron trichloride and water
d germanium(IV) chloride and water

2 Compare the extraction of iron from its ore, described in the Background reading in Topic 16, with the extraction of aluminium, described in this Topic, and answer the following questions.

a What compound forms the principal ore of **i** iron, **ii** aluminium?
b How is silicon(IV) oxide, an impurity in both ores, removed in each case?
c By reference to the Ellingham diagrams in Topic 16, estimate the minimum temperature needed for carbon to reduce aluminium oxide.
d Impure copper can be purified by electrolysis, using a cathode of pure copper and making the impure copper the anode. Why cannot aluminium be purified in this way?
e Find out the current prices of aluminium and iron, for example from the commodity index of the *Financial Times*. Why is aluminium so much more expensive than iron?

*3 Gallium is the element immediately below aluminium in the Periodic Table. Use your Periodic Table and the *Book of data* to compile a summary of information about the element, its oxide, and its chlorides.

You should indicate which information you obtain directly from the *Book of data* and which information you deduce from the chemistry of boron and aluminium.

4 From your experience of the behaviour of the oxides of the elements you have studied, classify the following as acidic, basic, or amphoteric oxides, briefly stating how you decide in each case:

BeO B_2O_3 GeO_2 SO_3 Tl_2O_3

5 Tin(IV) bromide can be prepared by direct combination of tin with a solution of bromine in an inert solvent. In a preparation of tin(IV) bromide, 2.5 cm³ of bromine were dissolved in 20 cm³ of an inert solvent in a small flask fitted with a reflux condenser. About 4 g of tin were added, a little at a time. The flask was cooled from time to time, by means of a beaker of cold water. When all the tin had been added, the flask was warmed gently until the colour of the bromine had disappeared. The apparatus was rearranged for distillation. The inert solvent was first distilled off and then the tin(IV) bromide (boiling point 202 °C). The tin(IV) bromide was collected in a specimen tube where it solidified to colourless crystals of melting point 33 °C which weighed 8.5 g.

(Sn = 119, Br = 80, density of bromine = $3.12\,g\,cm^{-3}$.)

a What can you say about the structure of tin(IV) bromide?
b How would you expect tin(IV) bromide to behave when put into water?
c Suggest a suitable inert solvent for use in this preparation.
d Calculate the percentage yield of tin(IV) bromide.
e What might you expect to happen if tin were reacted with pure bromine?
f Lead(IV) bromide cannot be made by a similar reaction. What generalization about the elements of Group IV does this illustrate?

6 Sodium thiosulphate, $Na_2S_2O_3 \cdot 5H_2O$, can be prepared by refluxing aqueous sodium sulphite solution with an excess of powdered sulphur for about 2 hours.

a Write an ionic equation for the reaction.
b What happens when sodium thiosulphate reacts with
i dilute hydrochloric acid
ii iodine solution?
In both cases say what you would see and give an equation for the reaction.

7 Copper reacts with nitric acid of various concentrations to give copper(II) ions, Cu^{2+}(aq), and a mixture of oxides of nitrogen. It is possible to adjust the concentration so as to make each oxide predominate.

a Using the data in figure 18.18, explain how electrode potentials show that a set of reactions is likely.

b Possible chemical reactions between nitric acid and copper include

$$3Cu(s) + 12H^+(aq) + 6NO_3^-(aq) \longrightarrow 3N_2O_4(g) + 6H_2O(l) + 3Cu^{2+}(aq)$$
$$3Cu(s) + 8H^+(aq) + 2NO_3^-(aq) \longrightarrow 2NO(g) + 4H_2O(l) + 3Cu^{2+}(aq)$$
$$3Cu(s) + 7.5H^+(aq) + 1.5NO_3^-(aq) \longrightarrow$$
$$0.75N_2O(g) + 3.75H_2O(l) + 3Cu^{2+}(aq)$$
$$3Cu(s) + 7.2H^+(aq) + 1.2NO_3^-(aq) \longrightarrow$$
$$0.6N_2(g) + 3.6H_2O(l) + 3Cu^{2+}(aq)$$
$$3Cu(s) + 7.5H^+(aq) + 0.75NO_3^-(aq) \longrightarrow$$
$$0.75NH_4^+(aq) + 2.25H_2O(l) + 3Cu^{2+}(aq)$$

What compound(s) might be formed with
i Concentrated (14M) nitric acid
ii 7M nitric acid
iii 2M nitric acid?

c Suggest a means of removing traces of the NO_2/N_2O_4 equilibrium mixture from an impure sample of dinitrogen oxide, N_2O, made by this reaction. (You should refer to the notes after figure 18.18 to help you with this answer.)

d According to figure 18.18, nitrous acid is also a possible product of the reaction of copper with nitric acid. Why do you think it unlikely that there is much nitrous acid among the products?

Index

References to specific salts are indexed under the name of the appropriate cation; references to substituted organic compounds are indexed under the name of the parent compound.

a

absolute electrode potentials, 168
accelerated wear testing, 268
acceptor impurity, *see p*-type impurity
acetic acid, ethylenediaminetetra-, *see* edta
acid–base equilibria, 22–31
  in human body, 42–8
Acid Blue 40, 291–2
acidic oxides/hydroxides, 356–7
acids, defined, 22
  strengths, 26–7
  see also amino acids; carboxylic acids
activated complex, 237–8
activation energy, 145, 146, 205
  effect of catalysts on, 147
acyl chlorides, preparation, 65, 67
  reactions, 73; with amines, 87, 324; with ammonia, 324; with water, 75–6
  reduction, 77
  synthesis, 324
addictive drugs, 306, 307, 309–10, 330–31
addition polymerization, 264–5, 272–3, 279–82
adenosine triphosphate, 211
adsorption, 238–9
aerobic respiration, 211
affinity chromatography, 109
air, hydrogen sulphide in, 394
  lead in, 382, 384, 385–6
alanine, 92
alanine, phenyl-, 92
alcoholism, 309
alcohols, dehydration, 323
  ester formation from, 2, 3–4, 9, 68, 73, 76, 324
  oxidation, 323, 324
  reactions: with acyl chlorides, 73, 76; with carboxylic acids, 63, 68; with halide ions, 322
  synthesis, 69, 77, 321, 322
aldehydes, identification, 314–15
  oxidation, 323
  reduction, 322
  synthesis, 324
alfalfa, 256

alkanes, halogeno-, dehydrohalogenation, 323
  hydrolysis, 138–42
  reactions: with alkalis, 322; with amines, 86–7; with ammonia, 323; with cyanide ions, 322
  synthesis, 322
alkenes, reactions: with halogens or hydrogen halides, 322; with sulphuric acid, 321
  synthesis, 323
aluminium, 357–9
  as reducing agent, 246
  extraction, 359–61
aluminium chloride, 359
aluminium hydroxide, 358
alveoli, 42
amethocaine, 330
amides, hydrolysis, 76
  reactions, 74
  reduction, 323
  synthesis, 324
amine oxidase, 256
amines, 81–7
  diazotization, 289–90, 294–5
  synthesis, 77, 323
amino acids, 87–96
  in proteins, 97–105
  separation, 94–6, 97
ammonia, 352–3
  catalysts for synthesis of, 239–41
  distribution between solvents, 4–6
  enzymic production, 209–10
  hydrogen–nitrogen–ammonia equilibrium, 12–13, 16, 18, 51–4, 55, 207
  in complex ions, 86, 219, 229, 236–7
  liquid, as ionizing solvent, 25–6
  reactions: in solution, 390; with carboxylic acids, 63; with ethanoyl chloride, 73, 76; with hydrogen chloride, 25; with water, 24–5
ammonia–ammonium chloride buffer solutions, 33, 35–6
ammonium chloride, ionization in ammonia, 25
ammonium ion, pK_a, 86
ammonium vanadate(v), 224–5

amphetamines, 308, 309
amphoteric compounds, 356
amylases, 107, 111, 112, 113
anaemia, 254, 256, 257
analgesic drugs, 305–7
angina pectoris, 113–14
aniline purple, 293
antibiotics, 112, 304–5
arginine, 93
Arrhenius equation, 146
ascorbic acid oxidase, 256
asparaginase, 115
asparagine, 93, 115
aspartic acid, 93, 115, 128
aspirin, 297–300, 301, 302–4, 307
atactic poly(chloroethene), 283, 284
ATP, *see* adenosine triphosphate
auxochromes, 296
azobenzene, 4-dimethylamino-, 295
azobenzene, 4-hydroxy-, 295
azo-dyes, 288–96

b

Bacon, Francis, 84
Bakelite, 262, 267
ballistic missiles, 365
barbiturates, 307
barley, nutrient requirements, 253, 257
bases, defined, 22
  strengths, 26–7
  see also acid–base equilibria
basic oxides/hydroxides, 356–7
bauxite, 360
benzamide, preparation, 76
benzene, 1,2-dihydroxy-, 229, 231, 232, 255
benzene, 4-nitromethyl-, oxidation, 332
benzenediazonium chloride, 289, 295
benzene-1,2,-dicarboxylic acid, 84
benzene-1,2-dicarboxylic anhydride, 266
benzidine, *see* biphenyl-4,4'-diamine
benzocaine, 330–34
benzoic acid, esterification, 65–6, 68
  K_a, to measure, 41
  reaction with phosphorus pentachloride, 67
  reduction, 69
benzoic acid, 2-ethanoylamino-, 83–4
benzoic acid, 2-hydroxy-, derivatives, as drugs,
    297–304
    complexes, 229, 231, 232
benzoic acid, 4-nitro-, esterification, 332
benzonitrile, hydrolysis, 76
  reduction, 77
benzoyl chloride, 67, 74
bidentate ligands, 232, 236–7
Bill of Mortality, 301
biological cells, energetics, 209–11

biphenyl-4,4'-diamine, 262
biuret test for peptide groups, 91
blackberries, 384
blast furnace, 249
blood, 42
  buffers in, 44–6
  carbon dioxide carriage in, 43–6, 47–8
  oxygen transport in, 42–3, 47–8
blood platelets, 303
blood serum, enzymes in, 113–14
boehmite, 360
boiling points, determination, 316
Boltzmann constant, 144–5
borax, 357
boric acid, 357–8
boron, 357–9
  as micronutrient, 257
  doping with, 377–8
boron trichloride, 355, 359
Brady's reagent, *see* hydrazine, 2,4-
    dinitrophenyl-
bread, 111
brewing, 111
bromine, bromine–bromide equilibrium, 182
  bromine–hydrogen–hydrogen bromide
    equilibrium, 10–11
  gas–liquid equilibrium, 1, 3, 21–2
bromophenol blue, 40–41
buffers, 32–7, 67
  in blood, 44–6
Burke, Edmund, 292
butane, 1-bromo-, hydrolysis, 138–9
butanedioic acid, 2-hydroxy-, 106–7
butanoic acid, pK_a, 67
trans-butenedioic acid, 106–7
butylamine, 82–3
butylammonium ion, pK_a, 86

c

caffeine, 308, 309
calcium carbonate, decomposition equilibrium,
    2, 21, 208
calcium carbonate–hydrogen chloride reaction,
    kinetic study, 129–31
calomel electrode, 181
cannabis, 309
canned food and drink, 384
carbamide, thio-, 236
carbon, as reducing agent, 246–8
  determination in organic compounds, 311
  in steel, 250
  oxidation, 244
  oxidation numbers, 362
  see also diamond; graphite
carbon dioxide, carriage in blood, 43–6, 47–8
carbonic anhydrase, 42, 44

carbonyl compounds, identification, 314–15
    reactions with cyanide ions, 322
carbonyls, 219
carboxylic acids, 63–72
    acyl chloride synthesis from, 324
    alcohol synthesis from, 321
    amide synthesis from, 324
    derivatives, 72–80
    nitrile synthesis from, 322
    synthesis, 323
Carothers, W.H., 263–4, 265
cast iron, 250
catalysts, effect on reaction rates, 125, 147–52
    for ammonia synthesis, 239–41
    for contact process, 395, 396
    for reduction of carboxylic acid derivatives, 77
    hydrogen ion, for esterification, 9
    transition metals as, 220, 237–42
    Ziegler, 272
    see also enzymes; initiators
cataracts, 303
    enzymic treatment, 115–16
cells, see biological cells; voltaic cells
cellulases, 112
cellulose ethanoate, dyeing of, 291–2
Chain, E.B., 305
chain transfer reactions, 281
cheesemaking, 111
chemical kinetics, see reaction rates
chemisorption, 238–40
chest infection, 48
chirality, in liquid crystals, 78–80
    of amino acids, 88, 91
chlordiazepoxide, 308
chloric(I) acid, pK_a, 41
chloride shift, 44–5
chlorides, Group IV elements, 362
    p-block elements, 353, 354–5, 359, 362
chlorophyll, 210
chlorosis, 253–4
chlorpromazine, 308
chocolates, soft-centred, 111
cholestene-3-one, 337, 340
cholesterol, 335, 337–8
cholesteryl benzoate, 74–5, 77–8
chromatography, enzyme purification by, 109
    see also ion-exchange chromatography; paper
        chromatography
chromium(II) ethanoate, preparation, 234–5
chromium oxide, reduction, 246
chromophores, 296
chymotrypsin, 115–16
cobalt, as micronutrient, 257
cocaine, 330–31
coconut oil, 72
codeine, 306

cold drawing of polymers, 271–2
collision theory of reaction kinetics, 144–5
colorimetry, reaction monitoring by, 131,
    149–52
colour, 295–6
combustion analysis, 310–13
complex ions, 81, 86, 219, 228–32
Conan Doyle, Arthur, 330–31
concentration, effect on electrode potentials,
    174–6, 184–5
    effect on reaction rates, 125, 126–42
condensation polymerization, 265, 274
conductance bridge, 133
conduction band, 367, 368, 370–71
confectionery industry, 111
conformation of proteins, 99
contact process, 11, 395–8
contraceptive pill, 335, 336
co-ordination compounds, 228
    see also complex ions
co-ordination number, 228
co-polymers, 285–8
copper, as micronutrient, 256
    copper–silver ion system, 22, 166, 167, 198–9
copper complexes, 86, 219, 228–32
copper(II) sulphate–zinc system, 166, 167
    see also Daniell cell
copper(I) thiocarbamide chloride,
    preparation, 236
Corey, R.B., 103
corrosion, 359
cotton, dyeing of, 291–2
coupling reactions, 289, 290
cretinism, 258
cross-linked polymers, 270
cryolite, 360
crystal pulling, 375–7
crystallinity in polymers, 271, 272, 274, 283
cyclohexane hazard, 262, 263
cysteine, 93

d

Dalton, John, 12
Daniell cell, 168–9, 170, 193–5, 201–3
d-block elements, see transition elements
decanedioyl dichloride, reaction with
    hexane-1,6-diamine, 84–5, 87
decanoic acid, 72
delocalized electrons, 63, 295–6, 352, 391
denaturing of enzymes, 110
derivatization of organic compounds, 310–13
desmosine, 256
detergents, 70–71
diabetes, 114, 303, 304
diamond, 366–7

diazepam, 308
diazotization, 289–90, 294–5
di(benzoyl) peroxide, initiator, 280
dichromate (VI), precautions in use, 234
di(dodecanoyl) peroxide, initiator, 265
diethylamine, 86
    infra-red absorption spectrum, 81
di-2-ethylhexyl benzene-1,2-dicarboxylate,
    plasticizer, 285
dihydrofolate reductase, 99
dilatometry, reaction monitoring by, 132
dinitrogen oxide, 388
dinitrogen tetraoxide, 388
    nitrogen dioxide–dinitrogen tetraoxide
    equilibrium, 207–8
dioctyl hexanedioate, plasticizer, 285
diosgenin, 337
Direct Red 23, 291–2
disodium tetraborate-10-water, *see* borax
Disperse Yellow 3, 291–2
'Dispersol' Fast Yellow G, 296
disproportionation reactions, 389–90, 391–3
dissociation constant, 29
    pK_a, 41
    to measure, 39–41
distillation of ethoxyethane, 338–9
distribution ratio, 6
dodecanoic acid, 72
donor impurity, *see* n-type impurity
doping, 378
drugs, 297–310, 319, 321
Durrer, R., 251
dust, lead in, 383, 386–7
dyes, 288–96
    see also indigo

e

Edman protein analysis, 97–9
edta, 229, 231, 232, 237
Ehrlich, P., 304
elastin, 256
elastomers, 270–71
    see also rubber
electrical conductivity, of metals, 367
    of semiconductors, 368, 369
    reaction monitoring by measuring, 133
electrode potentials, 168
    effect of concentration on, 174–6
    to calculate, 176–8
    to measure, 184
    used in catalyst selection, 241–2
    see also standard electrode potentials
electrodes, contribution to e.m.f., 170
    hydrogen, 170–72
    reference, 170, 181
    silver, 175–6

electromotive force, 169
    contributions of electrode systems, 170
    related to free energy, 201–2
    to calculate, 188–9
    to measure, 172–3; for determination of silver
        ions, 179–80; under non-standard
        conditions, 202–3
electrons, delocalized, 63, 295–6, 352, 391
    expansion of octet, 352–3
    in metals, 367
    in semiconductors, 365–72
Ellingham diagrams, 243–8
emulsion polymerization, 287
endothermic reactions, 164, 200
    effect of temperature change, 17
entropy changes, 163–5, 204–5
    in reactions of complexes, 236–7
    in voltaic cells, 196–8
    zero, at equilibrium, 49–55, 164
enzootic ataxia, 256
enzymes, 99, 105–8, 209
    in medicine, 113–16
    industrial uses, 109–12
    transition elements in, 255, 256, 257
    see also carbonic anhydrase; trypsin; urease
equilibria, 1–55
    acid–base, 22–31; in human body, 42–8; *see
        also* dissociation constant
    characteristics, 2–3
    effect of pressure and temperature, 12–19
    heterogeneous, 19–22
    homogeneous, 19
    in buffer solutions, 32–7
    in indicator solutions, 37–41
    relative concentrations, 3–11
    zero total entropy change in, 49–55, 164
    see also redox reactions
equilibrium constant, 8, 9, 10–11, 204–5
    and free energy, 54–5
    effect of pressure changes, 16, 18, 19; of
        temperature changes, 16, 18
    for oxide decomposition, 243–4
    in terms of partial pressures, 12–14
    to calculate, 207–9
    to find, 192–5
    see also stability constants
Equilibrium Law, 10–11
esters, reduction, 77, 321
    synthesis, 2, 3–4, 9, 68, 73, 76, 324
ethanal, decomposition, 147
ethanamide, formation, 76, 77
    reactions, 74, 76
ethanamide, N-methyl-, formation, 76
ethane, chloro-, 86–7
ethane, 1,2-diamino-, in complex ions, 236
ethane, ethoxy-, distillation, 338–9

ethane-1,2-diammonium ion, pK_a, 86
ethanedioic acid–manganate(VII) reaction,
    kinetic study, 148–52
ethanenitrile, preparation, 77
ethanoic acid, esterification, 2, 3–4, 9
    glacial, 64
    infra-red absorption spectrum, 63–4
    pK_a, 41, 67
    reactions, 64–5, 67
ethanoic acid–ethanoate buffer solutions,
    32, 33–5, 67
ethanoic acid, amino-, see glycine
ethanoic anhydride, 297, 299–300, 303
ethanol, ester formation, 2, 3–4, 9, 73
ethanoyl chloride, formation, 65
    reactions, 73; with amines, 83, 87; with
        nucleophiles, 76
ethene, polymerization, 272
ethene, chloro-, polymerization, 272–3, 279–82,
    283–4, 287
ethene, 1,1-dichloro-, co-polymers, 286
ethenyl ethanoate, co-polymers, 286
ethyl ethanoate, formation, 2, 3–4, 9, 73
ethyl 4-nitrobenzoate, reduction, 332–3
ethylamine, reaction with chloroethane, 86
exothermic reactions, 164, 198
    effect of temperature change, 17
extensive properties, 3

f

fats and oils, 69–72
fermentation, 109, 111, 112
'ferrous sulphate' tablets, see 'iron tablets'
fibres, 268–70, 272
ficin, 112
field effect transistor, 380
first order reactions, 131, 153–4, 156
fishing weights, 387
Fleming, Alexander, 305
Flixborough disaster, 262, 263
Florey, H.W., 305
flow in molten polymers, 279
foams, polymer, 264
food, lead in, 383–4
food industry, enzymes in, 110–11
free energy change, 55, 165–6, 200, 201–2
    see also standard free energy change
free radicals, 147, 272, 279
freezing, entropy change, 49–50
fumarase, 106–7

g

galena, 383
gallium arsenide, 382
gas constant, 145

gelatin, enzymic dissolution, 112
germanium, extraction, 373
    oxidation numbers, 362
    purification, 374–5
    semiconducting properties, 365
    transistor manufacture, 377
Gibbs free energy change, see free energy
    change
gibbsite, 360
glass electrode, 181
glucose, 210, 211
    detection in urine, 114
glucose oxidase, 114
glutamic acid, 90–91, 93
glutamine, 93
glycerol, see propane-1,2,3-triol
glycine, 87, 90–91, 92
glycine, ethanoyl-, X-ray diffraction study, 99,
    101
goitre, 258
Goldschmidt process, 246
gramophone records, polymer formulation for,
    276–88
graphite, reaction with water, 18
Griess, P., 294
Group IV elements, 361–7

h

haemocyanin, 256
haemoglobin, 42–3, 44–5, 89, 254–5
half-life time studies of reactions, 131, 154, 155,
    156–7
Hall effect, 372
hallucinogenic drugs, 309
'hard' drugs, 306, 309–10
hazards, in use of drugs, 303–4, 310; see also
    addictive drugs
    manufacturing, 262, 263
heart disease, 113–14, 303
heart surgery, 48
α-helix, 102–3, 104
Henderson–Hasselbalch equation, 46, 47
herbal remedies, 301–3
heroin, 306, 309–10
heterogeneous catalysis, 238–41
heterogeneous equilibria, 19–22
hexadecanoic acid, 72
hexadecanol, 72
hexadentate ligands, 232, 237
hexane-1,6-diamine, in nylon synthesis, 84–5,
    87, 274
hexanedioic acid, in nylon synthesis, 274
6-hexanolactam, polymerization, 273
histidine, 93
Hoffmann, F., 303
Hofmann, A.W. von, 293

holes, in semiconductors, 369, 370, 372
homogeneous catalysis, 238, 241–2
homogeneous equilibria, 19
hydrazine, 2,4-dinitrophenyl-, 314–15
hydrides, hydrolysis, 355
hydrochloric acid, *see* hydrogen chloride
hydrogen, as reducing agent, 246–8
    determination in organic compounds, 311
    hydrogen–bromine–hydrogen bromide
        equilibrium, 10–11
    hydrogen–iodine–hydrogen iodide
        equilibrium, 2, 6–8, 16, 17–18
    hydrogen–nitrogen–ammonia equilibrium,
        12–13, 16, 18, 51–4, 55, 207
    oxidation, 244, 245
hydrogen bonding, in amines, 85
    in carboxylic acids, 66
    in polymers, 271, 284
    in textile dyeing, 292
hydrogen bromide–hydrogen–bromine
    equilibrium, 10–11
hydrogen chloride, pH of solution, 28
    reactions: with ammonia, 25; with calcium
        carbonate, kinetic study, 129–31; with
        sodium thiosulphate, kinetic study, 143–4;
        with water, 23–4
hydrogen electrode, 170–72, 176
hydrogen iodide–hydrogen–iodine equilibrium,
    2, 6–8, 16, 17–18
hydrogen ion, 24, 26
    see also pH
hydrogen sulphide, 391, 394
    hydrogen sulphide–manganate(VII) reaction,
        190–92
hydrogencarbonate ion, in blood, 43–6, 47, 48
hydrolases, 111–12
hydronium ion, 26
hydroxonium ion, 26

i

immobilized enzymes, 110, 112
indicators, 37–41
indigo, 326–9
infra-red absorption spectra, carboxylic
    acids, 63–4
    diethylamine, 81
    for compound identification, 316–19, 320
    2-hydroxybenzoic acid derivatives, 298–9, 301
initiators of polymerization, 265, 266, 272,
    279–80
insulin, 89, 304
    amino acid composition, 97, 99, 100
    analysis of hydrolysate, 96
    role in metabolism, 114
integrated circuits, 365, 380–82

intensive properties, 3
intrinsic conductivity, 369, 372
invertase, 111, 112
iodide, iodide–iodine equilibrium, 183–4,
    189–90
    iodide–iron(III) reaction, 182–4, 189–90,
        208–9
    iodide–persulphate reaction, 241–2
iodine, as micronutrient, 254, 258
    catalyst for ethanal decomposition, 147
    hydrogen–iodine–hydrogen iodide
        equilibrium, 2, 6–8, 16, 17–18
    iodine–iodide equilibrium, 183–4, 189–90
    iodine–propanone reaction, kinetic study, 128
        134–7
ion-exchange chromatography, for amino acid
    analysis, 97
ionization constant, 23
iron, as catalyst: for iodide oxidation,
    242; in ammonia synthesis, 239–41
    as micronutrient, 254
    corrosion of, 359
    iron(II)–iron(III) equilibrium, 183–6, 189–90
    iron(III)–iodide reaction, 182–4, 189–90,
        208–9
    properties of pure, 250
    redox chemistry, 220–23
iron and steel industry, 248–53
iron(III) oxide, 245, 246
'iron tablets', analysis, 226–7
isoleucine, 92
isotactic poly(chloroethene), 283

k

keratins, 103, 112
ketones, identification, 315
    reduction, 322
    synthesis, 324
kinetics of reactions, *see* reaction rates
    and under specific reactions
Kolbe, H., 302
Kwolek, S.L., 84

l

lactase, 112, 256
Largactil, 308
lasers, 382
law of partial pressures, 12
LD steelmaking process, 251–3
Le Chatelier's principle, 16–17
lead, oxidation numbers, 362
    pollution by, 382–7
lead–magnesium cell, 188–9
lead(IV) oxide, 363–5
leucine, 92

leukaemia, 115
Librium, 308
life processes, energetics, 209–11
ligands, 219, 228–32, 236–7
light-emitting diodes, 382
lipase, 110, 112
lipids, 69–72, 110
liquid crystals, 74–5, 77–80
lithium tetrahydridoaluminate reductions, 69, 321, 323
Lowry–Brønsted theory, 22
lucerne, 256
lysergic acid diethylamide (LSD), 309
lysine, 93

m

macronutrients, 253
magnesium–lead cell, 188–9
maltase, 110
maltose, enzymic hydrolysis, 110
manganate(VII), reactions: with ethanedioic acid, kinetic study, 148–52; with hydrogen sulphide, 190–92
manganate(VII)–manganese(II) equilibrium, 182, 187
manganese, as micronutrient, 255
manufacturing hazards, 262, 263
manufacturing plant design, 286
Marco Polo, 326
margarine, 70
marijuana, 309
Marker, R.E., 336
mass spectrometry, for compound identification, 316–19, 320
for r.m.m. determination, 312
of proteins, 99
mauve, 293
mechanisms of reactions, *see* reaction mechanisms
mescalin, 309
metabolic pathways, 209
metal/metal ion systems, 166–74, 196–7
metals, electrons in, 367
methanoic acid, pH of solution, 29–30
pK_a, 67
methanol, reactions: with benzoic acid, 65–6, 68; with ethanoyl chloride, 76
methanol, phenyl-, preparation, 69, 77
methionine, 92
methyl benzoate, preparation, 65–6, 68
reactions, 73–4, 77
methyl 2-methylpropenoate, polymerization, 265
methyl orange, 37, 296

methylamine, reaction with ethanoyl chloride, 76
methylamine, phenyl-, preparation, 77
micelles, 287
microbial enzymes, 109
microelectronics, 365–82
micronutrients, 253–8
migraine, 303
Minute-man missile, 365
mitochondria, 211
molybdenum, as catalyst in ammonia synthesis, 239–40
as micronutrient, 254, 256, 257
monodentate ligands, 231, 236–7
Morgan, P.W., 84
morphine, 306, 309–10
myocardial infarction, 113–14
myoglobin, 89, 99
structure, 103, 105

n

naphthalen-2-ol, 289, 290
necrosis, 254
nematic liquid crystals, 78
twisted, 78–80
Nernst equation, 178–9, 186, 187, 203
nickel carbonyl, 219
nickel complexes, 237
nicotine, 308, 309
ninhydrin test for amino acids, 91
nitrate ion, 352–3, 388
nitric acid, 388
pK_a, 41
reactions, 389
nitriles, 76–7
hydrolysis, 323
reduction, 323
synthesis, 322
nitrogen, adsorption, 239–41
bonding, 352–3
determination in organic compounds, 311
hydrogen–nitrogen–ammonia equilibrium, 12–13, 16, 18, 51–4, 55, 207
nitrogen–oxygen–nitrogen monoxide equilibrium, 16
nitrogen compounds, 387–90
nitrogen dioxide–dinitrogen tetraoxide equilibrium, 207–8
nitrogen monoxide, 388
nitrogen–oxygen–nitrogen monoxide equilibrium, 16
nitrogenase, 209
nitrous acid, 388
reactions, 389–90; with amines, 288–9, 290

nomogram from Henderson–Hasselbalch
equation, 47
n-type impurity, 371, 372
nutmeg seed fat, 70, 72
nylon, 265, 271, 275
dyeing of, 291–2
manufacture, 273–4
physical properties, 267–8, 282
'rope trick', 84–5, 87

o

octadecanoic acid, 69, 72
octadec-9-enoic acid, 72
octadec-9-en-1-ol, 72
octanoic acid, 72
oestrogens, 336
oil of wintergreen, 297, 299, 300–301
olive oil, 72
opium alkaloids, 306
orders of reaction, 127–8, 130–31
from half-life times, 156–7
organic compounds, combustion analysis,
310–13
derivatization, 313–16
spectroscopic identification, 316–19
osmium, as catalyst in ammonia synthesis,
239–41
oxidation, 167
oxidation numbers, Group IV elements, 361–2
nitrogen, 387
sulphur, 391
transition elements, 218–19, 220–27
oxide masking process, 378–9
oxides, acid–base properties, 356–7
Group IV elements, 362–5
thermodynamic stabilities, 243–8
oxonium ion, 26
oxygen, determination in organic compounds,
311–12
nitrogen–oxygen–nitrogen monoxide
equilibrium, 16
sulphur dioxide–oxygen–sulphur trioxide
equilibrium, 11, 14, 15, 16
transport in blood, 42–3, 47–8
oxyhaemoglobin, 43

p

paint, darkening and restoration, 394
lead in, 383, 386–7, 394
palm oil, 72
pancreatitis, diagnosis, 113
papain, 112
paper chromatography of amino acids, 94–6
Parkesine, 262
partial pressure, defined, 12

partition coefficient, 6
Pauling, L., 103
p-block elements, 352–98
pectinases, 112
penicillins, 112, 305
peptide group, 88
detection, 91
dimensions, 100, 101–2
Perkin, W.H., 292–3, 294
pernicious anaemia, 257
peroxidase, 114
peroxides, initiators, 265, 272, 280
peroxodisulphate(VI)–iodide reaction, 241–2
Perspex, *see* poly(methyl 2-methylpropenoate)
pethidine, 306–7
petrol, lead in, 383, 385–6
pewter, 383
pH, 26–31
of blood, 45–8
pH meters, 27, 180–82
titrations using, 30–31
phase, defined, 19
phenol, coupling reaction, 295
phenol, 4-nitro-, 296
phenolic resin, 267, 275
phenolphthalein, 37, 38–9
phenyl isothiocyanate, 97–8
phenylamine, 82–3, 85–6, 87
diazotization, 289, 290, 295
in synthesis of mauve, 293
phenylamine, *N,N*-dimethyl-, coupling reaction
295
phenylammonium ion, pK_a, 86
Phillips, Peregrine, 395
phosphorus, bonding, 353
doping with, 377–8
phosphorus chlorides, 353, 355
photoresists, 378, 379
photosynthesis, 210–11
physical adsorption, 238–9
'pining', 257
plant, manufacturing, design, 286
plant lipids, 69–70
plant nutrients, 253–8
plasma etching, 382
plasticizers, 273, 284–5
plumbosolvency, 385
pollution, 70–71, 382–7
poly(benzyl glutamate), X-ray diffraction
study, 102–3
poly(2-chlorobuta-1,3-diene), 275
poly(chloroethene), 264–5, 274, 276
for gramophone record manufacture, 279–88
manufacture, 272–3
physical properties, 267–8, 282, 283, 284
polydentate ligands, 231

polyesters, dyeing of, 291–2
  resin, 266
poly(ethane-1,2-benzene-1,4-dicarboxylate), 275
poly(ethene), 276
  manufacture, 272
  physical properties, 267–8, 272, 273, 282–3
polymers, 262–88
poly(2-methylbuta-1,3-diene), 275
poly(methyl 2-methylpropenoate), 275
  physical properties, 267–8
polypeptides, 88, 99
poly(propenamide), 266
poly(propene), 274, 276
poly(propenenitrile), 274
polystyrene, 264, 274
polyvinyl chloride, see poly(chloroethene)
potassium dichromate(VI), precautions in use,
  234
potassium manganate(VII)–ethanedioic acid
  reaction, kinetic study, 148–52
potassium peroxodisulphate(VI), initiator,
  266, 272
pre-exponential factor, 146
pregnenolone, 337
procaine, 330
progesterone, 337
progestogens, 336
proline, 92
Pronase, 112
Prontosil, 304
propane, 2-bromo-2-methyl-, hydrolysis, 140–42
propane-1,2,3-triol, esters, see fats and oils
  in polyester synthesis, 266
propanoic acid, pK_a, 67
propanone, identification, 314–15
  propanone–iodine reaction, kinetic study,
  128, 134–7
propenamide, polymerization, 266
proteases, 112
proteins, 88–91
  blood plasma, 44
  chemical and structural studies, 96–105
  see also enzymes
'psychedelic' drugs, 309
p-type impurity, 371, 372
purine metabolism, 257
PVC, see poly(chloroethene)
pyrometallurgy, 242–53

q
quinine, 304

r
radiotracer studies, solid–solution equilibria, 2,
  20

rate constant, 127–8
  effect of temperature, 146
rate equation, 126–8
  integration, 153–5
rate-determining step, 136–7, 239
rates of reaction, see reaction rates
Rauwolfia, 302
reaction mechanisms, 126, 136–42
reaction rates, 125–6
  effect of catalysts, 125, 147–52
  effect of concentration, 125, 126–42
  effect of temperature, 125, 126, 142–6
  of pyrometallurgical processes, 242–3, 248
redox reactions, 182–211
  entropy changes in, 198
  in voltaic cells, 198–200
  metal/metal ion systems, 166–74, 196–7
  to find equilibrium constant, 192–5
  to predict likelihood, 189–92, 203–5
reduction, 167
reference electrodes, 170, 181
Reinitzer, Friedrich, 77
relative molecular mass, control in
    polymerization, 279–82
  determination, 312
reserpine, 302
resins, phenolic, 267, 275
  polyester, 266
respiration, aerobic, 211
respiratory acidosis, 48
reversible changes, 3
river pollution, 70–71
Roberts, I., 68
rubber, 275
  physical properties, 267–8
ruthenium, as catalyst in ammonia synthesis,
    239–41

s
salicin, 302
salivary α-amylase, 107
Salvarsan, 304
Sanger, F., 96–7, 99
'scouring', 256
Second Law of Thermodynamics, 163
second order reactions, 131, 154–5, 156
sedatives, 307–8
semiconductors, 365, 368–77
  see also transistors
serine, 88, 92
Seveso disaster, 262
sheepskin, wool removal from, 111–12
silane, 373–4
  hydrolysis, 355
silane, tetrachloro-, hydrolysis, 355
silane, trichloro-, 374

silicon, extraction, 373–4
  impurities in, 370
  oxidation numbers, 362
  purification, 374–7
  semiconducting properties, 365
  transistor manufacture, 377–9
silver, copper–silver ion system, 22, 166, 167, 198–9
  determination, by e.m.f. measurement, 179–80
silver/silver chloride electrode, 182
silver chloride, solid–solution equilibrium, 19–21
  solubility product, 20, 180
skin creams, enzymes in, 112
soap, 70
sodamide, ionization in ammonia, 25
sodium nitrate, solid–solution equilibrium, 1–2, 3
sodium tetrahydridoborate reductions, 322
sodium thiosulphate–hydrogen sulphide reaction, kinetic study, 143–4
soil, lead in, 382, 384
solder, 384
solid–solution equilibria, silver chloride, 19–21
  sodium nitrate, 1–2, 3
solubility product, of silver chloride, 20, 180
sperm whale oil, 72
stability constants, 228–9, 230, 231
standard electrode (redox) potentials, 172, 176, 186–8, 205
  applications, 188–95, 387–8, 391–3, 394
standard free energy change of formation, 205–9
standard free energy change of reaction, 55, 204–5, 207
  for photosynthesis, 210
  see also Ellingham diagrams
starch, enzymic hydrolysis, 107, 111
steel, see under iron and steel industry
steric factor, 146
steroids, 335–40
stimulants, 308–9
stoicheiometry of complexes, 232–3
Stone, Rev. Edmund, 302
strong acids, 28–9
substrate, defined, 105
subtilisin, 110
sulphadiazine, 321
sulphanilamide, 304, 321
sulphate ion, 391
sulphathiazole, 321
sulphite ion, 391
  reactions, 393–4
sulphonamides, 304, 321
sulphur compounds, 390–98
sulphur dichloride, 391

sulphur dioxide, 391, 393
  sulphur dioxide–oxygen–sulphur trioxide equilibrium, 11, 14, 15, 16
sulphur trioxide, 391
  absorption, 397–8
  sulphur dioxide–oxygen–sulphur trioxide equilibrium, 11, 14, 15, 16
sulphuric acid, manufacture, 11, 394–8
  pH of pure, 28
suspension polymerization, 287
swans, 387
'swayback', 256
syndiotactic poly(chloroethene), 283, 284
synthesis, to plan, 325–6
syphilis, 304

t

temperature, effect on electrode potentials, 185
  effect on equilibrium constant, 16, 18
  effect on reaction rates, 125, 126, 142–6
tetracyclines, 305
tetradecanoic acid, 72
textile industry, enzymes in, 111–12
textiles, dyeing of, 290–92
thalidomide, 262, 307–8
Thermit process, 246
thermolysin, 99
thermoplastic materials, 271, 282
thermosetting polymers, 270
thiosulphate–hydrogen ion reaction, kinetic study, 143–4
threonine, 93
thyroxine, 258
tin, oxidation numbers, 362
tin(IV) oxide, 362–3, 364
titrations, acid–base, 30–31
  reaction monitoring by, 131, 134, 140–41, 148–9
tooth powder, enzymes in, 112
trace elements, see micronutrients
tracer studies, of ester formation, 68
  of solid–solution equilibria, 2, 20
transistors, 365, 372, 377–9
  see also integrated circuits
transition elements, 218–57
triboluminescence, 84
triethylamine, 86
tritolyl phosphate, plasticizer, 285
trypsin, 89, 99
tryptophan, 92
tungsten, as catalyst in ammonia synthesis, 239–40
tyrosinase, 256
tyrosine, 93

u

uranyl(VI) nitrate hexahydrate, 84
urea, enzymic hydrolysis, 105–6
urease, 89, 105–6
Urey, H.C., 68
urine, detection of glucose in, 114

v

vacuum degassing, 251
valence band, 367, 368, 370–71
valine, 88, 92
Valium, 308
vanadium, oxidation numbers, 223–6
viscosity of molten polymers, 279, 281
vitamin B_{12}, 257
voltaic cells, Daniell, 168–9, 170, 193–5,
    201–3
  diagrams, 169–70, 184, 187
  redox reactions in, 183–8, 198–200
  see also electromotive force

w

washing powders, enzymes in, 110
waste disposal, 262
water, entropy change of freezing, 49–50
  in complex ions, 219, 228–9

ionization, 22–3
lead in, 383, 385
pollution by detergents, 70–71
reaction with graphite, 18
waterfowl, 387
weak acids, 28–30
willow extracts, 302, 303
wintergreen, oil of, 297, 299, 300–301
wool, enzymic dissolution, 111–12

x

xanthine oxidase, 257
X-ray diffraction studies, ethanoylglycine, 99,
    101
  poly(benzyl glutamate), 102–3

y

yeast, 111, 256

z

zero order reactions, 131, 136, 156
Ziegler, Karl, 272
zinc, as micronutrient, 255–6
zinc–copper(II) sulphate reaction, 166, 167
  see also Daniell cell
zinc oxide, reduction, 246–8
zone refining, 374–5, 376